Whaling in the Antarctic

Queen Mary Studies in International Law

Edited by

Malgosia Fitzmaurice
Phoebe Okowa

VOLUME 23

The titles published in this series are listed at *brill.com/qmil*

Whaling in the Antarctic

Significance and Implications of the ICJ Judgment

Edited by

Malgosia Fitzmaurice
Dai Tamada

BRILL
NIJHOFF

LEIDEN | BOSTON

Library of Congress Cataloging-in-Publication Data

Names: Whaling in the Antarctic : the ICJ Judgment and its implications (2014 : Kōbe-shi, Japan) | Fitzmaurice, M., editor. | Tamada, Dai, 1974- editor. | Kobe University Centre for International Law, sponsoring body.
Title: Whaling in the Antarctic : significance and implications of the ICJ judgment / edited by Malgosia Fitzmaurice, Dai Tamada.
Description: Leiden ; Boston : Brill Nijhoff, 2016. | Series: Queen Mary studies in international law ; volume 23 | Includes papers presented at a conference "Whaling in the Antarctic : The ICJ Judgment and its Implications" held at the Kobe University Centre for International Law, May 31-June 1, 2014 discussing the case "Whaling in the Antarctic (Australia v. Japan: New Zealand intervening)". | Includes bibliographical references and index.
Identifiers: LCCN 2016008189 (print) | LCCN 2016011658 (ebook) | ISBN 9789004313644 (hardback : alk. paper) | ISBN 9789004313828 (E-book)
Subjects: LCSH: Whaling--Law and legislation--Antarctica--Cases--Congresses. | Whaling--Law and legislation--Antarctic Ocean--Cases--Congresses. | Environmental law, International--Cases--Congresses. | Australia--Trials, litigation, etc.--Congresses. | Japan--Trials, litigation, etc.--Congresses.
Classification: LCC KWX797 .W43 2016 (print) | LCC KWX797 (ebook) | DDC 343.07/692809167--dc23
LC record available at https://lccn.loc.gov/2016008189

Typeface for the Latin, Greek, and Cyrillic scripts: 'Brill'. See and download: brill.com/brill-typeface.

ISSN 1877-4822
ISBN 978-90-04-31364-4 (hardback)
ISBN 978-90-04-31382-8 (e-book)

This book is printed on acid-free paper and produced in a sustainable manner.

Contents

Notes on Contributors

Theodore Christakis
is Professor of International Law at the Faculty of Law of the University Grenoble Alpes and a Senior Member of the Institut Universitaire de France. He also teaches at the Paris School of International Affairs. He is Director of the Centre for International Security and European Studies (CESICE) and director of the Master's degree in International and European Law of the University Grenoble Alpes. He has published or edited eight books and more than 45 articles and book chapters on International Law.

Malgosia Fitzmaurice
has a Chair of Public International Law at Queen Mary University of London. Her main research interests are: the law of treaties; international environmental law; whaling; Arctic law. She was invited by several universities as a Visiting Professor, such as Berkeley, Sorbonne-Pantheon; Kobe. Her most recent publication will be a monograph *Whaling and International Law* published in 2015 by Cambridge University Press.

Caroline Foster
is an Associate Professor at the University of Auckland, New Zealand, previously employed by the New Zealand Ministry of Foreign Affairs and Trade. Related publications include: *Adjudication, Arbitration and the Turn to Public Law 'Standards of Review': Putting the Precautionary Principle in the Crucible* [2012] 3[3] JIDS 525; *Diminished Ambitions? Public International Legal Authority in the Transnational Economic Era* (2014) 17(2) JIEL 440; and *Science and the Precautionary Principle in International Courts and Tribunals: Expert Evidence, Burden of Proof and Finality* (Cambridge University Press, 2011).

Shotaro Hamamoto
is Professor of the law of international organizations, Kyoto University. *Professeur invité* at Université de Paris I and Sciences Po de Paris. Counsel of Japan in *Whaling in the Antarctic* (ICJ, 2014) as well as in *Hoshinmaru* and *Tomimaru* (Japan v. Russia, ITLOS, 2007). Assistant to the Spanish Government in *Fisheries Jurisdiction* (Spain v. Canada, ICJ, 1998). Japanese Representative in the WG II of the UNCITRAL (2010–15) and in the Investment Committee of the OECD (2011).

Mika Hayashi
currently teaches international law in the Graduate School of International Cooperation Studies, Kobe University, Japan. Her previous career includes conference interpreter and court interpreter for English, French and Japanese. Mika's principal research interest relates to international security and the law of armed conflict. Her most recent article in these fields is 'Reacting to the Use of Chemical Weapons: Options for Third States', *Journal on the Use of Force and International Law*, Vol.1 (2014). Mika holds MA from the University of Tokyo and DESS from the INALCO, Paris.

Joji Morishita
is Director-General at the National Research Institute of Far Seas Fisheries. Japan's Commissioner to the International Whaling Commission. Served as Fisheries Attaché at the Japanese Embassy in the United States from 1993 to 1996. Involved in international fisheries issues since 1982 in the Fisheries Agency of the Japanese Government and also covered such areas as the Convention on Biological Diversity, CITES, and UN General Assembly Informal Consultations on the sustainable fisheries resolution. Master of Public Policy (MPP), Harvard University. BS, Kyoto University.

AJ (Tony) Press
is an Adjunct Professor at the Antarctic Climate and Ecosystems Cooperative Research Centre, and the Institute for Marine and Antarctic Studies at the University of Tasmania, Hobart. He was formerly CEO of the ACE CRC; Director of the Australian Antarctic Division; Australia's Commissioner to the Commission for the Conservation of Antarctic Marine Living Resources; and, Delegate to Antarctic Treaty Consultative Meetings. He recently provided a major report to the Australian government on Australia's Antarctic interests.

Donald R. Rothwell
is Professor of International Law, and Deputy Dean at the ANU College of Law, Australian National University where he has taught since July 2006. His research has a specific focus on law of the sea, law of the polar regions, and international law within Australia. He has acted as a consultant or been a member of expert groups for UNEP, UNDP, IUCN, the Australian Government, and advised the International Fund for Animal Welfare (IFAW) on Japanese whaling.

Hironobu Sakai
is Professor of International Law at Graduate School of Law, Kyoto University. His recent publications in English include: 'New Relationship between the United Nations and Regional Organizations in Peace and Security: A Case of

the African Union', in Shotaro Hamamoto, Hironobu Sakai & Akiho Shibata (eds.), *"L'être situé", Effectiveness and Purposes of International Law* (Brill, 2015), pp.165–189, '*La bonne administration de la justice* in the Incidental Proceedings of the International Court of Justice', *Japanese Yearbook of International Law*, Vol.55 (2012), pp.110–133.

Akiho Shibata
is professor of international law and Director, Polar Cooperation Research Centre, Graduate School of International Cooperation Studies (GSICS), Kobe University, Japan. His recent publications include: 'Japan and 100 Years of Antarctic Legal Order: Any Lessons for the Arctic?' *Yearbook of Polar Law*, Vol.7 (2015); 'ICRW as an Evolving Instrument: Potential Broader Implications of the *Whaling* Judgment', *Japanese Yearbook of International Law*, Vol. 58 (2015); and *L'être situé, Effectiveness and Purposes of International Law* (Co-editor, Brill, 2015).

Yuri Takaya
is a Research Fellow and a Lecture at the Kobe University, Japan, teaching international space law. Her recent publications include: 'Computer Network Attacks in Outer Space: The Case of Harmful Interference to Satellite-based Communications', Proceedings of the fifty-seventh Colloquium on the Law of Outer Space, Eleven International Publishing (2015) 519–532. Obtained her B.A. at the Waseda University, Japan, LL.M. at the Leiden University, The Netherlands, and Ph.D. at the Paris XI University, France.

Dai Tamada
is Professor of International Law at the Graduate School of Law, Kobe University. He holds LL.M. (Kyoto University in 2000) and Ph.D (Kyoto University in 2014). His research focuses on peaceful settlement of international disputes and international investment law, among others. His masterpiece is *Legal Effects of the International Courts' Judgments* (Yuhikaku, 2012, vii + 242 pp, in Japanese). He has been committee member in the government organs, such as Ministry of Economy, Trade and Industry.

Christian Tams
is Professor of International Law at the University of Glasgow, where he directs the Law School's LLM in international law. He is a qualified German lawyer (admitted in 2005) and holds an LLM and PhD from the University of Cambridge. His research focuses on legal aspects of dispute resolution, investment law, the law of treaties and UN law. Christian is an academic member of Matrix Chambers, London, and frequently advises States and other actors in proceedings before international courts and tribunals.

Introduction

Malgosia Fitzmaurice and Dai Tamada

The *Whaling in the Antarctic* case (*Australia and Japan. New Zealand intervening*) was decided by the International Court of Justice, on 31 March 2014.[1] Two months after the Judgment, on 31 May and 1 June 2014, Kobe University Centre for International Law held a Symposium on the case,[2] with the participation of 15 speakers, including academics, practitioners and government representatives. Speakers discussed a wide range of topics covering all the aspects of the Judgment; and, the result of this symposium is the present publication.

1 Legal Issues Contained in the Judgment and the Case

The Judgment, and the case itself, raise a variety of issues which need to be examined and analysed from both a legal and a political point of view. Many issues raised in the case will influence understanding of international law; and the Judgment of the Court will find its place in international law textbooks.

The following topics were analysed in the Judgment. (i) Preliminary issues before the ICJ: a jurisdictional objection, reservations to the optional clause declaration, *jus standi*, legal interest, standing, admissibility, no preliminary objection raised by Japan, no claim of provisional measures by Australia. (ii) Procedural issues before the ICJ: single round of written pleading, intervention of third country under Art 63 of the ICJ Statute, appointment of judge *ad hoc*, and cross-examination of experts. (iii) Issues on the Merits included the following: object and purpose of the International Convention on the Regulation of Whaling (ICRW), subsequent practice, subsequent agreement of parties, legal meaning of resolutions and recommendations of the IWC, discretion of the contracting parties under Art 8 of the ICRW, standard of review, reasonableness test, burden of proof, breach of a provision, reparation, cessation of unlawful acts. (iv) Impact and influence of the Judgment on the future behaviour of States: how to interpret the Judgment, impact of Judgment on the International Whaling Commission (IWC) and future scientific whaling.

---

1 *Whaling in the Antarctic* (*Australia v. Japan: New Zealand intervening*), Judgment of 31 March 2014. This judgment contained a jurisdictional aspect and a merit aspect altogether.

2 Website of the symposium is accessible at [http://www.edu.kobe-u.ac.jp/ilaw/en/whaling_sympo2014.html], which contains a list of case notes of the Whaling case.

 | DOI 10.1163/9789004313828_002

As is clear from the above, the Whaling Judgment contains a full set of legal issues of international law which should be discussed and analysed by academics and practitioners.

2 Background of the Case and Its Multilateral Nature

Whaling is a very controversial issue between the whaling countries (such as Japan) and the anti-whaling countries (such as Australia and New Zealand). Consequently, the *Whaling* case before the ICJ was, to some extent, a showcase of the views of both sides. It should be remembered, in this context, that whaling has been a multilateral issue, not confined to a bilateral dimension. In fact, whaling had been mainly discussed within the IWC, established under the ICRW. As is well known, the IWC failed to provide an effective and definitive solution in relation to Japanese whaling, since it has been very divisive on the whaling issue. Consequently, this issue, which is basically multilateral, has reduced, in fact, to a bilateral dispute between Australia and Japan, so that it could be submitted to the ICJ in conformity with the procedural requirements of the ICJ. However, the *Whaling* case, which is characterised by a multilateral aspect, cannot be entirely reduced to a simple bilateral dispute. The impact of the Judgment, therefore, has a multilateral rather than a bilateral influence, which is evident from discussions and debates within the IWC following the Judgment.

The bilateral aspect of the whaling issue is obviously a result of bringing it within the procedure of the ICJ, i.e. standing and legal interest of the Applicant. The ICJ, like other judicial bodies, is an institution for the settlement of bilateral disputes between sovereign States. This is undeniably clear from the adversarial nature of the procedure—i.e. the ICJ's procedure is basically designed as exchanges between the Applicant and the Respondent. This procedure is not suitable to solving multilateral issues, in which several States participate.[3] As will be examined in this book, the Whaling case, and the ICJ's Judgment, confirmed its previous position on the 'standing' of the Applicant State, shown in *Belgium v. Senegal*.[4] This then new approach seems to have been already established in the ICJ's jurisprudence; and this is why Marshal

3 In order to supplement the adversary nature of the ICJ procedure, the Court has invented some judicial notions and rules, e.g. intervention of third parties, Monetary Gold rules, and the doctrine of indispensable third parties.

4 *Questions relating to the Obligation to Prosecute or Extradite* (*Belgium v. Senegal*), Judgment of 20 July 2012, I.C.J. Reports 2012, p. 422.

Islands submitted applications against nine States alleging violations of Art 6 of the NPT[5] two months after the Judgement in the *Whaling* case.[6] Viewed from this point of view, the *Whaling* case can be evaluated as a cornerstone for admitting a wide range of standing where the applicable law contains obligations *erga omnes partes*. This admission of public interest litigation must have a huge impact on future cases before the ICJ, and on the litigation strategy of States. What is more, this trend has a significant impact in the sense that the ICJ's function can be extended from one of merely settlement of disputes, to one providing a monitoring mechanism for multilateral treaties, such as human rights treaties, environmental treaties, and disarmament treaties.[7] It is thus important to examine carefully the ICJ's pronouncements on this point.

3 Science and Law before the ICJ

The *Whaling* case raised another issue, i.e. how to discuss and evaluate the scientific results of Japanese whaling. The most important issue brought before the ICJ in this respect was whether the Japanese whaling program (JARPA II) could be seen as constituting 'scientific whaling' (scientific research or 'special permit whaling') or whether it constituted simply prohibited 'commercial whaling'. It was the role of the ICJ to examine whether JARPA II had actually contributed to scientific research on whales. The ICJ is composed of Judges[8] who as jurists know law, but are not scientists. Thus, we cannot expect from

5 *Obligations concerning Negotiations relating to Cessation of the Nuclear Arms Race and to Nuclear Disarmament.* Three cases are now in the general list of the ICJ, i.e. between Marshall Islands (Applicant) and India, Pakistan and the U.K. (Respondents).

6 On 24 April 2014, the Government of the Republic of the Marshall Islands simultaneously filed in the Registry of the Court separate Applications against nine States, i.e. (China, France, Russia, U.K., U.S.A., India, Israel, North Korea, Pakistan) accusing them of not fulfilling their obligations under Art 6 of the NPT. *See*, ICJ Press Release, No. 2014/18 (25 April 2014). Among the nine applications, three (against U.K., India and Pakistan) are entered into the general list of the ICJ and proceed to the procedure, since the three countries had declared the acceptance of optional clause of the ICJ Statute. Other applications, on the other hand, are not yet entered into the general list of the ICJ, since they are based on *forum prorogatum*. *See* Art 38 (5) of the Rules of the ICJ.

7 Those treaties should be regarded as examples which contain obligations *erga omnes partes*. So far, the ICJ has admitted the existence of those obligations in the Convention on the Prevention and Punishment of the Crime of Genocide and the Convention against Torture.

8 According to Art 2 of the ICJ Statute, the ICJ judges are of 'high moral character' who possess the 'the qualifications required in their respective countries for appointment to the highest judicial offices, or are jurisconsults of recognized competence in international law'.

Judges that they will fully examine and evaluate the scientific aspect of certain activities. As far as necessary, however, the Court is required to take a position towards the science. This kind of task is not unknown to the ICJ, which has been confronted with several environmental disputes, as well as other cases in which some special knowledge is needed. In the *Corfu Channel* case,[9] for example, special knowledge of valuation method was required to fix the amount of compensation. In many maritime delimitation cases and boundary cases, Judges are required to know the technology of large scale maps. Recently, the ICJ has tackled more directly environmental issues, in such cases as the *Gabčikovo-Nagyamros* case[10] and the *Pulp Mills* case,[11] in which the environmental impact evaluation was one of the questions raised before the Court.

The *Whaling* case was a good opportunity for the Court to clarify its position towards science in general. The litigating parties did have totally different viewpoints on the scientific results of Japanese whaling. However, on this point, the ICJ chose to reach its conclusions without answering a fundamental question—i.e. the Court did not touch directly upon the issue of what is 'scientific whaling', but rather whether Japanese whaling was 'for the purpose of scientific whaling'. It concluded that JARPA II cannot be regarded as 'for the purpose of scientific whaling'. A question may be posed; how can we regard a project as not 'scientific' without defining what *is* 'scientific'? The legal method used by the ICJ will be closely examined in several chapters of this book. It can safely be said here that the reasoning of the Judgment is quite difficult to understand. In other words, it contains a technical method of reasoning. In this regard, the concept of 'standard of review' has a fundamental impact and importance in the *Whaling* case Judgment and, consequently, requires a profound analysis. As is well known, the 'standard of review' has been adopted in the WTO-DSB and the investor-State dispute settlement (ISDS). But, the Court missed the opportunity to give an in-depth explanation why it applied and adopted the same criteria in the Whaling case. Thus, the *Whaling* case is important in the sense that it can influence the future role of the Court—i.e. whether it should exclusively rely on the discretionary interpretation of States or whether it should go further and apply objective criteria ('standard of review').

9 *Corfu Channel Case (Assessment of the Amount of Compensation Due from the People's Repbulic of Albania to the United Kingdom of Great Britain and Northern Ireland)*, Judgment of 15 December 1949, *I.C.J. Reports 1949*, p. 244.

10 *The Gabcikovo-Nagymaros Project (Hungary v. Slovakia)*, Judgment of 25 September 1997, *I.C.J. Reports 1997*, p. 7.

11 *Pulp Mills on the River of Uruguay (Argentine v. Uruguay)*, Judgment of 20 April 2010, *I.C.J. Reports 2010*, p. 14.

4 Reasoning Style of the Judgment

The substantive aspect of the Judgment indicates an important trend in the Court's reasoning and approach—i.e. approach based not on the civil law (continental law), but on the common law. This means that the French law tradition, or more generally the civil law tradition, has been losing its influence and persuasive power on the bench of the Court. In the *Whaling* case, this can be seen in the Court's reasoning, such as in relation to the burden of proof (burden to explain imposed on the Applicant), (see Judgment, para. 68) and the application of standard of review. This reasoning might have been completely different if the Court had adopted the civil law approach, an approach which was clearly adopted by two Judges, i.e. Judges Owada (Japan) and Abraham (France). Their arguments are based on the traditional civil law approach, e.g. the exercise of discretion, the abuse of discretion and the burden of proof imposed on the party which alleges an abuse of discretion. If the Court adopted the latter approach, it might have reached a different conclusion. It is thus possible to conclude that the discrepancy between the majority and the minority on the bench was based on the Judges' legal perception of international law. In this sense, the Judgement in the *Whaling* case should be closely analysed for a better understanding of a (possibly) new trend in the ICJ's reasoning.

5 Dispute Settlement by the ICJ

Both parties to the litigation, Australia and Japan, had a common expectation that the dispute between them could and should have been definitively and smoothly settled by the ICJ's judgment. This is shown by the fact that the Applicant (Australia) did not raised any claim regarding provisional measures and the Respondent (Japan) did not raised any preliminary objections. This means that both parties desired to enter the merit phase without any incidental procedure. This stance was based on both parties' expectations of winning the case. Notwithstanding whether their expectations were right or not, each party had a desire of settling the dispute by the ICJ's Judgment. As good citizens of the international community, fully committed to peaceful settlement of disputes, both parties have accepted the Judgment in the *Whaling* case without any complaint. After the Judgment, Australia, the winner, has not revelled in its victory and Japan, the loser, declared that it would accept the Judgment soon after its delivery. This case, it seems at first blush, is an ideal example of a definitive and friendly solution of a dispute between two countries by the ICJ's judgment. The *Whaling* case was between two developed countries which have established a good relationship between them. In addition, around the time of the Judgment, the two countries were looking to enhance their cooperative relationship, not only in the economic area (realised by the conclusion of an Economic Partnership Agreement between two countries just after the Judgment),[12] but also in the security area (as evidenced by the signature of a Defence treaty between the two countries).[13] It was the common intent of both parties not to make too much of an issue regarding the *Whaling* Judgment in their diplomatic relationship.

However, the reality is a bit different from this rosy picture. Japan, did at once stop its whaling (JARPA II) by revoking the special permit as was required by the Judgment; but declared that it would resume its whaling in the form of a new program, NEWREP-A. It is obvious that, while the ICJ's Judgment ordered Japan to revoke the special permit provided under the JARPA II, it did not say anything about other possible future whaling by Japan. It is also obvious that, once Japan modified its program of whaling, it could resume whaling which is permissible under Art 8 of the ICRW (scientific whaling). A question which arises here is whether NEWREP-A can be differentiated from JARPA II or not; and whether it does satisfy all the requirements indicated in the Judgment as to the 'standard of review'.

It is too early to give a more precise assessment of NEWREP-A. It is possible, however, to note some legal problems regarding this issue. Is there, at least potentially, a possibility of additional proceeding on the compliance of Japan with the Judgment? The ICJ Statute does not stipulate any particular procedure which allows parties to demand compliance with a Judgment, once rendered by the Court. In the history of the ICJ, however, some cases have been submitted to the Court asking to rule on the compliance with its Judgments.[14]

12 Japan-Australia Economic Partnership Agreement, signed on 8 July 2014. Text is available at [http://www.mofa.go.jp/files/000044322.pdf].

13 Agreement between the Government of Japan and the Government of Australia concerning the Transfer of Defence Equipment and Technology, signed on 8 July 2014. Text of it is available at [http://www.mofa.go.jp/files/000044447.pdf].

14 E.g., in *Haya de la Torre case* (*Colombia/Peru*) (Judgment of June 13th, 1951), the party asked the Court 'to determine the manner in which effect shall be given to the Judgment of November 20th, 1950', which had been rendered in *Asylum case* (*Colombia/Peru*). Nicaragua submitted an application against Colombia for alleging that the latter does not comply with the maritime delimitation indicated in the Judgment of 19 November 2012 in *Territorial and Maritime Dispute* (*Nicaragua v. Colombia*). *See*, Application Instituting Proceedings filed in the Registry of the Court on 26 November 2013, in *Alleged Violations*

If a similar action is taken in the *Whaling* case, then it will be apparent that there must be a significant problem with NEWREP-A, as not being much different from JARPA II.

Some of the answers to the questions raised in the Introduction will be found in the chapters of this book.

---

of Sovereign Rights and Maritime Spaces in the Caribbean Sea (*Nicaragua v. Colombia*). In this case, in addition, preliminary objections have been raised by the Republic of Colombia on 19 December 2014.

PART 1

The Law of Evidence and Standard of Review

CHAPTER 1

Methodologies and Motivations: Was Japan's Whaling Programme for Purposes of Scientific Research?*

Caroline E. Foster

This chapter studies the most novel yet pivotal aspect of the judgment by the International Court of Justice in *Whaling in the Antarctic* (*Australia v Japan*): the judicial methodologies applied to the legal issues relating to Japan's motivations for scientific permit whaling in the Antarctic.[1] In particular, the chapter focuses on how the Court hinged its judgment on the idea of a 'standard of review', which the Court considered to have been mutually supported by the disputing parties. This was a new decision-making technique for the Court.[2] Seen in a traditional light, this approach raises various tensions, not least in connection with how it moves away from the usual process of deciding cases based on an interpretation of the law and the application of rules about proof. An advantage of applying a 'standard of review' was that there was no need for the judgment to assess or confront directly the complex blend of reasons for Japan's whaling activities.

The Court's approach was in 'sync' to a degree with tentative steps in other contemporary adjudicatory *fora* involving the application of what are sometimes termed 'standards of review'. We can see this notably in the area of

* BA LLB (Cantuar) LLM PhD (Cantab) Associate Professor, Faculty of Law, University of Auckland, PO Box 92019, Auckland, New Zealand. Email:c.foster@auckland.ac.nz. Warm thanks to the organisers of the symposium for their kind invitation to participate. Dr. Caroline E. Foster is an Associate Professor at the University of Auckland, New Zealand. She gained her LLM and PhD from the University of Cambridge. She is also a graduate of the Academica Diplomática de Chile and holds a degree in French language and literature. Her book *Science and the Precautionary Principle in International Courts and Tribunals: Expert Evidence, Burden of Proof and Finality* (Cambridge University Press, 2011, reprinted 2013) was cited by Judges Simma and Al-Khasawneh in the *Case Concerning Pulp Mills on the River Uruguay*, and by Japan in *Whaling in the Antarctic*. Counter-Memorial of Japan, paragraph 9.16 and Annex 202, pages 351–358 (pages 1274–1276 of electronic document).

1 *Whaling in the Antarctic* (*Australia v Japan*), Judgment of 31 March 2014, [2014] ICJ Rep.

2 For background, see Chiara Ragni 'Standard of Review and the Margin of Appreciation before the International Court of Justice' in Lukasz Gruszczynski and Wouter Werner (Eds) *Deference in International Courts and Tribunals* (Oxford University Press, 2014) 319.

 | DOI 10.1163/9789004313828_003

investment treaty arbitration[3] and also to a degree in WTO dispute settlement.[4] In the medium term, it will be important to evaluate these new and experimental dynamics in the light of theoretical considerations, considering also comparisons and distinctions between the patterns appearing in these different *fora*. However there is not the scope here for that broader enquiry, and the present chapter's specific contribution is to analyse how the Court's use of the idea of a 'standard of review' interacted with proof of the central

3 For selected scholarly reflections regarding these developments, see the many contributions to Gruszczynski and Werner, above n 2; Caroline E. Foster, 'Diminished Ambitions? Public International Legal Authority in the Transnational Economic Era' (2014) 17(2) *Journal of International Economic Law* 355, 373–396; Valentina Vadi and Lukasz Gruszczynski 'Standard of Review in International Investment Law and Arbitration: Multilevel Governance and the Commonweal' (2013) 16(3) *Journal of International Economic Law* 613; Caroline Henckels 'Balancing Investment Protection and the Public Interest: The Role of the Standard of Review and the Importance of Deference in Investor-State Arbitration' (2013) 4(1) *Journal of International Dispute Settlement* 197; Caroline E. Foster 'Adjudication, Arbitration and the Turn to Public Law "Standards of Review": Putting the Precautionary Principle in the Crucible' (2012) 3(3) *Journal of International Dispute Settlement* 525; Stephan W Schill, 'Deference in Investment Treaty Arbitration: Re-conceptualizing the Standard of Review' (2012) 3(3) *Journal of International Dispute Settlement* 577; Rahim Moloo and Justin Jacinto 'Standards of Review and Reviewing Standards: Public Interest Regulation in International Investment Law', in Karl P Sauvant (ed), *Yearbook on International Investment Law and Policy* 2011–2012 (Oxford University Press, 2012); Caroline Henckels, 'Indirect Expropriation and the Right to Regulate: Revisiting Proportionality Analysis and the Standard of Review in Investor-State Arbitration' [2012] 15(1) *Journal of International Economic Law* 223; William W. Burke-White and Andreas von Staden, 'Private Litigation in a Public Law Sphere: The Standard of Review in Investor-State Arbitrations' [2010] 35 *Yale Journal of International Law* 283.

4 See, inter alia, Foster, 'Diminished Ambitions' above n 3; Tracey Epps, 'Recent Developments in WTO Jurisprudence: Has the Appellate Body Resolved the Issue of An Appropriate Standard of Review in SPS Cases?' (2012) 62(2) University of Toronto Law Journal 201; Michael Ming Du, 'Standard of Review under the SPS Agreement after EC-Hormones II' (2010) 59 International and Comparative Law Quarterly 441; Jan Bohanes and Nicolas Lockhart, 'Standard of Review in WTO Law' in Daniel Bethlehem and others (eds), The Oxford Handbook of International Trade Law (Oxford University Press, 2009) 378; Claus-Dieter Ehlermann and Nicolas Lockhart, 'Standard of Review in WTO Law', 7(3) *Journal of International Economic Law* (2004), 491; Catherine Button, The Power to Protect: Trade, Health and Uncertainty in the WTO (Hart Publishing 2004); Matthias Oesch, 'Standards of Review in WTO Dispute Resolution', (2003) 6(3) *Journal of International Economic Law* 635; Matthias Oesch, Standards of Review in WTO Dispute Resolution (Oxford University Press, 2003); Steven Croley and John H. Jackson, 'WTO Dispute Panel Deference to National Government Decisions: The Misplaced Analogy to the US Chevron Standard-of-Review Doctrine', in Ernst-Ulrich Petersmann (Ed.), International Trade Law and the GATT/WTO Dispute Settlement System (Kluwer, 1997), 187.

question whether or not Japan's scientific whaling programme in the Antarctic was 'for purposes of scientific research'.

The dispute before the Court focused specifically on the research programme operated by Japan from 1995 to 2014, known as 'JARPA II'. This was the second Japanese Whale Research Programme under Special Permit in the Antarctic, following a previous programme operating from 1987 to 1994 known as JARPA. The question was whether JARPA II fell within the applicable provision of the International Convention for the Regulation of Whaling 1946, Article VIII (1):

> ...Notwithstanding anything contained in this Convention any Contracting Government may grant to any of its nationals a special permit authorizing that national to kill, take and treat whales *for purposes of scientific research* subject to such restrictions as to number and subject to such other conditions as the Contracting Government thinks fit, and the killing, taking, and treating of whales in accordance with the provisions of this Article shall be exempt from the operation of this Convention... (emphasis added).

Looking at this provision, there was the possibility that the Court might determine whether JARPA II constituted 'scientific research' under Article VIII of the Convention. In the event, the Court did not find it necessary to define 'scientific research'.[5] The parties' experts had agreed that lethal methods could have a place in scientific research,[6] and the Court found that the JARPA II activities involving the lethal sampling of whales could broadly be characterised as 'scientific research'.[7] The Court also said it would not pass judgment on the scientific merit or importance of the stated research objectives for JARPA II.[8] Instead, the Court focused its decision on whether JARPA II was 'for purposes of' scientific research.

We tend perhaps to overlook that scientific disputes are not exclusively about questions that we can expect to be illuminated by science. They are often

5 See Judgment, paragraphs 127 and 172.

6 Judgment, paragraph 127, while distinguishing the experts' conclusions as scientists from the interpretation of the Convention, which was the task of the Court. Judgment, paragraph 82; and see also paragraph 131.

7 Indeed, in the Court's view even if there were non-lethal alternatives this would not necessarily mean that the grant of a special permit would fall outside the licence in Article VIII. Judgment, paragraph 137.

8 Judgment, paragraphs 88, 172.

also about the reasons why a State has undertaken or adopted a disputed activity, measure or programme. There are many kinds of scientific dispute where motivations are at issue.[9] We have seen cases concerning the adoption of import barriers to protect human animal and plant life and health,[10] export restrictions for environmental purposes,[11] environmental measures affecting foreign investments,[12] failure to complete the construction of large-scale projects on ecological grounds,[13] and the protection of fish stocks and marine life.[14] The Philip Morris investment treaty arbitration against Australia is a further contemporary example, with Philip Morris contending that plain packaging for tobacco products will actually undermine rather than support Australia's stated public health rationale, that it lacks demonstrable utility in relation to public health and that it is not, at the end of the day, 'for a proven public purpose related to the internal needs of Australia'.[15]

A State's reasons for pursuing an activity, measure or programme can be multiple, and they may certainly be difficult to establish directly, conclusively

---

9 Foster, 'Adjudication, Arbitration', above n 3.

10 For instance, consider, as referred to below, *European Communities: Measures Concerning Meat and Meat Products (Hormones)*, (WT/DS26) (WT/DS48), Report of the Appellate Body adopted 13 February 1998; *Australia-Measures Affecting Importation of Salmon*, Complaint by Canada (WT/DS18), Reports of the Panel and Appellate Body; *Australia-Measures Affecting the Importation of Apples from New Zealand* (WT/DS367), Reports of the Panel and Appellate Body adopted on 17 December 2010.

11 *China-Measures Related to the Exportation of Various Raw Materials*, (WT/DS394/)(WT/DS395)(WT/DS398), Reports of the Panel and Appellate Body adopted 22 February 2012; *China-Measures Related to the Exportation of Rare Earths, Tungsten and Molybdenum* (WT/DS431).

12 For example, *Methanex v US*, Final Award on Jurisdiction and Merits, 3 August 2005; *Chemtura Corporation v Canada*, Award of 2 August 2010.

13 *Case concerning the Gabčikovo-Nagymaros Project* (*Hungary/Slovakia*) [1997] ICJ Rep 7.

14 *Southern Bluefin Tuna Cases* (*New Zealand v. Japan; Australia v. Japan*) (Provisional Measures), Order of 27 August 1999, 38 ILM 1624. Consider also Reports of the Panel and Appellate Body in *EC-Measures Prohibiting the Importation and Marketing of Seal Products*, (WT/DS400 R) (WT/DS401/R) dated 25 November 2013, in which the Panel found that 'the objective of the EU Seal Regime is "to address the moral concerns of the EU public with regard to the welfare of seals"', paragraph 7.410.

15 *Philip Morris Asia Ltd v The Commonwealth of Australia*, Notice of Claim, 22 June 2011, paragraph 7.4. Australia claims that the plain packaging scheme is simply not expropriatory, based on the most modern conception of expropriation, according to which properly taken measures for the protection of human health and the environment that do not run counter to specific assurances will not constitute expropriation. See also Foster, 'Diminished Ambitions?' above n 3.

and objectively. As the Court observed in the *Whaling* case: 'a State often seeks to accomplish more than one goal when it pursues a particular policy…'.[16] As the Court took into account, individual government officials may have motivations that go beyond scientific research, but this will not 'preclude a conclusion that a programme is for purposes of scientific research within the meaning of Article VIII'.[17] Scientific disputes can therefore put international adjudicatory bodies in a difficult position. In *Whaling in the Antarctic* the Court's application of methodology revolving around 'standard of review' changed the way in which the Court might otherwise have approached the legal question of whether Japan's whaling programme was 'for purposes of scientific research' in terms of Japan's motivations and thus within the scope of Article VIII.

This chapter will be divided into two parts. Part I will deal with the cluster of methodologies adopted by the Court in its decision-making process, including employment of a standard of review, the Court's light touch approach to applying the traditional rules on burden of proof, and the Court's reliance on inference in finding that JARPA II was not for purposes of scientific research. Part II will make observations on how the Court's chosen methodologies enabled it to deal with the question of Japan's motivations in conducting scientific research whaling. These include observations on how employment of a standard of review enabled the Court to make an indirect assessment of Japan's motivations, and leave intact the presumption of good faith on the part of Japan.

Part 1 Methodologies

This part of this chapter considers the methodologies adopted by the Court in *Whaling in the Antarctic*, including (A) employment of a standard of review, (B) the Court's apparent non-application of the traditional rules on burden of proof, and (C) the Court's reliance on inference in finding that JARPA II was not for purposes of scientific research.

16 Judgment, paragraph 97. See also Report of the Appellate Body, *US-Measures affecting the Production and Sale of Clove Cigarettes*, WT/DS406/Appellate Body/R, adopted 24 April 2012: 'We further observe that measures often pursue a multiplicity of objectives, which are not always easily discernible from the text or even from the design, architecture, and structure of the measure'. Paragraph 113. In the end, the Appellate Body decided that determinations of products' 'likeness' should not be based on the regulatory purposes of technical regulations, although regulatory concerns would play a role. Paragraph 117.

17 Judgment, paragraph 97.

A Standard of Review

The idea of applying a 'standard of review' is novel in the International Court of Justice. The notion of 'standard of review' as a judicial decision-making tool in international courts and tribunals has taken on some currency in investment treaty arbitration, especially in the literature, and also to a degree in the World Trade Organisation (WTO). Applying a 'standard of review' facilitiates determinations on the legality of States' actions that do not directly call into question their motivations. The Court's adoption of a methodology based on standard of review in the *Whaling* case was specific to the case. The parties clearly envisaged that the Court might use such a methodology, and the Court was able to point to an articulation of the terms of a standard of review that it considered the parties had agreed would be applicable under Article VIII of the ICRC. However it is notable that the Court employed the terminology of standard of review as though it were already an established technique. This is perhaps evidence that the Court considers itself as belonging to the wider collective group of international courts and tribunals, within which the idea of standard of review has become an increasingly widely discussed concept.

The standard of review applied by the Court in the *Whaling* case as the test to address whether JARPA II was 'for purposes of' scientific research in accordance with Article VIII was whether 'in the use of lethal methods, the programme's design and implementation are reasonable in relation to achieving its stated objectives',[18] based on the evidence.[19] The Court found that the use of lethal sampling was not *per se* unreasonable in relation to JARPA II's research objectives.[20] However this standard of review did require the design and implementation of Japan's research programme to be related objectively to its identified scientific foundations.

The standard of review adopted in the *Whaling* case evokes resonances of the 'standard of review' articulated by the Appellate Body of the World Trade Organisation in relation to Article 5.1 of the WTO Agreement on Sanitary and Phytosanitary Measures in the cases *Canada-Continued Suspension of Obligations in the EC-Hormones Dispute, US-Continued Suspension of Obligations in the EC-Hormones Dispute* and *Australia: Measures Affecting the Importation of Apples from New Zealand.*[21] The Agreement on Sanitary and Phytosanitary

18 Judgment, paragraph 67.

19 Judgment, paragraph 98.

20 Judgment, paragraph 224.

21 Report of the Appellate Body, *Canada-Continued Suspension of Obligations in the EC-Hormones Dispute* (WT/DS320), and *US-Continued Suspension of Obligations in the EC-Hormones Dispute* (WT/DS321), paragraph 590, adopted on 14 November 1998. Report of the Appellate Body,

Measures is an agreement intended to help ensure that trade-inhibiting sanitary and phytosanitary measures, i.e. measures to protect human and animal plant life and health, are adopted only when necessary and not for protectionist purposes.[22] The Agreement's Article 5.1 requires that WTO Members' sanitary and phytosanitary measures be based on a risk assessment, which the jurisprudence has interpreted as requiring a rational or objective relationship between an SPS measure and a supporting risk assessment.[23] The Appellate Body has in recent years adopted the idea that Article 5.1 may be approached through the application of a standard of review, namely whether a measure is 'supported by coherent reasoning and respectable scientific evidence and is, in this sense, objectively justifiable'.[24] Whether this is truly the application of a 'standard of review' or more a matter of an advanced interpretation and application of Article 5.1 has to be questioned, but the Appellate Body has at least temporarily applied the language of standard of review.[25]

Will a court or tribunal still assess the science when it takes an approach like this? Practice in the WTO is illuminating. Although a WTO Panel will not review a Member's scientific work *de novo*,[26] assessing compliance with Article 5.1 has required adjudicators to look into how the science has been employed by a responding party.[27] As the Appellate Body has said, 'it is not possible to review

Australia-Measures Affecting the Importation of Apples from New Zealand (WT/DS367), adopted on 17 December 2010.

22 Agreement on Sanitary and Phytosanitary Measures (Marrakesh, 15 April 1994).

23 See Report of the Appellate Body, *European Communities: Measures Concerning Meat and Meat Products (Hormones)*, WT/DS26/AB/R; WT/DS48/AB/R, adopted on 13 February 1998, paragraph 193. Judge Keith agreed in general with the test as stated by the Court, but formulated it rather as: 'is the contracting Government's decision to award a special permit objectively justifiable in the sense that the decision is supported by coherent scientific reasoning?' Declaration of Judge Keith, paragraph 8.

24 Report of the Appellate Body in *Canada: Continued Suspension of Obligations in the EC-Hormones Dispute*, above n 21, paragraph 591; and *United States: Continued Suspension of Obligations in the EC-Hormones Dispute*, above n 21, paragraph 591; Report of the Appellate Body, *Australia: Measures Affecting the Importation of Apples from New Zealand*, above n 21, paragraph 213.

25 Caroline E. Foster 'International Adjudication: Standard of Review and Burden of Proof: Australia-Apples and Whaling in the Antarctic' (2012) 21(2) *Review of European Community and International Environmental Law* 80.

26 Report of the Appellate Body, *European Communities: Measures Concerning Meat and Meat Products (Hormones)*, above n 23.

27 In *Australia-Apples* the Appellate Body explained that there were two aspects to the standard of review under Article 5.1: 'scrutiny of the underlying scientific basis of the risk

the ultimate conclusions reached by the risk assessor in isolation from the reasoning and the intermediate conclusions that lead up to them'.[28] Indeed, in the WTO, especially under the SPS Agreement, the rules have been set up precisely so that by virtue of evaluating scientific dimensions of a Member's disputed measure we will seldom have to base legal findings on the direct evaluation of motivations. If trade disputes had frequently to be resolved on the basis of actually establishing protectionist motivations this could make WTO dispute settlement very difficult.

The likelihood that the Court would contemplate employing the notion of standard of review was foreseen by the parties to the *Whaling* case, particularly Japan, and it was at the very least partly in response to pleadings that the Court did so.[29] The Court considered that the parties had also identified common ground in relation to the articulation of an appropriate standard of review. The Court observed that Counsel for Japan had said, near the close of oral proceedings, that 'Japan agrees with Australia and NZ in regarding the test as being whether a State's decision is objectively reasonable, or 'supported by coherent reasoning and respectable scientific evidence and…, in this sense, objectively justifiable''.[30] The Court applied the standard of review to the question of whether the Japanese research programme was 'for purposes of' scientific research, i.e. the 'for purposes of' requirement.[31] The Court did this by considering the reasonableness of 7 elements, drawn by the Court from the parties' pleadings, of the programme's design and implementation in relation to its

assessment and scrutiny of the reasoning of the risk assessor based upon such underlying science'. Ibid, paragraph 215; see also paragraph 220.

28 Ibid, paragraph 226. As the Appellate Body said in *EC-Hormones*, total deference to national authorities could not ensure an objective assessment of the facts of a dispute. Report of the Appellate Body, *European Communities: Measures Concerning Meat and Meat Products* (*Hormones*), WT/DS26/AB/R; WT/DS48/AB/R, adopted on 13 February 1998, paragraphs 113, 114.

29 Written Observations of Japan on New Zealand's Written Observations, pages 27–31. See also Professor Lowe for Japan, CR 2013/15 17–20, paras 15–28; CR 2013/22 55–61, paragraphs 1–22. For Australia, see the submissions by Professsor Crawford, CR 2013/19 64–66, paras 17–24.

30 Professor Lowe for Japan, CR 2013/22 60, paragraph 21; Judgment, paragraph 66. Indeed, Japan had consistently emphasised the idea that the reasonableness of its actions was important.

31 Judgment, paragraphs 67, 127. For a critique of this step in the Court's reasoning see the Dissenting Opinion of Judge Owada at paragraph 35, quoting Counsel for Japan as indicating this test would apply as a way to delineate the Court's depth of enquiry into whether JARPA II was to be regarded as 'scientific research', rather than going to an assessment of whether JARPA II was 'for purposes of' scientific research.

stated objectives. The 7 elements were: Japan's decisions regarding the use of lethal methods; the scale of use of lethal sampling; the methodology used to select sample sizes; comparison of sample size to actual take; the programme's timeframe; the programme's scientific output; and the degree of coordination with related research projects.[32]

Standard of review offered practical benefits to the Court in exercising its adjudicatory function in the *Whaling* case. Introduction of a standard of review enabled the Court to decide the case through the medium of an objective process based on an expectation that an Article VIII research programme's design and implementation will be reasonable in relation to achieving its stated objectives. Reliance on purely interpretive methods might not have enabled the Court as easily to have carried out its task. Yet there was also within the Court a push back against the adoption of a standard of review. Judge Yusuf did not accept the Court's recourse in the *Whaling* case to an 'extraneous and undefined standard of review', which he was not persuaded was grounded in law and found to be 'obscure and debateable'.[33] The Judge favoured greater emphasis on the interpretation and application of terms of Article VIII of the Convention,[34] in the light of the increased emphasis on conservation in the International Whaling Commission since the adoption of the Convention in 1946.[35]

The idea that international courts and tribunals may engage in 'review' rather than 'adjudication' as we know it requires serious consideration and discussion. Will this change the nature of the adjudicatory process? We have witnessed in this case a move away from traditional approaches to proof. One way to view the Court's approach in this case is to draw the conclusion that when the Court applies a standard of review, the adjudicatory process is more a matter of assessing whether a State's conduct meets the 'test' set out in the standard of review, and less a matter of assessing whether requisite matters are 'proven' as such.

32 Judgment, paragraph 88.

33 Dissenting Opinion of Judge Yusuf, paras 3, 12, 15, 24, 28.

34 Ibid, paras 3, 11, 12, 28.

35 Ibid, paras 7, 26, 46–47, 60. Judge Xue also referred to interpretive methodology, placing an accent on the influence of the conservation efforts of the parties to the Convention. Separate Opinion of Judge Xue, paragraphs 12, 15, 28, although the Judge appeared also to accept the use of a standard of review in relation to Japan's decision to issue permits, see above. Judge Sebutinde took issue with the derivation of the standard of review, considering the Court should have elaborated the standard of review with reference to paragraph 30 of the Schedule, and the relevant IWC Guidelines. Separate Opinion of Judge Sebutinde, paragraph 6.

B A Light Touch Approach to Application of the Usual Rules on the Burden of Proof

In general, the allocation and discharge of the burden of proof are directly referred to only infrequently by international courts and tribunals. Established rules or practices there are, however.[36] In international law the burden of proof lies traditionally with the applicant, or the party invoking a particular legal rule or provision.[37] This practice is seen widely as the basis for the allocation of the burden of proof also in national legal systems. Its roots lie in the Roman law civil trial in which *ei incumbit probatio qui dicit non qui negat*, commonly expressed in the formula *actori incumbit probatio*. This will mean the burden of proof usually lies with the applicant. However a party relying on an exception or defence will became the *actor*, and assume the burden of proof in relation to the exception: *reus in exceptione fit actor*.[38] Either party may discharge its burden through evidence, and even potentially by resting its case successfully on a presumption in its favour or, occasionally, by benefiting from judicial inferences.[39]

A traditional application of the rules on burden of proof in the *Whaling* case would have placed the burden of proof on Australia, as applicant or *actor*. Australia would have had to establish the applicability of the legal provisions with which it asserted Japan to be out of compliance, and prove all the necessary associated facts. However, as the parties were aware, there was the possibility that the Court might regard Article VIII(1) as an exception, and thereby allocate the burden of proof to Japan under this provision.[40] As it happened, the Court made no finding on whether Article VIII (1) was an exception.[41] Possibly it was thought that this restricted methodology would less effectively

36 C. Brown, *A Common Law of International Adjudication* (Oxford University Press, 2007), 92.

37 Caroline E. Foster, 'Burden of Proof in International Courts and Tribunals', (2008) 29 *Australian Year Book of International Law* 27, emphasizing that the burden's allocation will generally pivot on the assertion of a particular legal rule or provision, although there have been occasions where international courts have found that it may also pivot on the assertion of a fact.

38 See Caroline E. Foster, *Science and the Precautionary Principle in International Courts and Tribunals: Expert Evidence, Burden of Proof and Finality* (Cambridge University Press, 2011), 185–186 and 210, including for further references. Also as noted in the Judgment, paragraph 52.

39 See below.

40 Australia viewed Article VIII (1) as an exception. Japan argued that Article VIII (1) was rather a free standing provision, an exempting provision.

41 Although Judge Cançado Trindade did consider Article VIII (1) to be an exception to the Convention and stated that the Party relying on this provision needed to demonstrate

communicate the Court's appreciation of the overall problems associated with JARPA II. Possibly, even, Article VIII (1) might not in fact have qualified as an exception. Article VIII(1) is after all phrased as a savings clause, although Article XX of the WTO General Agreement on Tariffs and Trade, which has long been regarded as an exception that attracts the burden of proof, is also phrased as a savings clause.[42]

There was also the distinct possibility in this case that the Court might allocate the burden of proof in the *Whaling* case with specific reference to the precautionary principle, although this was not expressly advocated by Australia or New Zealand.[43] The essence of the precautionary principle is that scientific uncertainty ought not be used as a reason to delay action to address potentially serious harm.[44] Consistent with this principle, the due administration of justice could be considered to require the Court to exercise its inherent powers and reverse the burden of proof. This would help ensure fairness between disputants, rather than perpetuating the disadvantage carried by the party with the burden of proof in a situation of scientific uncertainty.[45] Fairness underpins the concept of the burden of proof, and the burden of proof should

---

that its whaling programme was 'for purposes of scientific research'. Separate Opinion of Judge Cançado Trindade, paragraphs 21 and 22.

42 Stating that 'Nothing in this Agreement shall be construed…' Foster, above n 25, 87. For discussion on why Article VIII might not qualify as an exception while Article XX does so qualify, ibid, 89. See also Caroline E. Foster, 'The Right to Regulate: an Inherent Power enjoyed by a Government?', Guest Editorial, *European Society of International Law*, Quarterly Newsletter, December 2014. Available at <www.esil-sedi.eu/node/92>

43 See oral submissions of Professor Crawford for Australia, CR 2013/18 39 paragraph 1(4). However both Australia and Japan, as well as New Zealand, addressed the precautionary approach more broadly in their submissions. Judge Ad Hoc Charlesworth and Judge Cançado Trindade drew attention to the principle's relevance for the Convention's interpretation. Separate Opinion of Judge *Ad Hoc* Charlesworth, paragraphs 6–8; Separate Opinion of Judge Cançado Trindade, paragraphs 57 and 70, see also paragraphs 23, 54, 60–69. Judge Cançado Trindade observed also that expert writing increasingly examines the precautionary principle. Ibid, paragraph 71.

44 For a commonly accepted reference point, see Declaration of the United Nations Conference on Environment and Development 1992 (the Rio Declaration) UN Doc. A/CONF.151/26/Rev.1, Principle 15: 'In order to protect the environment, the precautionary approach shall be widely applied by States according to their capabilities. Where there are threats of serious or irreversible damage, lack of full scientific certainty shall not be used as a reason for postponing cost-effective measures to prevent environmental degradation'.

45 For the technical methods by which the Court might formally have done so, see Foster, above n 38, 272–280.

not work against fairness in circumstances of scientific uncertainty.[46] These considerations could have led the Court expressly to allocate to Japan the burden of demonstrating that JARPA II was indeed 'for purposes of scientific research'. Japan appears to have contemplated that the Court might do so, and made efforts in written and oral pleadings to arguments to dissuade the Court.[47]

Importantly, we must take into account that this possibility still remains open to the Court in future cases in the wake of the Court's 2010 judgment in the *Case Concerning Pulp Mills*. The Court's reasoning on this issue in that case appears to have been based only on the language of the Statute of the River Uruguay, as invoked by Argentina.[48] The *Pulp Mills* judgment did not address broader international legal arguments for the reversal of the burden consistent with the precautionary principle. Comments in the Separate Opinions of Judge *Ad Hoc* Charlesworth and Judge Cançado Trindade in the *Whaling* case offer a bridge to the future for the potential reversal of the burden in subsequent cases.[49]

In the event, the Court adopted neither of the two options described above. Rather, the Court's application of a standard of review in the *Whaling* case seems to have been connected with a relatively informal and malleable approach to allocation of the burden of proof. The Court did not refer to the burden. Yet the Court seemed to look to Japan to assist it with explanations on the history and design of JARPA II.[50] How is this to be explained? We could see this in the light of parties' general obligations to cooperate with and assist international courts and tribunals on evidentiary matters.[51] However the Court itself explained that it was turning to Japan for an explanation of the basis of

46 Ibid, 189–193.

47 Written Observations of Japan on New Zealand's Written Observations, pages 22–15, and, before the Court, Professor Boyle for Japan, CR 2013/16 29–37.

48 *Case Concerning Pulp Mills on the River Uruguay*, Judgment of 20 April 2010, [2010] ICJ Reports, paragraphs 160–164.

49 Stating that 'In relation to Article VIII...an implication of the precautionary approach is that lethal methods must be shown to be indispensable to the purposes of scientific research'. Separate Opinion of Judge *Ad Hoc* Charlesworth, paragraph 10. Judge Cançado Trindade expressed concern about the uncertainties surrounding the very activity of scientific research, and wondered, 'As to the *cas d'espèce*, would this mean that whales would keep on being killed, and increasingly so, for "scientific purposes" and amidst scientific uncertainty? I do not think so...'. Separate Opinion of Judge Cançado Trindade, paragraph 74.

50 Judgment, paragraphs 68, 141, 144.

51 Foster, above n 38, 187. We might additionally observe that placing the burden on Australia required proof of a negative proposition, and in circumstances where it was Japan who had the best access to all relevant evidentiary material. See, ibid, 206 and 207–208.

the decision to grant scientific research permits because it was a Japanese decision under Article VIII (1) that was at issue.[52] This could be considered to reflect rather the idea that the Court may have been inclined to frame its role in this case as being to 'review' Japan's decision, rather than to 'adjudicate' the question of Japan's compliance with the terms of the ICRW, and did in fact consider this to involve stepping away from the usual rules on burden of proof.

It was almost as though adoption of a standard of review had reversed the burden of proof.[53] At the pivotal point in the Court's reasoning, it was held that 'the evidence *does not establish* 'that JARPA II's design and implementation were reasonable in relation to achieving its stated objectives''.[54] At numerous points in the part of its judgment dealing with Article VIII (1) the Court inferred in various ways that it was incumbent on Japan to provide it with evidence.[55] There are ways to view these expectations of Japan consistently with established approaches to the rules on burden of proof, but the Court made no direct reference to such approaches – neither to the idea that Australia might have successfully established a *prima facie* case requiring to be rebutted by Japan, nor that the evidentiary burden could have swung to Japan even though the true burden of proof remained with Australia.[56]

---

52 'Inherent in such a decision is the determination by the State party that the programme's use of lethal methods is for purposes of scientific research. It follows that the Court will look to the authorizing State, which has granted special permits, to explain the objective basis for its decision'. Judgment, paragraph 68.

53 Indeed, Judge Owada considered that the Judgment 'appears to be applying the standard of objective reasonableness in such a way that it is the granting Party that bears the burden of establishing that the scale and the size of the lethal take envisaged under the programme is reasonable in order for the programme to be qualified as a genuine programme "for purposes of scientific research"'. Dissenting Opinion of Judge Owada, paragraph 43. The Judge believed the burden should lie instead with the Applicant. Ibid, paragraphs 44–45. See also Dissenting Opinion of Judge Abraham, paragraph 31: 'La Cour se montre en permanence particulièrement exigeante à l'égard du Japon, comme si c'était le défendeur qui devait prouver son bon droit'.

54 Judgment, paragraph 227.

55 See Judgment at paragraphs 137, 141, 144, 185, 193, 194, 206, 222, 226.

56 On the common law distintion between the evidential burden and the true burden, and for further references, see Foster, above n 38, 205. Cf Judge Yusuf appeared to allocate the burden to Australia in relation to the purposes for which permits had been issued. Dissenting Opinion of Judge Yusuf, paragraph 52. Judge Xue set out clearly the view that 'It is up to Australia to prove with convincing evidence' that Japan's activities did not involve scientific research, based on the well-established principle that *onus probandi incumbit actori*, with Japan having the right to rebuttal. Separate Opinion of Judge Xue, paragraph 15. However in relation to the broader matter of Japan's decision to issue

C Reliance on Inference

Employment of a decision-making methodology based on standard of review in this case does though appear to have involved the Court's reliance on *inference* in determining that the Japanese programme was not 'for purposes of' scientific research. Inference has been seen at work previously in relation to questions about States' motivations in World Trade Organisation dispute settlement in scientific cases. In *Australia – Measures Affecting Importation of Salmon* Canada successfully established an unrebutted *prima facie* case that distinctions in levels of protection made by Australia resulted in 'discrimination or a disguised restriction on international trade' under Article 5.5 of the SPS Agreement. The Panel reached this conclusion on the basis of three 'warning signals'[57] and three other 'factors' derived from the architecture and structure of the Australian measures.[58] Contrastingly, in *EC-Hormones*, the Appellate Body

---

permits under Article VIII (1) the Judge seemed at least to tolerate the idea of applying a standard of review. Indeed, the Judge also made various allusions to the need for Japan to have proven a point (paragraphs 19, 21, 26). One possibility is that we could read these as referring simply to Japan's task of rebutting the Australian case (as indicated in paragraph 27 of the Judge's Separate Opinion).

57 Report of the Panel, *Australia – Measures Affecting Importation of Salmon*, Complaint by Canada (WT/DS18). The three 'warning signals' were as follows. The first was the arbitrary distinction in Australia's levels of sanitary protection in relation to products presenting the same risk, prohibiting salmon products but tolerating imports of herring for bait and controlled import of live ornamental finfish. Ibid, paragraph 8.149. The second was that the arbitrary difference was a rather substantial difference. Ibid, paragraph 8.150. The Appellate Body has said that where there is an arbitrary or unjustifiable difference in a WTO member's level of protetion against different risks this may operate as a warning signal that the measure might be being applied in a manner that is discriminatory or a disguised restriction on international trade. Report of the Appellate Body, *European Communities: Measures Concerning Meat and Meat Products* (*Hormones*), above n 23, paragraph 215. The third 'warning signal' in the *Salmon* case was that the ban on salmon products was inconsistent with Articles 5.1 and 2.2 of the SPS Agreement – it was not based on a risk assessment and was maintained without sufficient scientific evidence. Ibid, paragraph 8.151. The Appellate Body accepted the Panel's reliance on all three 'warning signals'. Report of the Appellate Body, *Australia – Measures Affecting Importation of Salmon*, above n 11, paragraphs 159–166.

58 Report of the Panel, *Australia – Measures Affecting Importation of Salmon*, above n 11, paragraph 8.159. The first factor was Australia's application of substantially different implementing measures to products presenting the same risk, prohibiting salmon products but tolerating imports of herring for bait and controlled import of live ornamental finfish. Ibid, paragraph 8.153. The second factor consisted of the difference between the conclusions of Australia's Draft Report on the risks of imported salmon, which would have allowed entry of the product under specified conditions and its Final Report which

overruled the Panel's finding of 'discrimination or a disguised restriction on international trade' under Article 5.5, rejecting as an error of law the importance attached by the Panel to the supposed multiple objectives of the European Communities.[59] The Appellate Body was unable to share the inference apparently drawn by the Panel that the ban on growth promotion hormones was not in fact designed to protect the population from cancer.[60] The Panel's finding was not supported either by the architecture and structure of the EC Directives or by the evidence.[61]

Inference has also been seen at work in investment treaty arbitration in relation to questions connected with States' motivations, as in *Methanex v US*, a NAFTA investment dispute defended by the United States in relation to a Californian ban on the gasoline additive methyl tert-butyl ether (MTBE). To establish jurisdiction, the Tribunal in that case had to decide whether the measures in question 'related to' Methanex. Methanex argued that they did relate to Methanex, and questioned their alleged scientific basis. The Tribunal handled the issue by saying the question was whether the Californian scientific conclusions 'were so faulty that the Tribunal could reasonably infer that the science merely provided a convenient excuse for the hidden regulation of Methanol producers'.[62] The Tribunal found that the ban did not 'relate to' Methanex, and the Tribunal therefore lacked jurisdiction under Article 1101(1) of NAFTA, as well as finding each of Methanex's allegations against the US unfounded.[63] The arbitral tribunal considered that the ban was motivated by

recommended a prohition on the relevant products. Ibid, paragraph 8.154. The third factor was a doubt that Australia applied similarly strict standards to the internal movement of salmon products and in its internal monitoring and prevention systems for fish diseases. Ibid, paragraph 8.155–8.158. The Appellate Body endorsed the Panel's reliance on two of the three factors, excluding only the first factor, on the basis it was essentially the same as the first warning signal. Report of the Appellate Body, *Australia – Measures Affecting Importation of Salmon*, Complaint by Canada (WT/DS18), paragraph 8.151.

59 Report of the Appellate Body, *European Communities: Measures Concerning Meat and Meat Products (Hormones)*, above n 23, paragraph 245.

60 Idem. See also Caroline E. Foster, 'Public Opinion and the Interpretation of the World Trade Organisation's Agreement on Sanitary and Phytosanitary Measures' (2008) 11(2) *Journal of International Economic Law* 427, 447.

61 Ibid, paragraph 246.

62 *Methanex v US*, Final Award on Jurisdiction and Merits, 3 August 2005, Part IV Ch E, paragraph 19.

63 E.g. The Tribunal concluded that the US measure in that case, was 'made for a public purpose, was non-discriminatory and was accompanied with due process...the California ban was a lawful regulation and not an expropriation'. *Methanex v US*, Final Award on Jurisdiction and Merits, 3 August 2005 pt IV, ch D, paragraph 15. See also paragraph 7.

an 'honest belief, held in good faith and on reasonable scientific grounds, that MTBE contaminated groundwater and was difficult and expensive to clean up'.[64]

Inference appears to have been determinative in the Court's consideration of the 7 elements of the JARPA II's design and implementation. In the course of its analysis of these 7 elements, the Court identified a range of difficulties, doubts, concerns, and less than appropriate characteristics associated with the design and implementation of Japan's research programme which subsequently came together to form the foundation of the Court's finding that Japan's research programme was not 'for purposes of' scientific research.

Before reviewing these 7 elements, it is helpful to recall Japan's stated research objectives. The research objectives for JARPA II as identified in its Research Plan were fourfold. They included monitoring of the Antarctic ecosystem, modelling competition among whale species, elucidation of temporal and spatial changes in stock structure, and improving the management procedure for Antarctic minke whale stocks.[65] The Research Plan set annual sample sizes at 50 fin whales, 50 humpback whales and 850 Antarctic minke whales plus or minus 10%.[66] Japan considered lethal methods indispensable for the programme,[67] which required data from internal organs, including ear plugs (to determine specimens' age) and stomach contents (to determine eating habits), as well as measurements of blubber thickness to gauge changes in the conditions of prey.[68]

The first element of JARPA II's design and implementation considered by the Court was Japan's decision-making regarding the *use of lethal methods*. The Court considered that the JARPA II Research Plan should have included 'analysis of the feasibility of non-lethal methods as a means of reducing the planned

64 Ibid, Part III, ch A, paragraph 102(2).

65 Judgment, paragraph 113. See Government of Japan, 'Plan for the Second Phase of the Japanese Whale Research Program under Special Permit in the Antarctic (JARPA II) – Monitoring of the Antarctic Ecosystem and Development of New Management Objectives for Whale Resources', SC/57/O1, http://www.icrwhale.org/pdf/SC57O1.pdf See also Counter-Memorial of Japan, 143, listing: Objective 1 (Estimation of the biological parameters required for the stock management of the Southern Hemisphere minke whale); Objective 2 (Elucidation of the role of whales in the Antarctic marine ecosystem); Objective 3 (Elucidation of the effect of environmental change on cetaceans); Objective 4 (Elucidation of the stock structure of Southern Hemisphere minke whales to improve stock management).

66 Judgment, paragraph 123.

67 Judgment, paragraph 122.

68 Judgment, paragraphs 129, 133, 162.

scale of lethal sampling'.[69] There was little evidence of this, which the Court found 'difficult to reconcile'[70] with Japan's obligation to give due regard to recommendations in IWC Resolutions and Guidelines and Japan's statement that it used lethal methods only to the extent necessary to meet its scientific objectives.[71] The Court did not say so expressly, but rather appears to have drawn an inference from these circumstances that Japan's research programme might not be 'for purposes of' scientific research. The Court did not expressly assess whether this element of the programme's design and implementation was reasonable in relation to the programme's stated objectives,[72] although the Court did appeal to 'reason' in making the observation that 'It stands to reason that a research proposal that contemplates extensive lethal sampling would need to analyse the potential applicability of these advances in relation to a programme's design'.

The same observation may be made in relation to the Court's analysis of the *second* element of the programme's design and implementation: the *scale of the use of lethal methods* in JARPA II. The Court referred to weaknesses in Japan's explanation for proceeding with larger sample sizes in JARPA II without a pause following the conclusion of JARPA and prior to the IWC Scientific Committee's final review of JARPA. The Court considered these to 'lend support to the view' that the sample sizes and launch date were not driven 'by strictly scientific considerations'.[73] These weaknesses gave weight to the theory advanced by Australia that Japan was seeking as the priority to maintain whaling without any pause.[74] Reasonableness, although the place where it was referred to was in an earlier paragraph in the judgment, was again the pivot for inference: 'In particular, a State party may not, in order to find the research for which a special permit has been granted, use lethal sampling on a greater scale than is otherwise reasonable in relation to achieving the programme's stated objectives'.[75]

69 Judgment, paragraph 137.

70 Judgment, paragraph 144.

71 Judgment, paragraphs 137, 141, 144.

72 Judgment, paragraph 137. This inference appears additionally to be based on references in an unpublished paper cited by Japan to the desirability of lethal methods because they will recover research costs. Judgment, paragraphs 138, 140, 143, 144.

73 Judgment, paragraph 156.

74 Judgment, paragraph 156.

75 Judgment, paragraph 94, 142. The Court observed early in its reasoning that the sale of whale meat 'taken alone' was insufficient to establish that JARPA II was not for purposes of scientific research, and that the scale of lethal sampling etc had also to be considered. Judgment, paragraph 94.

The *third* element of JARPA II's design and implementation was *the way in which Japan determined sample sizes*. Here the Court delved deeper into scientific methodologies. The Court set out to evaluate whether the Japanese sample sizes were reasonable in relation to achieving the objectives of JARPA II.[76] Australia contended that Japan had designed the JARPA II programme around the predetermined sample size of approximately 850 minke whales a year, this being a figure that would meet Japan's funding and commercial objectives.[77] The Court addressed separately the JARPA II sample sizes set for (1) fin and humpback whales and (2) minke whales. Japan sought to obtain by lethal sampling information concerning the research items of interest for all three species, including pregnancy rate, age of sexual maturity and feeding patterns.[78] Japan explained how overall sample sizes were set by looking at the diverse sample sizes needed to study the research items of interest in each species relative to the period of time over which sampling would be conducted.[79] The Court found that sample sizes selected for fin and humpback whales (50 whales of each species per annum) were a factor that 'cast doubt on' the reasonableness of JARPA II's design because this was not large enough to provide statistically reliable information and to measure all the trends that the programme sought to cover.[80] Importantly, also, Japan's decision to set the sampling period for fin and humpback whales at 12 years was out of alignment with the 6 year high-catch sampling period for minke whales. Given that JARPA II had the stated objectives of assessing competition between species and of ecosystem analysis, the 12 year sampling period for fin and humpback whales was a factor that 'cast doubt on' the reasonableness of the minke whale sample sizes in relation to achieving the programme's objectives,[81] and therefore on the reasonableness of the overall design of the programme. The Court considered that the scant analysis and justification for the decisions that had generated JARPA II's sample sizes for minke whales 'raised further concerns' about whether the programme's design was reasonable in relation to achieving its stated objectives.[82]

76 Judgment, paragraphs 172, 185.

77 Judgment, paragraphs 158, 184.

78 Judgment, paragraph 161.

79 See illustration, paragraph 182 of Judgment; Counter-Memorial of Japan, Vol. 1, p. 261.

80 Judgment, paragraphs 179, 181, 194, 196.

81 Judgment, paragraphs 178, 181, 191, 193, 195, 196.

82 Judgment, paragraph 198. See also paragraphs 188, 193, and 195 where the Court refers to the lack of transparency seen in the evidence on Japan's decision-making and absence of a consistent explanation for the 6 year high-catch sampling period for minke whales.

The *fourth* element of JARPA II's design and implementation was the significant gap between the sample sizes envisaged in the programme and the *actual take*. For minke whales, the target sample size was 850 whales per annum plus or minus 10%. In the 2005–2006 season 853 minke whales were taken. Approximately 450 minke whales were killed each year on average, with figures of 170 in the 2010–2011 season and 103 in the 2012–2013 season.[83] Australia considered that actual take was a function of Japan's commercial market for whale meat rather than scientific considerations.[84] Japan referred to the sabotage carried out by the Sea Shepherd Conservation Society, with specific reference to events in the 2008–2009, 2009–2010, and 2012–2013 seasons, and to the fire on the *Nisshin Maru* in the 2006–2007 season. The Court noted that JARPA II's take of minke whales had already dropped to 505 and 551 whales in the 2006–2007 and 2007–2008 seasons, prior to this sabotage.[85] Japan also stated that the research undertaken in the JARPA II programme could still achieve scientifically useful results, although with decreased accuracy or a longer period of research.[86] The Court found this statement to raise 'further doubts' about the reasonableness of the sample size,[87] discussed above, and that it 'added force' to Australia's argument.[88] Japan's very low take of a total of 18 fin whales over seven seasons against a target sample of 50 fin whales per annum[89] was explained by Japan with reference also to the *Nisshin Maru*'s being too small to pull on board such large whales.[90] Japan had suspended its take of 50 humpback whales per annum in 2007, although continuing to issue permits for humpbacks, as a political decision in response to a request from the Chair of the IWC.[91] Combined with Japan's statement that a lower take of minke whales would not prejudice JARPA II's potential to obtain meaningful scientific results, the Court found the gap between target sample sizes and actual take for fin and humpback whales to 'undermine' Japan's reliance on JARPA II's stated research objectives relating to monitoring the Antarctic ecosystem and modelling competition among whale species as justification for the expanded target sample size for minke whales in JARPA II compared with JARPA I,[92] and

83 Judgment, paragraph 202.
84 Judgment, paragraph 205.
85 Judgment, paragraph 206.
86 Judgment, paragraph 208.
87 Judgment, paragraph 209.
88 Judgment, paragraph 209.
89 Judgment, paragraph 201.
90 Judgment, paragraph 203.
91 Judgment, paragraphs 201, 205, 206, 212.
92 Judgment, paragraph 210, 212.

'cast further doubt on' the characterisation of JARPA II as a programme for purposes of scientific research.[93] Japan's assertion that it could rely on non-lethal sampling for fin and humpback whales suggested there was no strict scientific necessity for using lethal methods.[94] That the actual take of fin and humpback whales was largely or even entirely the result of logistical and political considerations 'further weakened' Japan's assertions about the relationship between JARPA II's research objectives and its target sample sizes.[95] In the end, the Court took the view that JARPA II's target sample sizes were larger than reasonable in relation to achieving the stated objectives.[96]

The *fifth* element of JARPA II's design considered by the Court was its *open-ended timeframe*, with the Court observing that a timeframe with intermediary targets 'would have been more appropriate' for a programme for purposes of scientific research.[97]

The *sixth* element of JARPA II considered by the Court was its *scientific output*, which 'appeared limited' in light of the length of time since JARPA II had begun in 2005 and the lethal sampling of about 3,600 minke whales over that period.[98]

In relation to the *seventh* element of JARPA II's design and implementation, *cooperation with other research institutions*, the Court observed that 'some further evidence' of cooperation *could have been expected*, given the programme's emphasis on the Antarctic ecosystem and regional environmental changes.[99]

Having addressed each of the 7 elements one by one, the Court then looked at the overall range of difficulties, doubts, concerns, and less than appropriate characteristics associated with the design and implementation of Japan's research programme to find, applying the standard of review, that 'Taken as a whole…the evidence does not establish that the programme's design and implementation are reasonable in relation to achieving its stated objectives'.[100] The Court went directly to stating that 'The Court concludes that the special permits granted by Japan for the killing, taking and treating of whales in connection with JARPA II are not 'for purposes of scientific research' pursuant to Article VIII, paragraph 1, of the Convention'.[101] Concerns relating to all

93 Judgment, paragraph 212.
94 Judgment, paragraph 211.
95 Judgment, paragraph 212.
96 Judgment, paragraphs 212 and 224.
97 Judgment, paragraph 216, referring to Annex P.
98 Judgment, paragraph 219.
99 Judgment, paragraph 222.
100 Judgment, paragraph 227.
101 Judgment, paragraph 227 and see paragraph 247 (2) of the *dispositif*.

of the seven elements of JARPA II's design and implementation fed into this conclusion.[102]

In all of this we should bear in mind that inference is an established methodology for Courts dealing with factual situations where evidence is scarce. Both common law and civil law incorporate the practice,[103] as well as international law.[104] The Court itself has classically commented that where there are difficulties involved in gathering evidence, 'a more liberal recourse to inferences of fact and circumstantial evidence' may be permissible.[105] In considering the part played by inference in the Court's judgment in the *Whaling* case we may also want to bear in mind that issues relating to a State's motivations or purposes may be even more challenging than the sort of more definitively binary questions about a State's knowledge seen for instance in the *Corfu*

---

102 Judgment, paragraphs 225–226. The Dissenting Opinion of Judges Bennouna and Abraham expressed concern about the Court having reached its conclusion in this way, on the basis of 'queries, doubts and suspicions, based on a selection of indicators…'. Dissenting Opinion of Judge Bennouna, page 1, and see Dissenting Opinion of Judge Abraham, paragraphs 29, 37 and 41–44. Judge Abraham did not find there to be a convincing case in relation to any one of the 7 elements addressed by the Court and did not seee that *en masse* the reasons described by the Court became any more convincing. Ibid, paragraph 37. 'Mais une doute, et même une accumulation de doutes, suffisent-ils pour faire une preuve?' Ibid, paragraph 44. 'Les doutes ne sont pas des preuves'. Ibid, paragraph 46. The Judge referred to the evidence of Japan's expert, Professor Walløe, that the necessary sample size over a 6 year period was in the order of 900 minke whales a year. Ibid, paras 47–48.

103 Bin Cheng, *General Principles of Law As Applied by International Courts and Tribunals* (Cambridge: Grotius Publications Ltd., 1987) 323–325.

104 'In cases where proof of fact presents extreme difficulty, a tribunal may [thus] be satisfied with less conclusive proof i.e. *prima facie* evidence…the inference in every case must, however, be one which can reasonably be drawn'. Cheng, above n 102, 323 and 325, as cited by Durward V. Sandifer, *Evidence before International Tribunals* (University Press of Virginia, 1975) 173.

105 Concluding in the *Corfu Channel Case* that Albania must have known of the minelaying in the Corfu Channel that had led to the loss of two British ships. *The Corfu Channel Case (United Kingdom v Albania)* Order of 17 December 1948 I.C.J. Reports 1947–1948 p. 124; Judgment of 9 April, [1949] ICJ Reports 1949 pp. 2, 18. The circumstantial and indirect evidence relied upon included the geographical configuration of the relevant territory, the estimated time required for mine laying, the distance of the minefield from the coast, the absence of an Albanian investigation into certain events, and notes kept by the Albanian Government. Mojtaba Kazazi, *Burden of Proof and Related Issues: A Study on Evidence before International Tribunals* (Kluwer Law International, 1996) 261. The Court also took into account that the 'exclusive territorial control exercised by a state within its frontiers' affected disadvantageously the capacity of the complaining party to provide proof of its case. Judgment of 9 April, above, 18.

Channel case.[106] Motivations may be multiple and mixed, with different actors within the State pursing diverse purposes, as the Court has recognised. Motivations come in bundles, not always easily unpacked. This leads to the further point that the overall inference drawn by the Court in the *Whaling* case is arguably not an inference of fact, but an inference of law.

Part 2 Motivations

Part 2 of this chapter will consider how the Court's chosen methodologies enabled it to deal with the question of Japan's motivations in conducting scientific research whaling. These include observations on how employing a standard of review enabled the Court (A) to make an indirect assessment of Japan's motivations, and (B) to preserve the presumption of good faith on the part of Japan.

A *Indirect Assessment of Blended Motivations*

Employing the mechanism of the standard of review, combined with a preparedness to rely on inference, kept the Court from potentially needing to make a direct assessment of Japan's motivations to see if they met the 'for purposes of' requirement under Article VIII (1). Rather, the Court's analysis remained at one remove. This meant that the Court did not need to concern itself closely with the question of unbundling the blend of motivations that Japan might have held in issuing the special permits for JARPA II. We have seen above that the Court was tolerant of the likelihood that diverse State officials will have a range of aims in pursuing a given policy. However if the Court had not applied a standard of review and had had to assess directly whether Japan's permits were 'for purposes of scientific research' this could have drawn the Court into contemplating the extent to which diverse motivations could be permissible, and how motivations could be ranked or characterised in terms of their importance to an activity.

Should it be the case that permits must be characterisable as being issued solely for the purposes of scientific research?[107] Professor Crawford for Australia argued that this was consistent with the approach taken by the Court in *Dispute*

106 *Corfu Channel Case*, above n 104.

107 This was the view adopted in the Separate Opinion of Judge Cançado Trindade at paragraph 80. The Judge considered that the provision in Article VIII (2) for the processing of whale meat from scientific research programmes was simply there in order to avoid waste. Separate Opinion of Judge Cançado Trindade, paragraph 84. The Judge did not consider

Regarding Navigational and Related Rights (*Costa Rica v Nicaragua*).[108] Alternatively, should it be the case that purposes other than scientific research only become relevant when it is established by convincing evidence that they have played a decisive role and have been the true source of the motivation that has given rise to the activities?[109] Judge Owada took just such a view, being prepared to set aside as irrelevant in principle 'ulterior motives harboured by some individuals involved in the action, whatever their position may be'.[110] Or should we adopt the idea that scientific research must be the preponderant purpose, while allowing for the sale of whale meat consistently with Article VIII (2)?[111] There is also the view that it is perfectly acceptable for a State to take into account that it will be able to defray research costs through commercialisation, and to choose its research methods based on an assessment of the bottom line.[112] Certainly, Judge Greenwood took the view that statements suggesting science was not Japan's only consideration did not in themselves take JARPA II outside the scope of Article VIII (1).[113]

A stark contrast may be drawn between the Court's approach and the approach of Judge Bhandari. The Judge found that the evidentiary record demonstrated plainly that Japan's whaling programme was a commercial enterprise.[114] The Judge referred, among various considerations, to evidence that the programme's take of minke whales was responsive to market forces, and to

that Japan's permits would have met this sole purpose test. Ibid, paragraph 80. Nor did the Judge believe scientific research was JARPA II's main purpose. Ibid, paragraph 83.

108 *Dispute Regarding Navigational and Related Rights* (*Costa Rica v Nicaragua*) Judgment of 2009 (2009) ICJ Reports 241, paragraph 61. The Court in that case was considering the meaning of the phrase 'con objectos de comercio'. The Court said there that 'expressly stating the purpose for which a right may be exercised implies in principle the exclusion of all other purposes'. Idem. Cited also by New Zealand, CR 2013/17 27, paragraph 41, observing that 'Any whaling programme has the potential to deliver *some* scientific information', idem, and that 'if a programme of whaling is designed for, or directed towards, achieving commercial purposes, even in part, it cannot claim to be special permit whaling under Article VIII'. Ibid, paragraph 42, responding to Judge Gaja's question 'If a whaling programme is both for purposes of scientific research and for commercial purposes, what are the applicable rules acording to the ICRW?' CR 2013/16 63.

109 Judge Owada uses the term *'fons et origo'*. Dissenting Opinion of Judge Owada, paragraph 22.

110 Idem.

111 Dissenting Opinion of Judge Yusuf, paragraph 55.

112 Dissenting Opinion of Judge Abraham, paragraph 39.

113 Separate Opinion of Judge Greenwood, paragraph 22. See also Dissenting Opinion of Judge Abraham, paragraph 26.

114 Separate Opinion of Judge Bhandari, paragraphs 2, 21.

efforts by the Japanese Government to stimulate domestic demand for minke whale meat.[115] The Judge also quoted statements from Japan's Minister for Agriculture, Forestry and Fisheries, the Director-General of the Japan Fisheries Agency and the Head of the Oceans Department of the Fisheries Agency on efforts to obtain international understanding and to ensure that whaling would be able to continue 'in some form or another'.[116] In addition he referred to statements from the Director-General in 1997 and 2012 that scientific whaling in the Southern Ocean was the only method for passing on the whaling tradition and was necessary for achieving a stable supply of minke whale meat.[117]

B *Effect on Presumption of Good Faith*

The Court's decision to use a standard of review may also have had a further effect. At no point did the Court pronounce on the idea that Japan had acted in bad faith. Indeed, Judge Greenwood made clear his view that Japan had not violated the implicit obligation to act in good faith.[118] It was possible to find, as did the Court and Judge Greenwood, that the permits issued for JARPA II were not for purposes of scientific research without a finding of bad faith. It may even be the case that a determination not to pronounce on the issue of good faith fed into the Court's decision to veer away from a traditional application of the rules on burden of proof. The issues of compliance and good faith are distinct. True, one of the generally accepted underpinnings of the rule on burden of proof in international courts and tribunals is the idea that a State will be presumed to have acted in accordance with its international legal obligations.[119] Perhaps because of this the presumption of compliance undergirding the rules on burden of proof is on occasion elided with the presumption that States act in good faith. Yet the two are not one and the same and WTO Appellate Body has been clear about this. A State presumed to have acted in good faith may not necessarily have achieved compliance with its obligations.[120]

115 Ibid, paras 21 and 34.

116 Ibid, paras 26–27.

117 Ibid, paragraph 29.

118 Separate Opinion of Judge Greenwood, paragraph 11. Nor did he consider that a case of bad faith had been demonstrated. Ibid, paragraph 29.

119 This assumed compliance is found also in domestic legal systems and has been linked with respect for the equality and dignity of individuals. Foster, above, n 38, 189ff. The presumption of compliance in international law sits well likewise with the idea of the dignity of sovereign equal States.

120 Report of the Panel, *Canada-Continued Suspension of Obligations in the EC-Hormones Dispute, Complaint by the EC* (WT/DS321) adopted on 31 March 2008, paragraphs 7.312–7.323.

On the other hand, Judge Yusuf expressed the view that the Judgment's conclusions implicitly entailed a finding of bad faith, taking into account that JARPA II, having been found not to constitute scientific whaling, was taken by the Court to constitute commercial whaling.[121] A number of the other judges viewed the Court's preparedness to make its pivotal finding against Japan under Article VIII (1) on the basis of inference as inconsistent with the principle that bad faith is not to be presumed, but must be established.[122] This was twinned with the view that the Court's reliance on a standard of review was unconvincing as a methodology and took the Court too far from the ordinary meaning of the terms of Article VIII (1).[123] Japan's permits should have been presumed to have been issued consistently with Article VIII(1) 'jusqu'à preuve du contraire'.[124] Yet saliently Professor Crawford, for Australia, cautioned against the Court's basing its judgment on a strong presumption of good faith in a case dealing with resources in the public domain which were a matter of collective interest.[125]

The presumption of compliance and the expectation that States will act in good faith were conflated both by the EC in its pleadings in these proceedings (see for example the Closing Statement of the EC, 15 September 2005 pp. 2–3) and by the Panel itself at paragraphs 7.312–7.357. Similarly the Appellate Body has been understood to say that a disguised restriction on trade as referred to in Article XX of GATT need not actually be hidden or dissimulated. Report of the Panel, *Brazil – Measures Affecting Imports of Retreaded Tyres*, WT/DS332/R, adopted 17 December 2007, as modified by Appellate Body Report WT/DS332/AB/R, paragraps 7.318–7.319, 7.326. See the findings agaist Brazil at paragraphs 7.349, 7.355 and 7.356.

121 Dissenting Opinion of Judge Yusuf, paragraph 54.

122 Judge Abraham considered it clear that Australia's case rested fundamentally on the idea that Japan had acted in bad faith, hiding its pursuit of commercial interests behind the appearance of a scientific research programme. Dissenting Opinion of Judge Abraham, paragraph 28. The Judge emphasised that the presumption of good faith requires to be met with solid rebuttal rather than weak arguments, simple doubts, suppositions and approximations. Ibid, paragraph 29. Judge Owada, too, emphasised that good faith was to be presumed and charges of bad faith against a sovereign state would have to be established by conclusive and indisputable evidence. Dissenting Opinion of Judge Owada, paragraphs 21–22, citing the *Lac Lanoux Arbitration* (*France v Spain*) RIAA Vol 12 281, also cited by Judge Yusuf.

123 Dissenting Opinion of Judge Abraham, paragraph 32.

124 Dissenting Opinion of Judge Abraham, paragraph 34.

125 Professor Crawford for Australia. 'To operate on a strong presumption of good faith is, in effect, to allocate those resources to a State good at concealment'. CR 2013/19, 65, paragraph 22.

Conclusion

We can understand, given all the dimensions of the case, why the Court turned to the idea of applying a standard of review in this case. This protected Japan against allegations beyond the four corners of the identified standard, and can be seen as an important effort to bring about peaceful resolution of the parties' dispute. However, the idea that international courts and tribunals will engage in 'review' rather than 'adjudication' as we know it requires serious consideration and discussion. What will be the benefits and the disadvantages for international dispute settlement? When will adjudication in its more original form remain available, if a Court in its discretion may decide instead to adopt a review-based approach as we saw in the *Whaling* case? Does a standard of review exist objectively and remain fixed in relation to a particular treaty? Will it be the same again in any subsequent dispute between the parties? Will it be the same in any subsequent disputes concerning the same Convention between other parties? In this case the Court sought to ground the application of a standard of review in statements put forward by the parties. Can there be additional or other authority for application of an standard of review? More deeply, what are the implications for public international law if international courts and tribunals increasingly apply standards of review?

Or should we take something quite different from the *Whaling* judgment? There is, perhaps, one key potentially enduring element at the heart of the methodologies employed by the Court in the *Whaling* case which can be distinguished from the application of a standard of review and which raises fewer questions. This is found in the Court's focus on the 'design and structure' of Japan's research programme, already a longstanding element in WTO dispute settlement practice predating the contemporary prevalence of reference to standard of review in the WTO.[126] As the WTO Appellate Body reiterated in 2014 in the *China – Rare Earths* case, a focus on measures' design and structure may afford 'an objective methodology' that helps a panel or the Appellate Body to assess whether a measure does what it purports to do.[127] A focus on structure

126 Consider Appellate Body Reports, *Japan – Taxes on Alcoholic Beverages*, WT/DS8/AB/R; WT/DS10/AB/R, WT/DS11/AB/R, adopted 1 November 1996, page 29; Appellate Body Report, *United States –Import Prohibition of Certain Shrimp and Shrimp Products*, WT/DS58/AB/R, adopted 6 November 1998. Note more recently the use of reference to the 'architecture and structure' of measures in cases under the WTO Agreement on Sanitary and Phytosanitary Measures and the WTO Agreement on Technical Barriers to Trade.

127 Appellate Body Report, *China – Rare Earths*, paras 5.96, 5.112. The Appellate Body was here applying the requirement in Article XX(g) of GATT 1994 that a measure 'relate to' the conservation of exhaustible natural resources.

and design does not necessarily involve the application of a standard of review. For instance, the WTO Appellate Body did not rest its decision on the idea of a standard of review in *China – Rare Earths*. Assessments of the 'structure and design' or 'design and implementation' of States' policies may be the important addition to ICJ jurisprudence to have emerged from the *Whaling case*, rather than the employment of a 'standard of review'.

Significantly, too, it appears that international adjudicatory bodies have found a decision-making tool, in the form of design and implementation assessments, that can in a case like *Whaling* effectively reverse the burden of proof and give effect to the precautionary principle, without doing so expressly. We can see this in the way that the Court looked to Japan for evidence to establish that 'in the use of lethal methods, the programme's design and implementation are reasonable in relation to achieving its stated objectives'. Of course it is an open question whether the precautionary principle might have played a more direct role had Japan been able to present evidence that more nearly persuaded the Court that the programme's design and implementation were at least 'reasonable' in relation to the use of lethal methods.

CHAPTER 2

From the Requirement of Reasonableness to a 'Comply *and* Explain' Rule: The Standard of Review in the *Whaling* Judgment

Shotaro Hamamoto *

Introduction

In *Whaling in the Antarctic*, the Court found that 'the special permits granted by Japan in connection with JARPA II do not fall within the provisions of Article VIII, paragraph 1, of the International Convention for the Regulation of Whaling'.[1]

Article VIII(1) provides:

> Notwithstanding anything contained in this Convention any Contracting Government may grant to any of its nationals a special permit authorizing that national to kill, take and treat whales for purposes of scientific research subject to such restrictions as to number and subject to such other conditions as the Contracting Government thinks fit, and the killing, taking, and treating of whales in accordance with the provisions of this Article shall be exempt from the operation of this Convention. Each Contracting Government shall report at once to the Commission all such authorizations which it has granted. Each Contracting Government may at any time revoke any such special permit which it has granted.

Australia argued that 'Article VIII(1) does not give Contracting Governments a discretion of a unilateral and subjective character as for the issue of a special permit'.[2] New Zealand considered that '[i]t is a discretion to grant Special Permits for purposes of scientific research', though the words in Article VIII 'provide a limited discretion for Contracting Governments to issue Special

---

* Professor of International Law, Kyoto University, Japan. Counsel and advocate of Japan in *Whaling in the Antarctic*. The usual disclaimer applies.

1 *Whaling in the Antarctic* (Australia v. Japan: New Zealand intervening), Judgment, 31 March 2014, para. 247 (2).

2 CR 2013/8, p. 31, para. 17 (Crawford).

 | DOI 10.1163/9789004313828_004

Permits for the specific articulated purpose of scientific research'.[3] Japan argued, for its part, that 'a Contracting Government authorizing a special permit has discretion to do so',[4] though 'Article VIII of the ICRW does not establish a completely unreviewable and self-judging right to designate any whaling activity as whaling "for purposes of scientific research"'.[5]

On this abstract level, accordingly, there was not much difference between the Parties and the intervening State with respect to the discretion given to the State conducting special permit whaling under Article VIII. Unsurprisingly, the Court, while considering that 'Article VIII gives discretion', confirmed that 'whether the killing, taking and treating of whales pursuant to a requested special permit is for purposes of scientific research cannot depend simply on that State's perception'.[6] The real problem was therefore how the Court should determine whether Japan exercised the discretion within the limit set forth by the Convention. It is at this point that issues relating to the standard of review come into play.

1 The Genesis of the Concept of Standard of Review in the *Whaling* Case

The *Whaling in the Antarctic* is the first case in which the Court employed the term 'standard of review',[7] though it is widely used in the WTO dispute settlement procedure and investor-State arbitration. It is therefore necessary to see how the term appeared in the discussion in the *Whaling* case.

The Memorial of Australia does not employ the term. This is quite understandable, because the Applicant's basic position defended in the Memorial is that Contracting Governments to the ICRW has no discretion in the determination of whether a whaling operation was carried out 'for purposes of scientific research'.[8] The term 'standard of review' is usually used to describe the degree of scrutiny or level of deference afforded by a court

3 Written Observations of New Zealand, 4 April 2013, paras. 38–39.

4 Counter-Memorial of Japan, 9 March 2012, para. 50.

5 Written Observations of Japan on New Zealand's Written Observations, 31 May 2013, para. 9.

6 Judgment, para. 61.

7 The ICJ 'has hardly ever made express reference to concepts like standard of review or margin of appreciation'. Chiara Ragni, 'Standard of Review and the Margin of Appreciation before the International Court of Justice', *in* Lukasz Gruszczyinski and Wouter Werner, *Deference in International Courts and Tribunals*, Oxford, Oxford University Press, 2014, p. 319.

8 Memorial of Australia, para. 4.116. As indicated above, Australia softened its position at the oral pleadings. *See supra* note 2.

or tribunal to the primary decision-maker in relation to its evaluation of factual and legal issues.[9] It is thus practically meaningless and theoretically impossible to use the term when one considers that the court or tribunal should exercise no deference the primary decision-maker, which in the present case is Japan.

The Counter-memorial of Japan does not use the term either, but refers to 'the margin of appreciation'.[10] Japan's basic position is, as indicated above, that Japan has discretion in issuing special permits but that this discretion is not unlimited. The Counter-memorial states that 'Japan does not question the right to challenge a State's decision to authorize special permit whaling on the ground that the authorization is arbitrary or capricious'.[11] In other words, the question is, according to Japan, whether the JARPA II operations are 'within the margin of appreciation';[12] and the task of the Court is, as the WTO Appellate Body stated in the *United States – Continued Suspension of Obligations in the EC – Hormones Dispute* case,[13] not to determine whether the measures taken by Japan are correct, but rather whether such measures are 'supported by coherent reasoning and respectable scientific evidence and [were], in this sense, objectively justifiable'.[14] Considering that Japan's granting of special permits was 'very obviously' not arbitrary or capricious, Japan emphasizes that Australia should address the question 'what further power Australia believes the Court to have to decide whether or not it shares Japan's view that JARPA II has been designed in a manner that is completely consistent with Article VIII of the ICRW'.[15]

The Written Observations of New Zealand takes a position close, but not identical, to Australia's, as the intervening State argues that, while 'whaling for scientific purposes can be objectively demonstrated',[16] each Contracting Government has a 'discretion' 'to determine the number of whales to be killed under Special Permit under Article VIII', the exercise of which 'remains subject

9 Caroline Henckels, 'Indirect Expropriation and the Right to Regulate: Revisiting Proportionality Analysis and the Standard of Review in Investor-State Arbitration', *Journal of International Economic Law*, vol. 15, 2012, p. 223, p. 238.

10 Counter-Memorial of Japan, paras. 9.7, 9.16.

11 Counter-Memorial of Japan, para. 9.7.

12 Counter-Memorial of Japan, para. 9.7.

13 United States – Continued Suspension of Obligations in the EC – Hormones Dispute, WT/DS320/AB/R, 16 October 2008, para. 590, *quoted in* Counter-Memorial of Japan, p. 412, n. 1099.

14 Counter-Memorial of Japan, p. 412, n. 1099.

15 Counter-Memorial of Japan, para. 9.7.

16 Written Observations of New Zealand, para. 52.

to review to ensure that it is exercised properly in light of the central obligation of good faith'; in other words, 'in a reasonable way to achieve the object and purpose of the Convention as a whole'.[17] New Zealand thus considers that Contracting Governments have no discretion with respect to the determination whether special permits are issued for purposes of scientific research, but have some discretion in the determination of the number of whales to be caught.

In its Written Observations on New Zealand's Written Observations, in which the term 'standard of review' appears for the first time in the present proceedings, Japan criticizes the first part of the intervening State's argument, stating:

> Like Australia, New Zealand does not address the standard of review that is applicable by the Court. To take only two possible approaches to this question, does New Zealand consider that the Court should ask (i) if Japan's view that JARPA II is scientific research is a view that no reasonable Government could reach or was adopted in bad faith, or should it ask (ii) if Japan's view was 'correct' and substitute its own (the Court's) assessment for that of the Contracting Government? Those are very different questions; and there are many more formulations of the criterion that might be adopted. Even at this late stage in the proceedings, Japan does not know what standard of review Australia and New Zealand say the Court should apply.[18]

In the first round of the oral pleadings, Australia submits somewhat contradictory opinions. While Professor Crawford, Counsel for Australia, maintains the Applicant's original position, denying that a margin of appreciation test should be applied,[19] the Solicitor-General assumes a position similar to Japan's when he states that Australia's 'ultimate proposition' is that (Japan's) abuse (of the right to grant special permits) is 'measured by whether the right holder – Japan – has so departed from standards of reasonableness and *bona fides* in the exercise of the right…that the right holder has reduced its treaty obligations to mere facultative ones'.[20] Japan, for its part, reaffirms its position, arguing that, neither *de novo* review nor complete deference being acceptable, the Court should examine whether Japan acted in good faith or bad faith in granting

17 Written Observations of New Zealand, para. 68.

18 Written Observations of Japan, para. 55.

19 CR 2013/8, p. 45, paras. 63–65 (Crawford).

20 CR 2013/11, p. 38, para. 43 (Gleeson).

special permits.[21] New Zealand also maintains its original position: there is no discretion with respect to the determination whether special permits are issued for purposes of scientific research but some discretion in the determination of the number of whales to be caught.[22]

In the second round of the oral pleadings, Australia criticizes Japan's argument that the Court should examine whether Japan acted in good faith, and thus implicitly withdraws the statement that the Solicitor-General delivered in the first round. '[J]udicial review', says Professor Crawford, 'should not be regulated by the Court wholly or primarily on the basis of such fluctuating and subjective notions as bad faith. The normal criterion for breach of treaty is whether the terms of the treaty, or any obligations reasonably to be inferred from them, are to be applied fairly and objectively'.[23] Japan, for its part, reminds the Court of the standard of review formulated in its Counter-Memorial, *i.e.* 'the key question was...not whether a State's decision was "correct", but whether it was "supported by coherent reasoning and respectable scientific evidence and is, in this sense, objectively justifiable"'.[24] '[R]easonableness must be given a role', because 'while the definition of scientific research is not solely a matter for the determination of each Contracting Government or State, neither is there one uniquely correct definition, so that the slightest divergence from that definition means that the characterization of a program as "scientific" puts the State in violation of international law'.[25]

2 The Standard of Review Adopted by the Court: The Requirement of Reasonableness

The Court, when determining the standard of review to be applied to the present case, places much emphasis upon reasonableness, as do the Parties and the intervening State. In paragraph 67, which is of crucial importance in the

21 CR 2013/15, pp. 18–19, 22, 24, paras. 21, 27–28, 43, 54 (Lowe).

22 CR 2013/17, p. 25, paras. 32–33 (Finlayson); pp. 35–36, paras. 2–3 (Ridings).

23 CR 2013/19, p. 65, para. 22 (Crawford).

24 CR 2013/22, p. 58, para. 12 (Lowe). Given this clear statement quoting the Counter-memorial, one would wonder why the Court said that '[n]ear the close of the oral proceedings, ...Japan refined its position regarding the standard of review' (Judgment, para. 66). Professor Lowe did no more than reiterate what had already been stated in the Counter-memorial. *See supra* note 14.

25 CR 2013/22, p. 59, para. 17 (Lowe).

Judgment, the Court adopts a two-step approach,[26] according to which it will assess:

- first, whether the programme under which JARPA II activities occur involves scientific research; and
- secondly, if the killing, taking and treating of whales is 'for purposes of' scientific research by examining whether, in the use of lethal methods, the programme's design and implementation are reasonable in relation to achieving its stated objectives.

The Court then goes on to say that '[t]his standard of review is an objective one'[27] and that 'the Court will look to the authorizing State, which has granted special permits, to explain the objective basis for its determination'.[28] We will not discuss questions relating to the burden of proof in detail.[29] More relevant to the present study is the fact that the Court points out a close relationship between the requirement of reasonableness and that of explanation.

3 Obligation 'To Give Due Regard to Recommendations'

Before going on to apply the standard of review thus established, the Court indicates that there exists an important obligation incumbent upon the States parties to the ICRW: the obligation to give due regard to IWC recommendations, which are not obligatory as such.

In the present case, Australia and New Zealand heavily relied on resolutions adopted by the IWC. The Court considers that the resolutions need to be classified into two categories: those adopted by majority without concurrence of a number of Contracting Governments including Japan, and others adopted by consensus. The first set of resolutions, adopted by majority, recommend that the killing of whales 'should only be permitted in exceptional circumstances where the questions address critically important issues which cannot be answered

26 As the Court says (Judgment, para. 70), this two-step approach was originally proposed by Australia in its Memorial (para. 4.36). However, none of the Parties or the intervening State referred to it in the following proceedings and the meaning or significance of the two-step approach was not discussed.

27 Judgment, para. 67.

28 Judgment, para. 68.

29 For our brief comment, *see infra* note 65 and the corresponding text.

by the analysis of existing data and/or use of non-lethal research techniques'.[30] The Court considers that, since they are adopted by majority and not by consensus, 'such instruments cannot be regarded as subsequent agreement to an interpretation of Article VIII, nor as subsequent practice establishing an agreement of the parties regarding the interpretation of the treaty within the meaning of subparagraphs (a) and (b), respectively, of paragraph (3) of Article 31 of the Vienna Convention on the Law of Treaties'.[31] These resolutions are therefore found to be irrelevant to the present case.

The second set of resolutions, adopted by consensus, do not establish a requirement, contrary to the first set of resolutions adopted by majority, that lethal methods be used only when other methods are not available. They do, however, call for an assessment of the feasibility of non-lethal alternatives.[32] With respect to these resolutions, the Court observes that 'the States parties to the ICRW…should give due regard', as the States parties 'have a duty to co-operate with the IWC and the Scientific Committee'.[33]

No one would be surprised to learn that Contracting Governments have a duty to co-operate with the IWC and thus are under the obligation to give due regard to recommendations adopted by consensus. Japan itself repeatedly argued that it had a duty to co-operate with the IWC and to consider carefully its comments and recommendations.[34] What matters is, therefore, the precise content of this 'obligation to give due regard to recommendations'.[35]

---

30 *See* Resolution 1995–9, *quoted in* the Judgment as an example of the resolutions adopted by majority (Judgment, para. 78).

31 Judgment, para. 83.

32 Judgment, para. 83. The Judgment (para. 78) refers to Resolution 1986–2 as an example of such resolutions. Resolution 1986–2 states, in relevant parts,

> [T]he Commission,
>
> […]
>
> Recommends that Contracting Governments when considering proposed research permits and the Scientific Committee when reviewing such permits and when reviewing the results of research from permits previously issued in accordance with the procedures of the Convention should take into account whether:
>
> (1) the objectives of the research are not practically and scientifically feasible through non-lethal research techniques;

33 Judgment, para. 83.

34 CR 2013/15, p. 21, paras. 36–37 (Lowe); CR 2013/16, p. 35, para. 22 (Boyle); CR 2013/16, p. 43, para. 18 (Pellet). As these records indicate, Japan included recommendations adopted by majority within the scope of due consideration.

35 In paragraph 83, the Court simply states that Contracting Governments 'should' give due regard to recommendations but talks of 'the obligation' to do so in later paragraphs (Judgment, paras. 137, 144).

4 A Combination of the Requirement of Reasonableness and the 'Obligation to Give Due Regard to Recommendations': An Obligation to Provide Reasons

The precise content of the 'obligation to give due regard to recommendations' is discussed by the Court when it applies the second step of the standard of proof, *i.e.* the requirement of reasonableness.[36] When arriving at this point, the Court has already confirmed that Article VIII expressly contemplates the use of lethal methods[37] and that 'the fact that a programme uses lethal methods despite the availability of non-lethal alternatives does not mean that a special permit granted for such a programme necessarily falls outside Article VIII'.[38]

However, the Court finds that Japan is in violation of the 'obligation to give due regard to IWC recommendations'.[39] As noted above, the relevant resolutions called for the States parties to take into account whether the objectives of the research are not feasible through non-lethal methods.[40] Japan referred to a document that it submitted to the Scientific Committee in 2007, when JARPA II was already operational.[41] However, the Court dismissed the relevance of the document for two reasons. First, it relates to JARPA rather than JARPA II and thus does not bear evidence of an assessment of the feasibility of non-lethal alternatives for JARPA II. Secondly and more substantially, although the document states that non-lethal methods are impractical, cost ineffective and prohibitively expensive, '[n]o analysis is included in support of these conclusions. There is no explanation of the relative costs of any methods'.[42] The Court finds that 'this is difficult to reconcile with Japan's obligation to give due regard to IWC resolutions'.[43] It follows that, according to the Court, the obligation to

36 Judgment, paras. 127–144.

37 Judgment, para. 83.

38 Judgment, para. 137. This is a highly important point as it relates to para. 246 of the Judgment that recognizes the possibility of future special permit whaling. The issue of future special permit whaling is however outside the scope of the present chapter.

39 Judgement, para. 144. The Judgement refers to 'IWC recommendations and Guidelines'. The relevant Guidelines were also endorsed by the IWC by consensus. *See* Judgment, paras. 47, 58, 83.

40 *Supra* note 33.

41 Judgment, para. 140. CR 2013/22, pp. 28–29, paras. 66–67 (Boyle). Seiji Ohsumi, Mutsuo Goto and Seiji Otani, 'Necessity of combining lethal and non-lethal methods for whale population research and their application in JARPA', SC/59/O2.

42 Judgment, para. 143.

43 Judgment, para. 144.

'give due regard to resolutions' consists of *an obligation to explain with evidence* how the relevant State has considered recommendations contained in such resolutions.[44]

5 A Refinement: An Obligation to Provide Reasons in Detail

As for the determination of the concrete number of whales to be sampled in research activities, the Court confirms the standard of review that it previously provided: 'the Court seeks here only to evaluate whether the evidence supports a conclusion that the sample sizes are reasonable in relation to achieving JARPA II's stated objectives'.[45] Under this requirement of reasonableness, the Court finds an obligation incumbent upon Contracting Governments to set out detailed reasons for determining the number of whales to be sampled under special permits. It is not sufficient to set out 'sample sizes' in accordance with accepted scientific standards. Contracting Governments are under an obligation to 'demonstrate[] a reasonable basis for annual sample sizes' 'in light of' 'stated research objectives'.[46]

A good example is the Court's analysis of the selection of a particular value for a given variable in JARPA II. For example, the 2005 JARPA II Research Plan states that it is necessary to catch 1,288 minke whales to detect a change in the age at sexual maturity at the level of yearly rate of change of 0.1 year in a six-year research period.[47] It indicates, at the same time, sampling 336 minke whales would be sufficient if the level of yearly rate of change is set at 0.2 year. Plainly speaking, a less accurate research would require a smaller number of samples. The Court points out that 'there is no consistent effort to explain why…JARPA II is designed to detect one particular rate or degree of change over another that would result in a lower sample size' except with respect to the study of the age at sexual maturity.[48] Scientists involved in JARPA II might respond that it is scientific common sense that a change of 1.0–1.5% in the apparent pregnancy rate of minke whales needs to be detected.[49] However, in light of the requirement of reasonableness as understood by the Court, unless

44 *See also* CR 2013/20, p. 42, para. 31 (Gleeson), pointing out the lack of such evidence.

45 Judgment, para. 172.

46 Judgment, para. 185.

47 Plan for the Second Phase of the Japanese Whale Research Program under Special Permit in the Antarctic (JARPA II), SC/57/O1, p. 17, p. 65 (Appendix 6).

48 Judgment, para. 189.

49 *See* the 2005 JARPA II Research Plan, *supra* note 47, p. 17.

the Contracting Government granting the special permit explicitly explains that the proposed rate of change to be detected is an accepted scientific common sense, the absence of explanation 'lends support to [the] contention that a predetermined overall sample size has dictated the choice of…the rate of change to be detected, rather than the other way around'.[50]

6 A Further Refinement: An Obligation to Provide Reasons in a Short Period of Time

Researchers, of any field for that matter, are all too well aware that any research activity will not proceed as it was originally planned. At a certain stage of research, research programme needs to be reviewed in light of information and data obtained in the course of the research, so as to reconsider and, if necessary, restructure research activities to better achieve the research objectives.

In the present case, there was a gap between the JARPA II target sample sizes and the actual catches as a smaller number of whales than targeted in the 2005 JARPA II Research Plan had been caught.[51] Japan stated several reasons for the gap. As for humpback whales (50/year targeted; 0 caught), it decided not to take any in response to a request by the chair of the IWC, in the midst of its 'peace process'. Regarding fin whales (50/year targeted; 0–10/year caught), the difference is due to sabotage activities by anti-whaling non-governmental organizations such as Sea Shepherd. As to minke whales (850/year targeted; 103–853/year caught), a fire on board a research vessel and the sabotage activities caused the difference.[52] Since the first review of JARPA II by the Scientific Committee was scheduled in 2014, Japan repeatedly reminded the Court that it was ready to modify JARPA II research programme, if the 2014 review by the Scientific Committee was to reveal the necessity of any modification.[53] This argument is simply ignored by the Court, which states:

> despite the number of years in which the implementation of JARPA II has differed significantly from the design of the programme, Japan has not made any changes to the JARPA II objectives and target sample sizes, which are reproduced in the special permits granted annually. In the

50 Judgment, para. 195.

51 Judgment, paras. 199–202.

52 Judgment, para. 203.

53 CR 2013/13, p. 38, paras. 70–71 (Hamamoto); CR 2013/22, p. 26, para. 54 (Boyle).

Court's view, two conclusions can be drawn from the evidence regarding the gap between the target sample sizes and actual take.[54]

> Japan's continued reliance on the first two JARPA II objectives to justify the target sample sizes, despite the discrepancy between the actual take and those targets, …cast further doubt on the characterization of JARPA II as a programme for purposes of scientific research.[55]

Why not wait for the Scientific Committee's review, which was to be held only a few months after the rendering of the judgment? Although the Court does not expressly refer to the obligation to give due regard to the relevant IWC resolutions in this context, this obligation seemingly played a decisive role. As mentioned above, the Court, when examining Japan's decisions regarding the use of lethal methods, affirmed the obligation to give due regard to such resolutions. In light of this obligation, a Member of the Court asked Japan what analysis it had conducted of the feasibility of non-lethal methods prior to setting the sample sizes *for each year* and the Court found Japan's reply unsatisfactory.[56] The fact that this question is quoted and given a significant importance in the judgment suggests that the Court considers that the requirement of reasonableness, or the obligation to give due regard to recommendations of the IWC, requires Japan to reconsider the original research programme every year in light of research results of the preceding year, to see whether it needs to be modified. Since such reconsideration must be carried out every year, it cannot be postponed until the Scientific Committee conducts its own review. Or, the Court might have found a contradiction in Japan's argument, since JARPA II was launched in 2005 without waiting for a review of JARPA by the Scientific Committee:[57] i.e. why, if Japan was able launch a new research programme without a review of the previous one by the Scientific Committee, could it not modify (or explain reasons for not modifying) the current programme without a review by it?

Conclusion: Obligation to Provide Reasons, or to 'Comply *and* Explain'

The Court, in its quintessentially important paragraph 67, states that it will examine 'whether, in the use of lethal methods, the programme's design and

54 Judgment, para. 209.

55 Judgment, para. 212.

56 Judgment, paras. 138–144.

57 Judgment, paras. 154–156.

implementation are reasonable in relation to achieving its stated objectives'. As previously noted, this is no surprise, as both of the disputing Parties as well as the intervening State placed emphasis upon the requirement of reasonableness. For example, Professor Lowe, Consel for Japan, summarized Japan's position in this respect as follows:

> Views may differ over the size of the sample that is needed, over whether priority should be given to the determination of simple abundance numbers or to the detailed understanding of changes in sexual maturity, pregnancy rates and stock structures. But these are not debates as to whether the questions are or are not scientific. These are debates about scientific questions. And Japan's view is that there is no uniquely correct answer to such questions. Uniquely correct answers to such questions do not exist, so there is no point in the Court looking for them.
>
> Yes: the Court can ask, could a reasonable State regard this as a properly-framed scientific inquiry. But it can no more impose a line separating science from non-science than it could decide what is and what is not 'Art'. In Japan's view, the correct question is, could a State reasonably regard this as scientific research?[58]

From this point, one can take either of the two ways to see whether Japan acted reasonably or not. One could understand, as Judge Owada states in his dissenting opinion, that the term 'reasonableness' would here be used as 'a yardstick for ascertaining whether a decision or an action is or is not "arbitrary" or patently "out of bounds".'[59] '[O]n ne saurait mettre en cause la parole d'un Etat à la légère et sa bonne foi doit être présumée jusqu'à preuve du contraire', says Judge Abraham.[60] In this case, there will be only two possible situations where the Court would be able to find Japan's violation of the ICRW, as pointed out by Judge Abraham:

> le premier cas est celui où il apparaît qu'il n'existe manifestement aucune relation raisonnable entre les objectifs annoncés et les moyens mis en œuvre, de telle sorte que ces moyens sont manifestement inaptes à atteindre les objectifs, ce dont on peut déduire que ceux-ci ne sont pas réellement recherchés; le second est celui dans lequel la taille des échantillons fixée par le programme est manifestement excessive au regard des

58 CR 2013/22, p. 60, paras. 19–20 (Lowe).

59 Dissenting Opinion of Judge Owada, para. 39.

60 Opinion dissidente de M. le juge Abraham, par. 34.

> nécessités de la recherche compte tenu des objectifs annoncés, ce dont on peut déduire que pour partie au moins le nombre de baleines dont la capture est autorisée a été fixé sur la base de motifs, ou pour la poursuite d'objectifs, non scientifiques (donc vraisemblablement commerciaux).[61]

If these criteria had been applied, it would have been difficult to find Japan's violation.[62] This conception of 'reasonableness' corresponds to the 'sens subjectif de raison' as explained by Olivier Corten in his masterpiece on the conceptions of the 'reasonable' employed by international courts and tribunals.[63]

But the Court adopts the 'acception objective de raison'[64] and understands the requirement of reasonableness as an obligation to state reasons. Whether Japan stated reasons or not is a purely objective question, which the Court can examine without placing burden of proof on either side: Australia does not have to indicate each point where Japan did not provide reasons.[65] It is true that 'to be reasonable' is often understood as 'to be able to provide reasons', particularly by the European Court of Human Rights but also by other instances.[66] However, the *Whaling* judgment is arguably the first one in which an international court or tribunal found that a State is obliged to provide reasons when conducting activities relating to issues dealt with in non-binding recommendations adopted by a treaty organ.

The Court's reasoning in *Whaling in the Antarctic* reminds us of the 'comply or explain' rule, which has been introduced in company law.[67] The 'comply or explain' rule gives the addressee of the rule a choice between complying with it

61 Opinion dissidente de M. le juge Abraham, par. 35.

62 Opinion dissidente de M. le juge Abraham, par. 36.

63 Olivier Corten, *L'utilisation du 'raisonnable' par le juge international*, Bruxelles, Bruylant, 1997, p. 446, para. 380.

64 Corten, *supra* note 63, p. 447, para. 381.

65 Telesetsky et al., consider that '[i]In creating the reasonableness test, the International Court ultimately shifts the burden of proof'. Anastasia Telesetsky et al., 'ICJ's Decision in Australia v. Japan; Giving up the Spear or Refining the Scientific Design?', *Ocean Development & International Law*, vol. 45, 2014, p. 328, p. 336. For a detailed analysis of the issues relating to the burden of proof in the *Whaling* judgment, *see* Chapter 1.

66 Corten, *supra* note 63, pp. 382–390, paras. 334–337. *See also* Jarrod Hepburn, 'The Duty to Give Reasons for Administrative Decisions in International Law', *International and Comparative Law Quarterly*, vol. 61, 2012, p. 641; Paul Craig, *EU Administrative Law*, 2nd ed., Oxford, Oxford University Press, 2012, pp. 340–355.

67 *See* Anne-Laure Boncori & Isabelle Cadet, 'Le *comply or explain*, un avatar de l'accountability', *Revue française de gestion*, t. 237, 2013, p. 35; Björn Fasterling & Jean-Christophe Duhamel, 'Le *comply or explain*: la transparence conformist en droit des sociétés', *Revue internationale de droit économique*, t. 23, 2009, p. 129.

and explaining why not complying with it.[68] Its use has been spreading due to its flexible nature in that it is 'perfectly adaptable to any kind of regulatory framework',[69] whether binding or non-binding. Precisely for this reason, the 'comply or explain' rule in domestic law is often critically analysed from a perspective of democratic theories, as the rule gives a binding character to norms that are non-binding in themselves.[70] To translate into international law parlance, the rule needs to be reconciled with sovereignty. However, in *Whaling in the Antarctic*, the Court goes even further, as it imposed a 'comply *and* explain' rule. It would be already difficult to establish a 'comply or explain' rule with respect to non-binding recommendations adopted by a treaty organ on the basis of a duty to co-operate with such an organ. How, then, is it possible to found a 'comply *and* explain' rule under such circumstances? Is it sufficient to refer to a duty to co-operate with a treaty organ? Although Japan expressly recognized that the applicable test would be a standard of reasonableness[71] and accepted a duty to co-operate with the IWC,[72] it has not accepted anything like a 'comply and/or explain' rule with respect to IWC resolutions. Rolland points out that 'the majority endorsed a reasonableness standard without ever explaining why it was appropriate'.[73] It is true that, while the Court indicates that Japan adopted a reasonableness standard, it does not explain why the standard amounts to a 'comply *and* explain' rule.

Could this 'comply and/or explain' rule be extended to other fields of international law? Since it does not depend on the wording of Article VIII of the ICRW in the present case, nothing seems to prevent such extension.[74]

68 Sridhar Arcot et al., 'Corporate Governance in the UK: Is the Comply or Explain Approach Working?', *International Review of Law and Economics*, vol. 30, 2010, p. 193.

69 Konstantinos Sergakis, 'Deconstruction and Reconstruction of the "Comply or Explain" Principle in EU Capital Markets', *Accounting, Economics and Law*, advanced publication, DOI: 10.1515/ael-2014-0007, November 2014, p. 1, pp. 20–21. *See also* Andrew Keay, 'Comply or explain in corporate governance codes', *Legal Studies*, vol. 34, 2014, p. 279, pp. 279–280; David Seidl et al., 'Applying the "comply-to-explain" principle: discursive legitimacy tactics with regard to codes of corporate governance', *Journal of Management and Governance*, vol. 17, 2013, p. 791, p. 780.

70 Peter O. Mülbert, 'Comply or Explain im deutschen Aktienrecht und EU-Hintergrund: § 161 AktG und Deutscher Corporate Governance Kodex', 16 December 2013, <http://www.securities.j.u-tokyo.ac.jp/w-papers/2014-5_Muelbert.pdf>, pp. 25–27.

71 Judgment, para. 66.

72 *See Supra* note 34.

73 Sonia E. Rolland, 'Whaling in the Antarctic (Australia v. Japan: New Zealand Intervening)', *American Journal of International Law*, vol. 108, 2014, p. 496, p. 500.

74 Smith considers that the Whaling judgment will strengthen (the treaty organs created under) marine resource treaties. p. 317. Jeffrey J. Smith, 'Evolving to Conservation?: The

That would, however, create a tremendous impact upon the exercise of State authority where it relates to issues and questions dealt with in non-binding recommendations adopted by a treaty organ. Each time, in each situation and with respect to each action, States would need to publicly provide detailed reasons in a timely manner; otherwise, they might be held internationally responsible for the failure to do so. May the judgment in *Whaling in the Antarctic* be remembered as the starting point of a profound evolution of international law regarding the authority and power of treaty organs through the introduction of a 'comply and/or explain' rule? And who wishes to replace this question mark with an exclamation mark?

International Court's Decision in the Australia/Japan Whaling Case', *Ocean Development & International Law*, vol. 45, 2014, p. 301.

PART 2

Substantive Law Aspects: The Law of Treaties

CHAPTER 3

The Whaling Convention and Thorny Issues of Interpretation

Malgosia Fitzmaurice

Introduction

The Judgment of the International Court of Justice (the 'Court' or the 'ICJ') in the *Whaling in the Antarctic* case[1] raised several issues regarding the law relating to the interpretation of treaties that will be analysed in this chapter. In Section 1, all the main points of the case concerning the law of treaties will be outlined and a short summary of the Judgment will be presented, highlighting particular points of the law of treaties that arose in this case. More detailed study of the issues relating to interpretation of treaties, and in particular their treatment by the Parties to the Whaling case, and by the Court and Judges, will be the subject of subsequent sections. Without doubt, the most important of these, which was discussed and analyzed both by the Parties to the case and in the Judgment, was the nature of the object and purpose of both the 1946 International Convention on the Regulation of Whaling (the 'ICRW') and in particular its Article VIII (special permits for scientific whaling) (Section 2). Other questions concerning the interpretation of treaties were the issues of 'evolutionary' (or 'evolutive' or 'dynamic') interpretation of treaties (Section 3) and the principle of systemic integration (Section 4). A further question of considerable importance (though not discussed extensively in the Judgment) was the legal nature of 'subsequent practice' in relation to the interpretation of treaties (Section 5).

1 General Overview of the Issues of the Law of Treaties in the *Whaling in the Antarctic* Case (the '*Whaling Case*')

As observed above, by far, the most important question on the law of treaties in this case was the interpretation of the 'object and purpose' of the ICRW.

1 *Whaling in the Antarctic* (*Australia v. Japan: New Zealand intervening*), ICJ, Judgment of 31 March 2014, http://www.icj-cij.org/docket/index.php?p1=3&p2=1&case=148, last accessed August 2015 (cited hereafter as 'The Judgment').

 | DOI 10.1163/9789004313828_005

In relation to the 'object and purpose' of the Convention, the pleadings of the Parties to the dispute were mostly based on the analysis of the Preamble to the ICRW and the contributions of the Resolutions of the International Whaling Commission (the 'IWC' or 'Commission') to the interpretation of the Convention.

The Parties to the dispute analysed various paragraphs of the 1946 Convention's Preamble in order to identify its 'object and purpose'. Australia and New Zealand emphasized the Preamble's conservation approach and the value of whales to future generations, while Japan emphasized in particular the last paragraph, which refers to 'the proper conservation of whale stocks' and to making 'possible the orderly development of the whaling industry'. To some extent, the Parties to the dispute relied on the *travaux préparatoires* to the ICRW, which according to Article 32 of the 1969 Vienna Convention on the Law of Treaties (the '1969 VCLT') is one of the of the 'supplementary means of interpretation'.

The most important statement in the Judgment concerning the 'object and purpose' of the Convention is contained in paragraph 56, which states: '[a] mendments to the Schedule and recommendations by the IWC may put an emphasis on one or the other objective pursued by the Convention, but cannot alter its object and purpose'. Paragraph 45 of the Judgment alludes to the issues of the evolutionary interpretation of treaties, where it says as follows in relation to the role of the IWC: '[t]he Commission has amended the Schedule many times. The functions conferred on the Commission have made the Convention an evolving instrument'. (As to which, see further below in Section 3.)

The Court also analyzed the 'object and purpose' of Article VIII of the ICRW (special permits regarding scientific whaling). The most important observation in the Judgment on the interpretation of Article VIII of the ICRW is that Article VIII forms an integral part of the Convention. For that reason it has to be interpreted in light of the object and purpose of the ICRW as a whole, and also taking into consideration other provisions of the Convention, including the Schedule. The Court also stated and that

> whaling conducted under a special permit which meets the conditions of Article VIII is not subject to the obligations under the Schedule concerning the moratorium on the catching of whales for commercial purposes, the prohibition of commercial whaling in the Southern Ocean Sanctuary and the moratorium relating to factory ships (para. 55).

The Court also made a statement relating to 'subsequent practice', which is one of the most complex issues of treaty interpretation. It said as follows:

> [a]rticle VI of the Convention states that "[t]he Commission may from time to time make recommendations to any or all Contracting Governments on any matters which relate to whales or whaling and to the objectives and purposes of this Convention. These recommendations, which take the form of resolutions, are not binding. However, when they are adopted by consensus or by a unanimous vote, they may be relevant for the interpretation of the Convention or its Schedule" (para. 46).

All of these issues are complex and contentious and their legal nature is not fully agreed upon in the theory and practice of the law of treaties. There are a lot of divergent views as to how they should be understood (such as the views relating to evolutionary interpretation of treaties and the principle of systemic integration, and the relationship between these two concepts).

In further sections of this chapter all the issues that have been signaled above, will be analysed within the context of the Judgment and Judges' Opinions; and against the broader background of the views of publicists and the jurisprudence of international courts and tribunals other than the ICJ.

2 The Object and Purpose of a Treaty

2.1 *General Considerations*

In this part of the chapter the role of the 'object and purpose' of a treaty will be presented in very broad brushstrokes.[2] The 'object and purpose of a treaty' is a frequently used phrase in the 1969 VCLT. It is used eight times in the text of the Convention.[3] Many of the most important questions raised by the 1969 VCLT are based on the notion of the 'object and purpose' of a treaty, such as treaty

---

2 There are many excellent publications on this subject, such as Richard Gardiner, *Treaty Interpretation* (Oxford: Oxford University Press, 2015, 2nd ed), 211 (cited hereafter as 'Gardiner'); Isabelle Buffard & Karl Zemanek, 'The "Object and Purpose" of a Treaty: An Enigma?' *Austrian Review of International and European Law*, 3 (1998): 311 (Cited hereafter as Buffard & Zemanek); David S. Jonas and Thomas N. Saunders, 'The Object and Purpose of a Treaty: Three Interpretive Methods', *Vand. J. Transnatl L* 43 (2010): 565 (cited hereafter as 'Jonas & Saunders'), http://www.vanderbilt.edu/jotl/manage/wp-content/uploads/Jonas Saunders-cr_v2.pdf; Jan Klabbers, 'How to Defeat the Treaty's Object and Purpose Pending Entry into Force: Toward Manifest Intent', *Vand. J. Transnatl.* L. 34 (2001): 283 (cited hereafter as 'Klabbers How to Defeat...'); Jan Klabbers, 'Some Problems Regarding the Object and Purpose of Treaties', *Fin. Y.B. Int'l L.* 2 (1997): 139–142 (cited hereafter as 'Klabbers Some Problems...'). *Y.B. Int'l L.* 2 (1997)...'.>

3 Articles 18; 19(c); 20(2); 31; 33; 41; 58(1); 60.

interpretation, reservations to treaties and the obligation not defeat the object and purpose of a treaty prior to its entry into force (Article 18 of the 1969 VCLT).

Notwithstanding such extensive use of this term, its meaning and content remain an enigma which, '[i]nstead of reducing the potential of future conflicts... [,] plants the seed of them'.[4] Jonas and Saunders are of the view that '[b]eyond this general idea, scholars have failed to create a definition with adequate clarity and detail to serve lawyers who must apply the term in practice'; and that '[t]he ambiguity of the term object and purpose creates problems in a range of current international law contexts'.[5] The concept of the object and purpose of the treaty is notoriously considered as 'vague',[6] but also as 'flexible'[7] and elusive. As Aust explains, the role of the object and purpose of a treaty, is not so much a search for the ordinary meaning of words in their context, but more for the purpose of confirmation of the interpretation—i.e. if the interpretation is incompatible with the treaty's object and purpose, it may be incorrect.[8] Bowman recommends caution is defining the object and purpose of a treaty, for while it is commonly regarded as obvious, in many cases it may be deceptive, as the whole concept is very elusive.[9]

There are a great number of judgments of various courts (international and national) which used the term 'object and purpose', including the ICJ.[10] In English law, the term 'object and purpose' is treated as a composite term.[11] However, in French public law there is a difference between '*l'objet*' of a legal instrument and '*le but*', which denotes the reasons for establishing *l'objet*.[12]

It is understood that ultimately French terminology (*l'objet et le but du traité*) was the source of the terms 'object and purpose' used in the 1969 VCLT.[13] Nevertheless, it may be said that the 1969 VCLT, unlike the French binary

4 Buffard & Zemanek, note 2, 343.

5 Jonas & Saunders, note 2, 565.

6 Frank Horn, *Reservations and Declarations to Multilateral Treaties* (Amsterdam, North Holland, 1988), 115.

7 Klabbers, Some Problems..., note 2, 138.

8 Anthony Aust, *Modern Law of Treaties and Practice* (Cambridge: Cambridge University Press, 3rd ed. 2013), 209.

9 Michael Bowman, '"Normalizing" the International Convention for the Regulation of Whaling', *Michigan Journal of International Law* 29 (2008): 300 (cited hereafter as 'Bowman').

10 E.g. in one of the most influential cases before the ICJ, *Reservations to Convention on Prevention and Punishment of Crime of Genocide* Advisory Opinion, 1951 ICJ; *LaGrand* case (*Germany v. United States*). Judgrnent, I.C.J. Reports, 2001, p. 466.

11 Gardiner, note 2, 212–213.

12 Buffard & Zemanek, note 2, 326.

13 Gardiner, note 2, 213.

concept of the object and purpose, adopted a unitary approach to the object and purpose. Notwithstanding this, the Court identified in the *Genocide Convention* Advisory Opinion two objects of this Convention: one was to 'safeguard the very existence of certain human groups'; and the other 'to confirm and endorse the most elementary principles of morality'.[14] However, in the same case the Court also established that the Genocide Convention had one purpose, which was identifiable with the intentions underlying the Convention, namely, the condemnation and punishment of genocide as a crime under international law, which encompasses a denial of existence of entire human groups, a denial which shocks the conscience of mankind and results in losses to humanity.[15] However, it has been noted that the Court had failed to explain why it understood the words object and purpose as two separate meanings, or how these two ideas are distinct. Equally, the Court did not provide an explanation how it arrived at the conclusion that the Convention had two objects and one purpose.[16] But the Court's general approach has been to treat 'object and purpose' as a composite concept, and this practice was followed by the Court in the Judgment in the Whaling case, in which no substantive distinction is made between 'object' and 'purpose'.

In this chapter it is of particular interest how the term 'object and purpose' is used in Article 31 of the 1969 VCLT ('General Rule of Interpretation'), which provides that: 'A treaty shall be interpreted in good faith in accordance with the ordinary meaning to be given to the terms of the treaty in their context and in light of its object and purpose'.

The object and purpose of a treaty is most frequently found in the treaty's preamble. This was confirmed by the Court in the *Oil Platform* case, where it was stated that the provisions of the preamble 'may be relevant and important guides to the manner in which the Treaty should be interpreted, and in order, as it were to "situate" it in respect of its object and purpose'.[17]

In order, however, to define the object and purpose of a treaty with precision, the interpreter must take into account all its provisions, which according to the dominant textual approach to interpretation adopted by the 1969 VCLT,

---

14 *Reservations to the Convention on the Prevention and Punishment of the Crime of Genocide, Advisory Opinion*, 28 May 1951, I.C.J. Rep.18, 23, http://www.icj-cij.org/docket/files/12/4285.pdf, accessed 4 August 2015.

15 See Jonas & Saunders, note 2, 379.

16 Jonas & Saunders, note 2, 580.

17 *Oil Platforms (Islamic Republic v. Iran v. United States of America), Preliminary Objections, Judgment, I.C.J. Reports 1996*, p. 803, at p. 813 (para. 27), http://www.icj-cij.org/docket/files/90/7287.pdf, accessed 4 August 2015.

should be based on the ordinary meaning of the treaty's provisions.[18] According to Marsden '[r]eference to the object and purpose is a secondary or ancillary process in the application of the general rule of interpretation'.[19] As Bowman explains, such an approach can be 'regarded as an attempt to strike a balance between the "textual" and "teleological" approaches to interpretation, while according a degree of primacy to the former', an approach which, as be notes, may in part be due to the vagueness of the concept of the 'object and purpose'.[20]

The value of the Preamble for the interpretive purposes can be overestimated; but on the other hand, as Bowman puts it:

> [f]or the purposes the preamble is intended to serve— typically, to provide some background detail regarding the motivations underlying the treaty's adoption and the objectives it seeks to advance—it is no less *authoritative* than any other aspect of the treaty.[21]

It must be noted, however, that Rosenne, for example, doubted whether multilateral treaties can have an object and purpose of their own at all.[22] Fitzmaurice, however, accorded some interpretative value the Preamble and stated that:

> Although objects of a treaty may be gathered from its operational clauses as a whole, the preamble is the normal place in which to embody, and the natural place to look for, an express or explicit general statement of the treaty's object and purposes. Where these are stated in the preamble, the latter, will, to that extent, govern the treaty.[23]

Bowman also suggests that the interpretative value of the Preamble is in its spelling out policy objectives which parties wished to include, while the treaty itself evidences specific legal purposes.[24]

---

18 Gardiner, note 2, 218–219.

19 Geoffrey Marston, 'United Kingdom Materials in International Law', *Brit. Y.B. Int'l L.* 43 (1992): 709.

20 Bowman, note 2, 318.

21 Bowman, note 2, 320–321.

22 Shabtai Rosenne, *Development in the Law of Treaties* 1945–1985 (Cambridge: Cambridge University Press, 1989), 83.

23 Sir Gerald Fitzmaurice, 'The Law and Procedure of the International Court of Justice 1951–54: Treaty Interpretation and Other Treaty Points', *BYIL*, 33 (1957): 228 (cited hereafter as 'G. Fitzmaurice').

24 Bowman, note 2, 321.

Despite its vagueness, the 'object and purpose' of a treaty is one of the three elements which are taken into consideration in order to establish the meaning of the treaty, the other two being the treaty's terms and the context of these terms.[25] According to Article 32(2) of the 1969 VCLT, context includes 'any agreement relating to the treaty which was made between all the parties in connection with the conclusion of the treaty' as well as 'any instrument…made by one or more parties…and accepted by the other parties'.

In the principles of interpretation set out by Gerald Fitzmaurice, the subsidiary principle of effectiveness relies on the interpretation of the treaty with reference to its object and purpose. However, according to Fitzmaurice, the group of subsidiary principles is only applicable in cases where the main principles of interpretation do not lead to a satisfactory result (a situation which was later formulated in Article 32 of the 1969 VCLT).[26] According to the principle of effectiveness (*ut magis valeat quam prereat*), treaties are to be interpreted by reference to their declared or apparent object and purpose and particular provisions are to be interpreted so as to give them the fullest effect consistent with the normal sense of the words and with the text as a whole, in such a way that a reasonable meaning can be attributed to every part of the

25 Jonas & Saunders, note 2, 577.

26 G. Fitzmaurice, note 23, 230. The main principles of interpretation (paraphrased) were as follows: Principe 1: of actuality or textuality (treaties are to be interpreted as they stand, on the basis of their actual text); Principle 2: the natural and ordinary meaning (that particular words and phrases are to be given their normal, natural and unstrained meaning in the context in which they occur). This principle can only be excluded by direct evidence that the terms used are to be understood in a manner different to their natural and ordinary meaning, or if such an interpretation would lead to an absurd and unreasonable result. Further, this principle may be subject to the principle of contemporaneity. Principle 3: of integration (treaties are to be interpreted as a whole). This principle is of fundamental importance and means that individual parts, chapters or sections of a treaty are not to be interpreted outside of their overall context. The remaining principles take effect subject to the previous group of principles, i.e. are only applicable if the main principles did not lead to satisfactory interpretative results. Principle 4 of effectiveness. According to the principle of effectiveness (*ut magis valeat quam prereat*) treaties are to be interpreted with a reference to their declared or apparent object and purpose and particular provisions are to be interpreted as give them the fullest effect consistent with the normal sense of the words and with the text as a whole, in such a way that a reasonable meaning can be attributed to every part of the text. Principle 5 of subsequent practice (recourse may be had to the subsequent practice of parties relating to the treaty practice of States parties after the entry into force of the treaty); Principle 6 of contemporaneity (terms of the treaty must be interpreted in the light of the linguistic usage current at the time when the treaty was concluded).

text. Fitzmaurice's principles of the interpretation were in part reflected in Articles 31 and 32 of the 1969 VCLT.

Broadly speaking, there are three main schools of (or approaches to) interpretation which are not necessarily exclusive of each other, but frequently are combined—as in the 1969 VCLT—but with a dominant approach of the textual school. The choice of a school means, rather, that primacy is accorded to a particular aspect of interpretation which is its dominant aspect, and after which the schools are named. They are: the intentions of the parties (or the founding fathers) school; the actual wording or text (or textual or ordinary meaning) approach; and the aims and objects of the parties (or teleological) approach.[27]

Jonas and Saunders observe that in practical application of the interpretation of the object and purpose of a treaty, a subjective approach would be more appropriate, as this concept refers to the goals which motivated its drafters and its ratification. However, there are two problems with the subjective method. The first is how to reconcile the multiple intentions of drafters with Article 31 of the 1969 VCLT, which is a general rule of the interpretation; and the second is how to reconcile the dominant textual approach to interpretation adopted by the 1969 VCLT with the subjective intentions of the drafters.[28] The same authors also note the circularity of the interpretative process: a treaty's object and purpose is ascertained through its text; but the text is understood when interpreted in light of its object and purpose. Jonas and Saunders suggest a four-step procedure, which is based on a process of moving back and forth between the specific to the general interpretative techniques. Step one involves the review of specific provisions of a treaty in order to ascertain whether common themes and ideas can be reconciled, or whether they conflict with each other. Step two is a result of the step one investigation, on the basis of which a tentative object and purpose of the treaty can be formulated. Step three is the return to the specific articles of the treaty, reexamining them in the light of the tentative conclusion on the object and purpose, noting conflicts and anomalies. Step four is the return to the general themes and—based on step three (conflicts and anomalies)—results in a revised and refined statement as to what is the object and purpose of the treaty. However, such a procedure does not ensure a clear, single, result, as different interpreters will come to diverse results, depending on their method of conducting the analysis.[29] It may also be

---

27 G. Fitzmaurice, note 23, 230. See also Ian Sinclair, *The Vienna Convention on the Law of Treaties* (Manchester: Manchester University Press, 2nd edn. 1984) 114–115.

28 Jonas & Saunders, note 2, 581.

29 Jonas & Saunders, note 2, 582.

mentioned that a purely teleological approach to treaty interpretation (based only on the object and purpose of a treaty), which may result in overriding the text of the treaty, is a rather contentious matter.

Equally obscure is the principle of 'effectiveness' which, in relation to the teleological approach, means that the objective is to produce an outcome that strengthens achievement of the aims of the treaty, which are identified by reference to its object and purpose. Such an approach should not result in the overriding of the text of the treaty but rather in the modification of the ordinary meaning of the term which is being interpreted, i.e. 'in the sense that that ordinary meaning is to be identified in their light' [the object and purpose].[30]

The above considerations all go to indicate that the concept of the object and purpose of a treaty, and its role in treaty interpretation, is rather obscure. The lack of the uniformity of views on the role of the object and purpose of the treaty in treaty interpretation was very evident in the pleadings in the *Whaling* case in relation to the interpretation the ICRW.

2.2 *The 'Object and Purpose' of the ICRW*[31] *in the Pleadings, Judgment and Judges' Opinions in the* Whaling *Case*

This section is concerned principally with analysis of the treatment of the issue of the object and purpose of the ICRW as a whole in the Pleadings and in the Judgments in the Case. This necessarily involves some reference to Article VIII of the ICRW (scientific whaling); but detailed analysis of the interpretation and effect of Article VIII in the Case will be presented below in Section 6.

2.2.1 Object and Purpose of the ICRW: The Memorial[32]

According to Australia, the ICRW embodies a significant change in its object and purpose by comparison to earlier whaling conventions: the 1931 Convention of the Regulation of Whaling[33] and the 1937 Agreement for the Regulation of Whaling.[34] According to Australia this change was exemplified by the wording of the Preamble to the ICRW which recognizes 'the interest of all the nations of the world safeguarding for future generations the great natural

30 Gardiner, note 2, 221–222.

31 The 1946 International Convention on the Regulation of Whaling, Washington: 2 December 1946, 161 *UNTS* 72 (entered into force 10 Nov 1948).

32 Memorial of Australia, 9 May 2011 (cited hereafter as 'the Memorial'), para. 2.16, http://www.icj-cij.org/docket/files/148/17382.pdf, accessed 5 August 2015.

33 The 1931 Geneva Convention on the Regulation of Whaling, Geneva: 24 September 1931, 155 *LNTS* 349 (entered into force on 16 January 1935).

34 The 1937 London Agreement for the Regulation of Whaling, London: 8 June 1937, 190 *LNTS* 79 (entered into force 7 May 1938).

resources represented by whale stocks'. Australia further suggested that a shift from industrial profitability as the sole objective of international regulation of whaling is evident in the second paragraph of the Preamble, where it says: 'the history of whaling has seen over-fishing of one area after another and of one species of whale after another to such a degree that it is essential to protect all species of whales from further over-fishing'. The Memorial also highlighted the provision of the Preamble to the effect that the system of international regulation of whale fisheries should ensure the proper and effective conservation and development of whale stocks, and the resultant decision to conclude a convention 'to provide for marine conservation and of whale stocks and the orderly development of the whaling industry'.[35] Therefore, Australia submitted, the ICRW had two fundamental objectives underlying the system it established. Firstly, it provided for the proper and effective conservation and recovery of all whale stocks; and secondly, 'relying on and flowing from' the first, to make possible the 'orderly development' of the whaling industry. Australia further noted the significance of the placing of the conservation of whale stocks for future generations as the first objective, prior to any reference to the whaling industry. It was further suggested that through the use of the conjunction 'thus' in the final paragraph of the Preamble, 'the orderly development of the whaling industry' is explicitly made dependent upon effective conservation of the whaling industry.[36] In conclusion, Australia in its Memorial stated unequivocally that while the ICRW is an instrument for the regulation of whaling 'its object and purpose was to establish a comprehensive regime to provide for the proper and effective conservation of all whale stocks'.[37] It was also argued that such an object and purpose is to be achieved 'in evolving and continuing fashion on the basis of best scientific advice';[38] and that 'the legal regime for the regulation of whaling has evolved from a system primarily designed to manage the exploitation of a natural resource to an increasingly conservation oriented regime' on the basis of Resolutions adopted by the IWC.[39]

35 The Memorial, note 32, paras. 2.17 and 2. 18.

36 The Memorial, note 32, paras. 2.19 and 2.20; Australia supported its view by reliance on the views of Patricia Birnie, 'Legal Aspects of Non-consumptive Use of Cetaceans' (1983), unpublished paper in which she said that the primary purpose of the ICRW was the conservation and the development of whale stocks and the secondary was the enabling of the whaling industry to operate in an orderly fashion.

37 Idem, para. 2.20.

38 Ibid.

39 Idem, para. 2.125.

2.2.2 Object and Purpose of the ICRW: The Counter-Memorial[40]

Japan argued that despite the assertions of Australia, concerns relating to the depletion of whale stocks have not changed the object and purpose of the ICRW.[41] Further, Japan interpreted the last sentence of the Preamble differently from Australia, and denied that the 'orderly development of the whaling industry' was contingent on the conservation of whale stocks. On the contrary, 'the conservation of whale stocks' was the means of achieving the goal of 'orderly development of the whaling industry', which according to Australia was only the initial goal of the Convention. Therefore, the ICRW's object and purpose is to enable the development of the whaling industry and to secure the sustainable use of whales.[42] The assertion of Australia that the object and purpose of the Convention has changed cannot be upheld, as there was no official amendment to the Convention.[43]

In interpreting the object and purpose of the ICRW, Japan relied on the *travaux préparatoires* of the ICRW and also noted that the Preamble to the ICRW refers to whaling as a 'resource'.[44] Japan very strongly supported the view that the 'orderly development of the whaling industry' 'is the key and final aim of the ICRW', which can be drawn from the natural meaning of the Preamble.[45] The interpretation of the ICRW as a wildlife preservation instrument, Japan argued, is to ignore its object and purpose as set out in its Preamble, which gives no basis for such an interpretation.[46] Moreover, Japan contended, that neither the object and purpose, nor the nature, of the ICRW have changed since its adoption in 1946.[47] Further, Japan, set out to identify the object and purpose of the ICRW by comparison with other international conservation treaties (such as the 1992 Convention on Biological Diversity), and came to the conclusion that the objective of sustainable use of whale resources is compatible with modern environmental treaty practice.[48] Japan argued that conventions the purpose of which was only conservation and preservation of species used a different language from the ICRW, its object and purpose being rather

40 Counter-Memorial of Japan, 9 March 2012, http://www.icj-cij.org/docket/files/148/17384.pdf, accessed 4 August 2015 (cited hereafter as 'The Counter-Memorial').

41 The Counter-Memorial, note 40, para. 6.2.

42 Idem, para. 6. 3.

43 Idem, para. 6.4.

44 Idem, para. 6.5.

45 Idem, para. 6.11.

46 Idem, para. 6.12.

47 Idem, para. 6.13.

48 Idem, para. 6.30.

the management of natural resources in order to secure their sustainable use.[49] Therefore, in conclusion, Japan stated that the primary purpose of the ICRW is the 'orderly development of the whaling industry' through proper conservation of whale stocks, which latter is not an objective and purpose of the Convention *per se* but only 'an appropriate means to insure a sustainable whaling'.[50]

2.2.3 Object and Purpose of the ICRW: New Zealand's Written Observations[51]

New Zealand also based its pleadings, in this respect, on the Preamble to the ICRW. It surmised from the Preamble that the adoption of the Convention resulted from the admission by the negotiating governments of their common interest in whale stocks and their recognition of the threat to whales caused by unregulated whaling, and from their desire to establish a collective regime for the regulation of all aspects of whaling, as a reaction to 'rampant uncontrolled whaling'.[52] This was the main object and purpose of the ICRW.[53]

New Zealand, similarly to Australia, focused its analysis of the Preamble to the ICRW on those of its paragraphs which referred to the preservation of whales in the interest of future generations, and the common interest in the restoration of whale stocks—i.e. on its ecological aspects.[54] By comparison to Australia, however, New Zealand adopted a more extensive approach to interpretation of the object and purpose of the ICRW. It did not confine itself to analyzing the Preamble, but also took into consideration the whole structure of the Convention—i.e. no aspect of whaling was left outside the Convention's rules. It regulates all activities linked to whaling, wherever and however conducted. Therefore in order to define the object and purpose of the ICRW, New Zealand also took into consideration the Schedule (an integral part of the Convention) which contains all substantive rules, such as open and closed seasons, the moratorium on whaling, etc. The ICRW is open to all States, whaling and non-whaling.[55] New Zealand argued that this indicates that the primary object and purpose of the Convention was not, contrary to the statement of

49 Idem, para. 6.34.

50 Idem, para. 6.35.

51 Written Observations of New Zealand, 4 April 2013, http://www.icj-cij.org/docket/files/148/17386.pdf, accessed 4 August 2015 (cited hereafter as 'Written Observations of New Zealand').

52 Idem, para. 23.

53 Idem, para. 25.

54 Idem, para. 24.

55 Idem, para. 26.

Japan, the 'orderly development of the whaling industry' but the safeguarding of whale stocks.[56]

According to New Zealand, the objective of the ICRW is collective regulation achieved through a process of collective decision-making, through the mechanism established in the IWC.[57] Therefore, according to New Zealand, the Preamble to the ICRW, as well as its scheme and structure, clearly indicate that its object and purpose flows from the 'interest of the nations of the world in safeguarding for the future generations the great natural resources represented by whale stocks'. The Parties to the Convention have agreed to work collectively to establish a system of international regulation to ensure the proper conservation of whale stocks and also the orderly regulation of the whale industry. Therefore, the object and purpose of the Convention was to replace unregulated, unilateral whaling by States with collective regulation as a mechanism to secure proper conservation and management of whales. The object and purpose of the ICRW, so interpreted, should, according to New Zealand, provide the background against which Article VIII it to be understood.[58]

2.2.4 Object and Purpose of the ICRW: Japan's Written Observations[59]

Japan's position, as already set out in the Counter-Memorial (above), was further strengthened in its Written Observations on New Zealand's Written Observations. In this document Japan again reiterated that the emphasis on the preservation of whale stocks, as the only objective and purpose of the ICRW, is misleading as it ignores the 'orderly development of the whaling industry', which was the purpose of the coming of the existence of the Convention.[60] Japan also submitted that notwithstanding the individual intentions of any particular Party when adhering to the Convention, there is not legal principle which would allow for a change of the object and purpose of the convention which would be adverse to the established text of the Convention, and to the will of its originators. In particular, States which at a later date accede to a convention cannot change its object and purpose. They accept the instrument as it stands, and 'to pretend otherwise would encourage an unprecedented

56 Idem, para. 27.

57 Idem, para. 28.

58 Idem, para. 33, note 34.

59 Written Observations of Japan on New Zealand's Written Observations, 31 May 2013, http://www.icj-cij.org/docket/files/148/17388.pdf, accessed 4 August 2015 (cited hereafter as Written Observations of Japan).

60 Idem, paras. 19–20.

form of hijacking of existing conventions'.[61] According to Japan, the object and purpose of the ICRW was never subject to an open and explicit amendment[62] and it could not be 'changed silently on the basis of tenuous inferences unsupported by evidence'.[63]

2.2.5 Object and Purpose of the ICRW: Oral Pleadings by Australia

During oral pleadings, Professor Boisson de Chazournes in very vivid language described the object and purpose of the ICRW from the Australian point of view.[64] In particular, she challenged the approach of Japan to the ICRW, which she described as treating the ICRW as a 'still born instrument' disappearing into a sort of 'black hole'.[65] She relied on the development of international law in the field of environmental law which affected and influenced the approach to the interpretation of the object and purpose of the ICRW. She argued that the 1946 ICRW constitutes 'a comprehensive and dynamic regime all elements of which must be taken into account in order to appraise the behavior of Japan and the lawfulness of JARPA and JARPAII'.[66]

Boisson de Chauournes further argued that the object and purpose of the ICRW relates to the conservation and recovery of whale stocks. She also referred to the subsequent practice of the International Whaling Commission which 'in the same way as development of general international law, confirms the con siderable importance of conservation' under the framework established by the ICRW.[67] Further, she pleaded that:

> mechanisms, tools and legal means for achieving the object and purpose of the ICRW are "not frozen in time". Means and mechanisms may change in the light of the challenges and needs of cetacean species conservation. And it is in the collective framework of the 1946 Convention that they must be specified in order to facilitate achievement of the object and purpose of the 1946 Convention.[68]

61 Idem, para. 22.

62 Idem, para. 23.

63 Idem, para. 19–25.

64 Laurence Boisson de Chazournes CR 2013/7 translation, Wednesday, 26 June 2013 at 10 am, http://www.icj-cij.org/docket/files/148/17440.pdf, accessed 4 August 2015 (cited hereafter as 'Boisson de Chazournes').

65 Idem, para. 44.

66 Idem, para. 69.

67 Ibid.

68 Ibid.

She also said that it is evident that the ordinary meaning of terms such as conservation and recovery implies that whatever 'use' there may be, it can only be 'sustainable use'.[69]

The objectives relating to conservation and recovery of whale stocks must be defined within the multilateral framework instituted by the ICRW, and therefore these legal questions cannot be defined 'unilaterally and on an *ad hoc* basis'.[70]

2.2.6 Object ad Purpose of the ICRW: Oral Pleadings by Japan

Professor Alan Boyle argued in his presentation the issues of object and purpose of the ICRW on behalf of Japan.[71] He started by criticizing the oral presentation of Australia on the object and purpose of the 1946 ICRW as confused, in particular in relation to sustainable use (which included use of natural resources): '[i]s it [Australia] for or against sustainable whaling'.[72] Boyle further analysed the wording of the Preamble of the Convention and concluded that conservation of whale stocks is not intended to be an 'end in itself'.[73] He also analysed the negotiating background of the Convention in order to define its object and purpose with precision. Analysing the development of the 1946 ICRW, he further argued that neither the establishment of Southern Ocean sanctuaries, nor the imposition of the moratorium on whaling (which were strictly of a utilitarian character), not the subsequent resolutions of the IWC had amended or altered the Convention.[74] He strongly stressed that there is not any dichotomy between the object and purpose of the conservation and sustainable utilisation of resources.[75]

Boyle, in analysing the object and purpose of the ICRW, relied on Article 31(3)(c) of the 1969 VCLT, the principle of systemic integration (see below). However, it is not quite clear whether he equated this principle with so-called 'evolutionary' interpretation of treaties (the status of which is in fact not quite clear-cut, as it will be explained further below), or whether the principle of systemic integration can be viewed as a substantively different institution from evolutionary interpretation. Be as it may, Boyle argued that, notwithstanding the

69 Idem, para. 36.

70 Idem, para. 69.

71 Alan Boyle, 3 July 2013, morning siting, CR 2013/13, http://www.icj-cij.org/docket/files/148/17426.pdf, accessed 4 August 2015 (cited hereafter as 'Boyle').

72 Idem, para. 6.

73 Idem, para. 13.

74 Idem, paras. 20 & 21.

75 Idem, para. 23.

principle of systemic integration, which is only a part of the interpretative process, the first task of the interpreter is to analyse the plain meaning of words in the their context and in the light of the object and purpose of the provision.[76] He also claimed that even evolutionary interpretation, applied to the ICRW, could not change fundamentally its object and purpose contrary to the plain meaning of the text, to the extent that a treaty about sustainable whaling would become a convention on elimination of whaling. According to Boyle, existing international law does not support the interpretation of Australia on the evolutionary reading of the ICRW, but rather the interpretation of Japan.[77]

Boyle further analysed the meaning of the term 'conservation' which in his view supports the coexistence of conservation and optimum and sustainable utilisation. Such an understanding of the term 'conservation' is, according to him, supported by case-law, such as *Fisheries Jurisdiction*[78] case and *Bluefin Tuna* cases.[79]

Boyle's final argument was based on the role of the IWC which he strongly denied had changed from the management of whaling to the protection of whales. The evidence against such an assertion is the development of the Revised Management Procedure which is specifically directed at the management of whaling.[80]

2.2.7 Object and Purpose of the ICRW: Oral Pleadings by New Zealand

The representative of New Zealand, Christopher Finlayson, also addressed the issue of the object and purpose of the ICRW.[81] Like all previous speakers, he referred to the Preamble of the ICRW. He rejected the interpretation by Japan of the sentence 'decided to conclude a convention to provide for the proper

---

76 Idem, para. 26.

77 Idem, paras. 27 & 29.

78 *Fisheries Jurisdiction (United Kingdom v. Iceland), Merits,* Judgement, *I.C.J. Reports* 1974, p. 31 para. 72, http://www.icj-cij.org/docket/files/55/5749.pdf, accessed 5 August 2015. *Fisheries Jurisdiction (Federal Republic of Germany v. Iceland), Merits, Judgment, I.C.J. Reports* 1974, p. 200, para. 64, http://www.icj-cij.org/docket/files/56/10713.pdf, accessed 4 August 2015.

79 *Southern Bluefin Tuna Cases (New Zealand v. Japan; Australia v. Japan)*, order of 27 August 1999, ITLOS. Nos. 3 & 4, paras. 28(2) and 29(2), https://www.itlos.org/fileadmin/itlos/documents/cases/case_no_3_4/Order.27.08.99.E.pdf, accessed 5 August 2015. Boyle, note 72, paras. 31 & 32.

80 Idem, para. 55.

81 Christopher Finlayson, morning sitting 8 July 2013, http://www.icj-cij.org/docket/files/148/17444.pdf, accessed 4 August 2015 (cited hereafter as 'Finlayson').

conservation of whale stocks and thus make possible the orderly development of the whaling industry', saying that the reliance by Japan on this paragraph of the Preamble alone distorted the sense of the Preamble as a whole. Reading the Preamble as a whole, by contrast, indicates that the objective of the Parties was to 'establish a system of international regulation of the whale fisheries to ensure proper and effective conservation and development of whale stocks' not 'the whaling industry'.[82] According to Finlayson, the purpose of the Convention was to provide a system through which the individual interests of the Parties in the future of whale stocks could be managed for the greater common interest of all the Parties through a comprehensive system of collective regulation.[83] The interpretation of the ICRW as an instrument expressing collective interest, is according to him, reflected by the structure and functions of the IWC, where all the Parties are present and which adopts all decisions collectively by a vote of its members, and the regulations so adopted are binding on them.[84]

2.2.8 Object and Purpose of the ICRW: Judgment and Judges' Opinions

The object and purpose of the ICRW in the Judgment is closely related to the interpretation of Article VIII of the Convention (see further below). However, para. 56 of the Judgment refers only to the object and purpose of the ICRW in a general sense. It reads as follows:

> The preamble of the ICRW indicates that the Convention pursues the purpose of ensuring the conservation of all species of whales while allowing for their sustainable exploitation. Thus, the first preambular paragraph recognizes 'the interest of the nations of the world in safeguarding for future generations the great natural resources represented by the whale stocks'. In the same vein, the second paragraph of the preamble expresses the desire 'to protect all species of whales from further overfishing', and the fifth paragraph stresses the 'need to give an interval for recovery to certain species now depleted in numbers'. However, the preamble also refers to the exploitation of whales, noting in the third paragraph that 'increases in the size of whale stocks will permit increases in the number of whales which may be captured without endangering these natural resources', and adding in the fourth paragraph that 'it is in the common interest to achieve the optimum level of whale stocks as rapidly

82 Idem, paras. 11 & 12.
83 Idem, para. 13.
84 Idem, para. 16.

> is possible without causing widespread economic and nutritional distress' and in the fifth that 'whaling operations should be confined to those species best able to sustain exploitation'. The objectives of the ICRW are further indicated in the final paragraph of the preamble, which states that the Contracting Parties 'decided to conclude a convention to provide for the proper conservation of whale stocks and thus make possible the – orderly development of the whaling industry'. Amendments to the Schedule and recommendations by the IWC may put an emphasis on one or the other objective pursued by the Convention, but cannot alter its object and purpose.

Thus, despite lengthy arguments by the parties both written and oral and the obvious importance accorded to the issue of the object and purpose of the ICRW generally by them, the Court did not analyse the point at great length, containing just the above succinct statement. The Court did, however, make further pronouncements concerning particular aspects of the general object and purpose of the Convention which are covered in later sections of this chapter.

The issue was, in fact, more extensively dealt with in Separate and Dissenting Opinions of the Judges.

Judge Owada in his Dissenting Opinion presented the following main overview of the different approaches by the Parties to the dispute relating to the object and purpose of the ICRW:

> It is argued on the one hand that there has been an evolution in the economic-social vista of the world surrounding whales and whaling over the years since 1946, and that this is to be reflected in the interpretation and the application of the Convention. It is argued, on the other hand, that the juridico-institutional basis of the Convention has not changed since it was drafted, based as it was on the well-established principles of international law relating to the conservation and management of fishing resources, including whales, and that this basic character of the Convention should essentially be maintained. This to my mind is the fundamental divide that separates the legal positions of the Applicant, Australia, and New Zealand as an intervener under Article 63 of the Statute, and that of the Respondent, Japan.[85]

85 Judge Owada, Dssenting Opinion, para. 4, http://www.icj-cij.org/docket/files/148/18138.pdf, accessed 4 August 2015 (cited hereafter as 'Judge Owada').

In order to identify the object and purpose of the ICRW, Judge Owada first analysed the Preamble of the ICRW, and further the structural and functional framework of the Convention in order to state its object and purpose. He took into account the object and purpose of the ICRW as it stood in 1946 at its signing and later, at the time of the imposition of the Moratorium on commercial whaling in 1982, and concluded that

> It is clear from this that the object and purpose of the Convention is to pursue the goal of achieving the twin purposes of the sustainability of the maximum sustainable yield ('MSY') of the stocks in question and the viability of the whaling industry. Nowhere in this Convention is to be found the idea of a total permanent ban on the catch of whales.[86]

Further, the concept of 'conservation of fisheries resources' has an element of 'maximum/optimum sustainable yield' as an integral part which is included in the Convention.[87] This object and purpose of the ICRW should be analysed within the framework of the essential characteristics of the regime established under the Convention, which should be the starting point to the understanding of the exact nature and structure of the regulatory regime contained in the concrete provisions of the Convention, and the legal nature of the rights and obligations prescribed for scientific activities under Article VIII as its central element.[88] Judge Owada is of the view that the Court failed to analyse the essential characteristics of the ICRW's regime in order to state its object and purpose, except for the brief statement in para. 45 that '[t]he functions conferred on the Commission have made the Convention an evolving instrument'.[89] (For a more detailed analysis of issues covered in Paragraph 45 of the Judgment, see below in Section 3.) Having analysed the structure of the ICRW, the IWC and its regulatory powers (including the system of adopting regulations and amendments to the Convention which is not absolutely binding on the parties—i.e. the system of tacit acceptance or opting out—) Judge Owada concluded that the ICRW has created a kind of self-contained regulatory regime on whales and whaling which may be comparable to the self-contained system of an intergovernmental international organization which has its own administrative autonomy with its regulatory regime for matters within the scope of its jurisdiction. Under such a system, while there is autonomy of the Parties *inter se*,

86 Idem, para. 9, note 84.
87 Idem, para. 10, note 84.
88 Idem, para. 11, note 84.
89 Idem, para. 12, note 84.

this is still subject to judicial review by the Court 'in accordance with the power given to it for interpreting and applying its constitutional document, namely, the Convention'.[90]

Interpreting the actions of Japan—i.e. stepping up its scientific programme—from the point of view of the entire regulatory regime, and accepting Japan's assertion that they were aimed at the provision of more evidence to the IWC in order to assist lifting the moratorium on commercial whaling, which was imposed as a temporary measure, Judge Owada was of the view that '[i] t would seem difficult to see anything wrong in the Respondent's course of action'.[91]

Judge Cançado Trindade made several comments concerning the object and purpose of the ICRW as a background to the process of establishing of the object and purpose of its Article VIII. According to this Judge the adoption of a Convention structure like that of the ICRW, with its own supervisory organ, is a proof that 'goal' of conservation is integral to its object and purpose, which is not 'limited to the development of the whaling industry'. Had that been the case, the Convention would have been structured differently.[92] Having analysed the Preamble to the ICRW, in particular the paragraphs relating to the conservation of whale stocks, he concluded that 'it appears that the primary object and purpose of the ICRW Convention can be found in the conservation and recovery of whale populations'.[93] The ICRW also evolved to face new challenges through Recommendations of the IWCR.

Judge Cançado Trindade set out very clear parameters in order to establish the object and purpose of the ICRW, which is to be construed in light of its text, its supervisory mechanism, and its nature as a multilateral treaty encompassing both whaling and non-whaling States. Accordingly the object and purpose of the Convention indicate, as a guiding principle, the conservation and recovery of whale stocks. The sustainable development of the whaling industry or the protection of commercial whaling is not accorded the same importance. A State Party cannot act unilaterally in order to decide whether its programme is fulfilling the object and purpose of the ICRW Convention, or the objective of conservation.[94]

Judge Cançado Trindade added an interesting aspect to the issue of the identification of the object and purpose of the ICRW—i.e. he observed its character

90 Idem, para. 14, note 84.

91 Idem, paras. 17 & 18.

92 Separate Opinion of Judge Canaçado Trindade, paras. 2 & 3, http://www.icj-cij.org/docket/files/148/18146.pdf, accessed 5 August 2015 (cited hereafter as 'Judge Canaçado Trindade').

93 Idem, para. 5.

94 Idem, para. 7.

as an 'evolving' or 'living' instrument, similar to other international instruments with regulatory bodies (see further below). To this effect, in the interpretation of the object and purpose of the ICRW, he compared it to other Conventions which adopted a conservation-oriented approach to marine mammals, such as the Convention on Biological Diversity, as the Whaling Convention is not stand-alone instrument, but 'is part and parcel of a system of collective guarantee and collective regulation oriented towards the conservation of living species'.[95]

Judge Greenwood also commented on the interpretative techniques concerning the object and purpose of the ICRW.[96] Like other Judges, he distinguished the interpretation of the object and purpose of the ICRW as provided, on the one hand, by Australia—based on the references to conservation in the Preamble, and on the resolutions of the IWC to evidence the development of the Convention; and, on the other hand, by Japan—relying in particular on the last paragraph of the Preamble.[97]

Judge Greenwood did not agree wholly with either of these approaches. Australia's approach, he maintained, is not fully reconcilable with the language of the Preamble. The language of the Convention and its *travaux préparatoires* undoubtedly indicated that one of the objectives of the ICRW was 'to ensure a future for the whaling industry by making sustainable whaling possible'. On the other hand, Japan's argument that the Convention treats conservation as wholly subordinate to the development of whaling is also untenable. The Preamble shows that both conservation and ensuring a future for sustainable whaling were considered purposes of the Convention.[98] In Judge Greenwood's view, that balance between the two purposes was not influenced by the Resolutions of the IWC, in the way that was submitted by Australia. Judge Greenwood further recalled that the IWC has the power to make two types of Resolution: Regulations to amend the Schedule (which in fact are amendments to the IWRC itself and are binding on States unless opted out from); and non-binding Recommendations. It was, according to Judge Greenwood, the Regulations adopted by the IWC which, in view of the Court, had made the ICRW 'an evolving instrument'. However, the question to be investigated was, rather, the role of Recommendations in the interpretative process.[99] Judge Greenwood,

---

95 Idem, paras. 58 & 59.

96 Separate Opinion of Judge Greenwood, http://www.icj-cij.org/docket/files/148/18150.pdf, accessed 6 August 2015 (cited hereafter as 'Judge Greenwood').

97 Idem, para. 3.

98 Idem, para. 4.

99 Idem, para. 5.

having analysed the Resolutions adopted by the IWC, concluded that those which were taken by the agreement of all parties could be treated as subsequent practice in the meaning of Article 31(3) (b) of the 1969 VCLT (see further below); but the Resolutions which lacked such agreement in their adoption 'cannot, therefore, be relied on to sustain an interpretation of the Convention which can bind Japan'.[100] Further, the difference in the procedural requirements concerning the adoption of, respectively, Regulations and Recommendations indicate that Recommendations cannot produce the same effect as Regulations. Likewise they have a less pronounced effect as the basis of an evolutionary interpretation of the provisions of the Convention.[101]

2.3 *Analysis of the Object and Purpose of the ICRW in Light of the Parties' Pleadings and of the Judgment and Judges' Opinions: Interim Conclusions*

The approach of the Parties to the issue in the Whaling case is a very good example of the view that the object and purpose of a treaty means whatever States want it to mean. Thus, in general, while the Parties to the dispute all looked for the object and purpose of the treaty mostly in its Preamble, each relied on its various parts, depending on their particular argument: Australia and New Zealand emphasizing the Preamble's conservation approach and the value of whales to future generations, and Japan stressing the last paragraph, which refers to 'the proper conservation of whale stocks and thus make possible the orderly development of the whaling industry'. However, some Judges noted adversely this focus of the Parties mostly on the relevant provisions of the Preamble to the 1969 VCLT, instead of taking into consideration the whole Convention in order to identify its object and purposey—as was suggested by Jonas and Saunders (see above). Such an approach would also have complied with the principle of integration, i.e. that treaties are to be interpreted as a whole, a principle which is of fundamental importance and means that individual parts, chapters or section of a treaty are not be interpreted outside of their overall context. The Parties did, however, rely in their Pleadings upon the *travaux préparatoires* to the ICRW, which according to Article 32 of the 1969 VCLT are a supplementary means of interpretation.

Infact, the quest relating to the object and purpose of the IWC did not emerge as a completely new issue during the *Whaling* case. It had been analysed a long time even before the Pleadings in the case.

100 Idem, para. 6.

101 Idem, para. 7.

Already in 1970s there was a discussion regarding the object and purpose of the ICRW. Views were roughly divided into two groups—those which assumed that conservation was the primary objective of ICRW (the secondary being the orderly development of the whaling industry);[102] and those who saw the development of the whaling industry as the ultimate aim—thus mirroring the Pleadings in the *Whaling* case.[103]

In 2006, the object and purpose of the ICRW was the subject-matter of one of the most important and controversial Resolutions adopted within the IWC, namely, Resolution 1-2006, the St Kitts and Nevis Declaration, which was adopted by thirty three votes to thirty two, with one abstention. This Resolution noted that that '...the purpose of the 1946 International Convention for the Regulation of Whaling (ICRW) is to "provide for the proper conservation of whale stocks and thus make possible the orderly development of the whaling industry" (quoted from the Preamble to the Convention) and that "the International Whaling Commission (IWC) is therefore about managing whaling to ensure whale stocks are not over-harvested rather than protecting all whales irrespective of their abundance"'.[104] Thereafter, the issue of the object and purpose of the ICRW continued to occupy the attention of the IWC.[105]

Bowman refers to the foundational principles of interpretation in order to find a correct manner in which to interpret the ICRW. Therefore, he approaches this task on the basis that a legal instrument should be interpreted responsibly, in good faith and in accordance with established legal criteria, in two phases.[106] The first phase, he suggests, is that it is necessary to extract from the preamble, together with the text as a whole and the surrounding circumstances, a more reliable and comprehensive sense of the original motivations behind the adoption of the ICRW, and the extent in which they have been *included* in its objectives and purposes. Secondly, it is of fundamental importance to ascertain how

102 J.E. Scarff, 'The International Management of Whales, Dolphins and Porpoises', *Ecology L.Q.* 6 (1977), 353–354.

103 S.V. Scott, *The Political Interpretation of Multilateral Treaties* (Leiden: Brill Academic Publishers, 2004), 125.

104 International Whaling Commission, St Kits and Nevis Declaration, Normalizing the International Whaling Commission, IWC/58/12, Agenda item 19 (2006), http://www.unesco.org/csi/smis/siv/Forum/SKNdeclaration06.pdf, accessed 4 August 2015.

105 The in depth analysis of Michael Bowman, 2008, represented a 'revised version of a paper that was made available to the International Whaling Commission in connection with the current deliberations concerning the future of that organization', Bowman, note 2, 293.

106 Idem, 312–314.

these elements have evolved during the almost seventy years of the Convention's existence.[107]

Regarding the first of these steps, Bowman suggests that an important element is the very adoption itself of the ICRW, which is the result of a single conference (not extended negotiations) and based almost entirely on the US draft. Therefore, although the final text of the Convention reflects the joint drafting effort, it cannot be ignored that in cases such as the ICRW, the final text may reflect particular interest of the main drafting State, the US.[108] Bowman turns his attention, in order to interpret the object and purpose of the ICRW, to the part of its Preamble where it is stated that the Parties desire to establish a system of international regulation for the whale fisheries to ensure proper and effective conservation and development of whale stocks on the basis of the principles embodied in the provisions of the 1937 International Agreement for the Regulation of Whaling. Notwithstanding that the 1937 Convention was based on the principle of the profitability of whaling industry, taking into account the whole text of the 1946 Convention, however, it may be said that the 1946 Convention, was aimed at establishing a different regulatory system.[109] Even so, Bowman, having analysed the whole text of the Preamble concluded that certain passages therein such as

> [r]ecognizing that the whale stocks are susceptible of natural increases if whaling is properly regulated, and that increases in the size of whale stocks will permit increases in the number of whales which may be captured without endangering these natural resources

indicate that at the time of the adoption of the ICRW it was not a part of the objectives of the Convention to preclude all whaling. However, the term 'orderly' indicates that some enhanced degree of organisation and regulation would be imposed on the whaling industry.[110]

Bowman next conducts a rigourous analysis of the text of the ICRW in order to further identify its object and purpose. According to him the key provision in this respect is Article V(2) of the ICRW which sets out the parameters of its further development. A close analysis of the wording of this article indicates that both the conservation of the whaling stocks and the interests of the whaling

107 Idem, 386–398.

108 Idem, 385.

109 Idem, 390.

110 Idem, 400, 402.

industry were taken into consideration.[111] Despite the new stress on conservation, it was not an intention to subjugate the whaling industry completely.[112] The examination of preparatory work of the ICRW is inconclusive—although it perhaps has a slight slant towards conservation over the interests of the whaling industry. It is really quite difficult, however, to state that the clear intention was to stop whaling in its entirety. Despite the wording of the Preamble, which may give the impression of whales being more than just a natural resource, there is little evidence to substantiate the argument that, at least originally, the whaling industry was intended to be obliterated. In support of this, it may be noted that that the original US draft of the Preamble had expressly included a statement that the 'ultimate objective should be to achieve and maintain the stocks at the level which will permit a sustained capture at maximum number of whales'.[113]

It may be stated that it appears that the original object and purpose of the ICRW was to balance the interests of the whaling industry and to conserve whale stocks (with perhaps a slight slant towards conservation) which thus did not exclude *per se* sustainable use of whaling resources. However, it may be disputed, as stated by Australia, that the primary object and purpose of the IWCR was only the preservation of whale stocks. Even from the point of the view of post-war economics and the food situation, whaling was viewed as a source of protein. Therefore, apart from the analysis based on the law of treaties, it is implausible that the intention of the drafters of the ICRW was a total ban on whaling.

3 The Object and Purpose of the ICRW and the Concept of Evolutionary Interpretation

The possibility that the concept of evolutionary interpretation might be applied to the interpretation of the object and purpose of the ICRW is possibly

111 Article V(2) 'These amendments of the Schedule (a) shall be such as are necessary to carry out the objectives and purposes of this Convention and to provide for the conservation, development, and optimum utilization of the whale resources; (b) shall be based on scientific findings; (c) shall not involve restrictions on the number or nationality of factory ships or land stations, nor allocate specific quotas to any factory ship or land station or to any group of factory ships or land stations; and (d) shall take into consideration the interests of the consumers of whale products and the whaling industry'. Bowman, note 2, 410.

112 Idem, 411.

113 IWC, United States Draft of the Preamble to the International Convention on the regulation of Whaling (Nov. 20–Dec. 2) (1946), cited in Bowman.

more challenging than the process of identifying its original object and purpose in accordance with more traditional means of interpretation. This issue will be analyzed in this chapter in two parts: the first part (Section 3.1) will be devoted to some general observations on the legal nature of the concept of evolutionary interpretation; second part (Section 3.2) will deal with the possibility of applying the concept of evolutionay interpretation to the ICRW's object and purpose in particular, pre-Judgment and post-Judgment.

Before coming to this, however, one should note that the issues concerning evolutionary interpretation, subsequent practice and systemic integration are at times very difficult to separate. These are all very complex issues which are concerned with treaties over time (though there are some disagreements concerning systemic integration in this respect)—i.e. the inter-temporal dimension of treaty interpretation (see on these issues further below).[114]

The issue of the temporal dimension of the law of treaties was further raised by the Working Group of the International Law Commission (ILC), within the framework of the fragmentation of international law, as a question of treaty interpretation.[115] In 2009, the ILC established a Working Group to work on the topic 'Treaties over Time', which work was commenced with analysis of subsequent agreement and practice (Professor Nolte as Special Rapporteur).[116]

---

114 The International Law Commission (hereinafter the 'ILC') attempted to address this dimension in the 1969 VCLT. but abandoned such an initiative. 1964 *Yearbook of the International Law-Commission* Vol. I, 33 *et seq*; 1964 *ILC Yearbook* Vol. II 8 et seq; 1966 Yearbook of the ILC, Vol. 1, pt.2, 184 et seq.1966 *ILC Yearbook*, Vol 2, 91. See also, *Island of Palmas* (the *Netherlands v. USA*) (1928) *RIAA* 829. Judge Max Huber made a statement that: 'A judicial fact must be appreciated in the light of the law contemporary with it, and not of the law in force at the time such a dispute in regard to it arises or falls to be settled. The existence of the right, in other words, its continued manifestation, shall follow the conditions required by the evolution of the law'.

Third Report, Waldock offered an article (draft article 56) appropriately entitled *The Inter-Temporal Law*, which provided:

1. A treaty is to be interpreted in the light of the law in force at the time when the treaty was drawn up.
2. Subject to paragraph 12, the application of a treaty shall be governed by the rules of international law in force at the time when the treaty is applied.

Read more: http://www.answers.com/topic/inter-temporal-law#ixzz36tXG1fjf, accessed 4 January 2015.

115 Report of the study Group on Fragmentation of International Law: Difficulties Arising from the Diversification and Expansion of International Law, finalised by Martti Koskenniemi (13 April 2006), UN Doc. A/CN. 4/L682, 242f, para. 478.

116 ILC 'Report of the International Law Commission on the work of its 60th Session' (5 May–6 June and 7 July–August 2008), UN Doc. A/63/10, Annex A, para. 352.

The role of the evolutionary (or dynamic) interpretation and subsequent practice in treaty interpretation was succinctly summed up by Nolte:

> A proper assessment of a treaty's object and purpose, in the light of unforeseen circumstances and its text, may even yield an 'evolutive' interpretation of certain rules in specific cases. However, the abstract possibility to interpret a treaty in a more or less dynamic fashion does not absolve interpreters from inquiring into a continuing role of the parties in the application of the treaty which can provide a sense of direction.[117]

3.1 *General Issues Relating to Evolutionary Interpretation of Treaties*

The evolutionary interpretation of treaties is one of the subjects of international law which has been extensively written about and analysed, in particular in relation to the practice of the European Court of Human Rights (ECtHR) (see further below).[118] Therefore only the main outline of the issues involved will be presented in this chapter.

In the first place, as has been argued, the term 'evolutionary' (sometimes 'evolutive' or 'dynamic') interpretation is more useful than the term 'teleological' as it is more inclusive of the notions involved.[119] Arato distinguishes two possible situations in which the intentions of the parties regarding treaty evolution can be identified. One is where development of the treaty over time was somewhat an original intention of the Parties to the treaty (evolutionary interpretation); and the second category, whereby the treaty is interpreted developmentally on the grounds of the later behavior of the Parties

---

117 Georg Nolte, 'Introduction', in *Treaties and Subsequent Practice*, Georg Nolte (ed.) (Oxford: Oxford University Press, 2013), 2 (cited hereafter as 'Nolte').

118 The recent publications on this subject are as follows: Pierre-Maria Dupuy, 'Evolutionary Interpretation of Treaties: Between Memory and Prophecy', in Enzo Canizzaro (ed.,) *The Law of Treaties Beyond the Vienna Convention* (Oxford: Oxford University Press, 2011), 123–137; Julian Arato, 'Subsequent Practice and Evolutive Interpretation: Techniques of Treaty Interpretation Over Time and Their Diverse Consequence', *The Law and Practice of International Courts and Tribunals*, 9 (2010): 443–449 (cited hereafter as 'Arato, Subsequent Practice'); Malgosia Fitzmaurice, 'Dynamic (Evolutive) Interpretation of Treaties'(Part I) *The Hague Yearbook of International Law* 21 (2008) (cited hereafter as 'M Fitzmaurice, Dynamic Interpretation, Part I'), 101; & Dynamic (Evolutive) Interpretation of Treaties Part II, *The Hague Yearbook of International Law* 22 (2009) (cited hereafter as M Fitzmaurice, Dynamic Interpretation, Part II), 3; Eirik Bjorge, *The Evolutionary Interpretation of Treaties* (Oxford: Oxford University Press, 2014).

119 M Fitzmaurice, Dynamic Interpretation, Part I, note 119, 102.

(interpretation based on subsequent practice).[120] However, in the view of the author of this chapter, the intentions of the Parties at times may be presumed very liberally, based on a presumption of such intentions made after the conclusion of the convention. An example of this may be found int the European Convention on Human Rights (ECHR), where the judicial practice of the ECtHR, (see further, below) which interpreted the ECHR in an evolutionary (it took several stages), was subject to a harsh criticism by several authors and Judges, such as Sir Gerald Fitzmaurice, who denied the existence of such an intention, thus insisting on the application of the principle of 'contemporaneity'.[121] According to Fitzmaurice, the principle of effectiveness (see above) is the *only* doctrine pursuant to which it may be legitimately assumed that the Parties intended their treaty to be capable of evolving over time.[122]

There is no absolute definition of what is evolutionary interpretation of treaties, and certain points are still unclear and contentious—one being the status of Article 31(3)(c) (embodying so called principle of systemic integration) of the 1969 VCLT, in relation to evolutionary interpretation, which states that in interpreting a treaty, account may be taken of 'any relevant rules of international law applicable in the relation between the parties'. There are views which appraise this article as embodying the concept of evolutionary interpretation, such as the Fragmentation Report;[123] and views which approach evolutionary interpretation and interpretation according to Article 13(3)(c) as different in substance.

The view also is held is that there are two types of the interpretation with the presence of the temporal element, both of them embracing the idea of the evolution of a treaty over time.[124] In broad brushstrokes, the difference is based on the origin of these two evolutionary approaches. One is subsequent practice, or, as Distefano named it 'endogenous evolutionary interpretation', which is due to new facts which emerged and can be adapted through adapting the treaty; and the other, 'exogenous evolutionary treaty interpretation',

120 Arato, Subsequent Practice, note 119, 433, 445.

121 G. Fitzmaurice, note 23, 203, 225 (1957).

122 Idem, 223.

123 ILC Study Group Report on Fragmentation, note 116.

124 Giovanni Distefano, 'Subsequent Practice of the Parties to a Treaty: Interpretation, Modification and Termination (La Pratique Subséquente des Etats Parties à un Traité)', *Annuaire Français de Droit International*, 40 (1995), 41, http://papers.ssrn.com/sol3/papers.cfm?abstract_id=1687643, accessed 6 August 2015.

originating externally to the treaty, due to changes in the normative environment outside the treaty itself, which occurred since the conclusion of the treaty—an interpretative approach which is in fact the embodiment of Article 31(3)(c).[125]

However, the view expressed by other authors suggest that '[e]volutive interpretation is conceptually independent from Article 31(3)(c)'.[126] Helemersen explains that:

> The difference is simply that Article 31.3.c is about interpretation in light of other law, while evolutive interpretation is about interpretation in light of some current meaning. This means that the range of relevant arguments to determine the evolution of an evolving term will often be much broader than just the 'rules of international law' that Article 31.3.c mentions.[127]

Helemersen maintains that interpretation in the light of Article 31(3)(c) is not based on the intentions of the parties, which are largely the basis of the evolutionary interpretation, Furthermore, Article 31(3)(c) may be invoked in cases where interpreting evolutively is not permissible, when the term being interpreted was not intended to be evolutive. However, there may be some overlap between these two approaches. The terms within the ambit of Article 31(3)(c) may have an evolutionary character and consequently the terms of a treaty subject to the interpretation may be given an evolutionary meaning.[128] Such an understanding of the relationship between these two concepts, in the view of the present author, is more persuasive than to unequivocally treat both of these interpretative techniques as identical.

However, despite many publications, the complex relationship between evolutionary interpretation, systemic integration (Article 31(3)(c)), and subsequent practice (Article 31(3)(b)) has not been solved in all respects and still awaits more elucidation.

The International Court of Justice relied in a number of cases on the evolutionary concept of the interpretation of the treaties. To mention some: the 1970

---

125 Ibid.

126 Sondre Torp Helmersen, 'The Evolutive Treaty Interpretation: Legality, *Semantics and Distinction*', *European Journal of Legal Studies*, 6 (2013): 126, 147, http://www.ejls.eu/12/150UK.pdf, accessed 6 August 2015 (cited hereafter as 'Helmersen').

127 Idem, 147.

128 Ibid.

Namibia advisory opinion;[129] the 1978 *Aegean Sea*;[130] the 2009 *Dispute Regarding Navigational and Related Rights*.[131]

The evolutionary interpretation of treaties based on intentions of the parties to a treaty (even presumed) was discussed above. The second type of such an interpretation is based on the treaty's object and purpose, which is particularly relevant to the *Whaling* case in which many of the of the interpretative arguments were made on the basis of the analysis of its object and purpose, as stated in its Preamble.

This approach, as stated by Arato, takes object and purpose as its starting point, rather than its outer limit. This type of evolutionary interpretation does not rely on the intentions of the parties, but determines a treaty or treaty provision to be evolutionary on the basis of its object and purpose.[132] The interpretative technique was applied in the *Iron Rhine Arbitration*.[133] As Arato explains, on the basis of such an evolutionary interpretation (based on the treaty's object and purpose), two questions may be asked: '(1) whether it is necessary to give the treaty an evolutive reading to make the agreement *effective* in terms of its object and purpose; and (2) how the demands of object and purpose substantiate the term's evolved meaning at the time of application'.[134] As he correctly states, however, due to the notorious difficulties linked to defining the object and purpose (or even objects and purposes) of a treaty, the general evidentiary standards of this mode of evolutionary interpretation are relatively insignificant.[135] In light of such complexities in the determination of the object

129 *Legal Consequences for States of the Continued Presence of South Africa in Namibia (South West Africa) notwithstanding Security Council Resolution 276 (1970)*, [1971] ICJ *Reports* 16, para. 53, http://www.icj-cij.org/docket/files/53/5595.pdf, accessed 7 August 2015.

130 *Aegean Sea Continental Shelf* [1978] ICJ *Reports* 3, para. 77, http://www.icj-cij.org/docket/files/62/6245.pdf, accessed 7 August 2015.

131 *Dispute regarding Navigational and Related Rights (Costa Rica v. Nicaragua)* [2009] ICJ *Reports* 213, paras. 64, 66, http://www.icj-cij.org/docket/files/133/15321.pdf, accessed 7 August 2015.

132 Arato, Subsequent practice, note 119, 473.

133 *Iron Rhine Arbitration (Belgium. v. Netherlands.), 27 R.I.A.A.* 35, 73 (para. 80) 'In the present case it is not a conceptual or generic term that is in issue, but rather new technical developments relating to the operation and capacity of the railway. But here, too, it seems that an evolutive interpretation which would ensure an application of the treaty that would be effective in terms of its object and purpose, will be preferred to a strict application of the intertemporal rule…' Reports of Arbitral Awards, http://legal.un.org/riaa/cases/vol_XXVII/35-125.pdf, accessed 5 August 2014.

134 Arato, Subsequent Practice, note 119, 473.

135 Ibid.

and purpose of a treaty tribunals must determine it on a case-by-case basis.[136] According to Arato there is a problem whether we deal with the effect of multiple objects and purposes, or with various strands of a single complex object and purpose, in substantiating an evolutive term. This observation is also relevant for the interpretation of the ICRW, as it has at least two identified (and seemingly contrasting) objects and purposes, namely, the conservation of whales and the orderly development of the whaling industry. Arato is of the view that in cases like that there are no clear standards for deciding what has the priority. The best way forward is to analyse the issue on a case-by-case basis, by the usual means of interpreting a treaty on the basis of its object and purpose—i.e. a tribunal should look at the preamble and broader treaty text, as well as its context, in order to make a determination on whether and in what way a particular treaty should evolve.[137] Finally, he states that the most important evidentiary standard for evolutionary interpretation 'concerns the relationship between a treaty's object and purpose, on the one hand, and a determination of both a treaty's evolutionary capacity and how it should be reinterpreted, on the other'.[138] A treaty is to be considered evolutionary in the light of its object and purpose only if such an analysis is *necessary* to give effect to its object and purpose.[139] Likewise, the substance of its reinterpretation must be similarly necessary. A 'necessary relation' constitutes the most important limit on superfluous application of evolutionary interpretation on the basis of object and purpose.[140]

The complexities of the evolutionary interpretation of a treaty are best illustrated by the interpretation of the ECHR by the ECtHR. In the view of the present author, treaty interpretation by human rights courts and bodies merit being categorised separately, especially in the light of the judicial practice of the ECtHR.[141] Even though its method of treaty interpretation is considered as belonging to the category of evolutionary interpretation of treaties, it has particular characteristics which, in the view of the present author, set it apart from other forms of such an interpretation, as a very singular kind in which (especially in later developments in evolutionary interpretation the ECtHR)

136 Idem, 475.

137 Ibid.

138 Ibid.

139 Idem, 476.

140 M Fitzmaurice, Dynamic Interpretation, Part II, note 119.

141 Malgosia Fitzmaurice, 'Interpretation of Human Rights Treaties' (Dinah Shelton ed.), *The Oxford Handbook of International Human Rights* (Oxford: Oxford University Press, 2013), 739–772.

law and morality combine to constitute together a very distinctive approach to interpretation. As is well known, the approach to the ECHR as a 'living' ('evolving' or dynamic) instrument has been present in the ECtHR's case law since the 1978 *Tyrer v United Kingdom* case.[142] In this Judgment the Court stated famously that:

> The Court must also recall that the Convention is a living instrument which, as the Commission rightly stressed, must be interpreted in the light of present-day conditions. In the case now before it the Court cannot but be influenced by the developments and commonly accepted standards in the penal policy of the member States of the Council of Europe in this field.[143]

In several later cases, such as *Marckx v Belgium*,[144] the ECtHR had refined and further developed its approach towards evolution of the Convention's provisions. In the *Marckx* case, the Court referred to two conventions, the Brussels Convention on theEstablishment of Maternal Affiliation of Natural Children and the European Convention on the Legal Status of Children Born out of Wedlock, in order to demonstrate the existence of 'commonly accepted standards'. However, these Conventions had not yet been ratified by the majority of the Contracting States at the time of the Judgment. The Court justified its approach by stating that the very existence of such conventions evidences that there is a 'a clear measure of common ground in this area amongst modern societies', and further that Belgian law itself showed signs of this 'evolution of rules and attitudes'.[145]

Letsas makes interesting comments on these early cases of the ECtHR regarding the ECHR as a 'living instrument'. The Court was asked to rule on morally sensitive issue in the respondent state, such as corporal punishment, children born out of wedlock and homosexuality. The Court took notice of the relevance of the prevailing moral climate within the respondent to the inter-

---

142 *Tyrer v. United Kingdom* (Appl. No. 5856/72) Judgment of 25 April 1978, Series A no. 26.

143 *Tyrer v. United Kingdom*, para. 31.

144 *Marckx v Belgium* (Appl. No. 6833/74) Judgment of 13 June 1979, Series A no. 31.

145 *Marckx v Belgium*, para. 41. See also, *Dudgeon v United Kingdom* (Appl. No. 7525/76) Judgment of 24 February 1983, at para. 60. See George Letsas 'The ECHR as a Living Instrument: Its Meaning and its Legitimacy', in: *Constituting Europe: The Europe an Court of Human Rights in a National, European and Global Context. Studies on Human Rights Conventions*, edited by Andreas. Follesdal, Birgit Peters and Geir Ulfstein, Cambridge: Cambridge University Press, 106, 111, http://dx.doi.org/10.1017/CBO9781139169295.005, accessed 6 August 2015 (cited hereafter as 'Letsas').

pretation of the Convention. Most importantly, the Court appealed to broader developments in the Council of Europe 'as a counterweight to the prevailing moral climate in the respondent state'.

Letsas explains that evolutionary interpretation by the ECtHR is linked to two other doctrines of the Court, which complement it, i.e. autonomous concepts and the margin of appreciation and analyses its positive and negative aspects.[146] According to him the 'old Court' granted a wide margin of appreciation to the respondent State in cases where there was no consensus among Contracting States on the moral issues raised by the applicant's case, particularly in relation to restrictions based on public morals. Until the late 1990s, the Court would rely on present-day standards among Council of Europe members as a counterweight to the moral climate prevailing in the respondent State, a climate prejudiced against the applicant's claim (*Tyrer*, *Marckx*, *Dudgeon*).

Having done so, it would examine the substance of the right in question and either conclude that there has been a violation or it would not be willing to recognise that a common standard exists, due to the lack consensus amongst contracting States, thus granting a wide margin of appreciation to the respondent State. The doctrine of the margin of appreciation was seen as a stumbling block to evolutionary interpretation.[147]

The 'new Court' relied on the existence of a trend of evolution, as a present-day common standard, similarly to the *Marckx* case loosely applied, in order to outweigh the respondent State's appeal to a margin of appreciation. This approach was further expanded by the new Court as it appealed to evidence of common ground and trends of evolution in international law materials, notwithstanding whether such materials are binding and whether most Contracting States have ratified them.[148] In *Demir and Baykara v Turkey*, the Court, we may say, relied on the principle of systemic integration, as it referred to taking into account other international instruments and, according to the standard of the *Marckx* case, also relied on common European standard.[149]

146 Letsas, note 145, 113.

147 Idem, 115.

148 Idem, 115–119. See *Goodwin v United Kingdom* (*Appl. No. 17488/90*) Judgment of 27 March 1996, at para. 74.

149 The Court said that 'has never considered the provisions of the Convention as the sole framework of reference for the interpretation of the rights and freedoms therein' and that 'in defining the meaning of terms and notions in the text of the Convention, it can and must take into account elements of international law other than the Convention, the interpretation of such elements by competent organs, and the practice of European States reflecting their common values', *Demir and Baykara v Turkey* (Appl. No. 34503/97) Judgment of 12 November 2008, paras. 69–86.

Letsas explains that the 'new Court' relied on evolving trends and emerging consensus in international law ('present-day conditions') 'as a justificatory basis for finding a practice (or policy) to be in breach of the Convention, even if the practice is followed by the majority of contracting states'.[150] There are, of course, some exceptions; but in broad brushstrokes, it may be said that the new Court:

> has moved away from placing decisive weight on the absence of consensus amongst Contracting States and from treating it as the ultimate limit on how far it can evolve the meaning and scope of Convention rights. The new Court treats the ECHR as a living instrument by looking for *common values* and *emerging* consensus in international law.[151]

This short introduction to evolutionary interpretation indicates that this is an area of international law still awaiting more study and more in-depth analysis. The method of 'evolving' of a treaty can be effected through its organs, parties' practice, reference to other international instruments, and judicial organs. In the view of the present author, there are similarities, but also great differences, between all these methods (see above the observations on systemic integration). Moreover, the practice of the ECtHR which is an autonomous body whose practice cannot be compared with that of regular organs established by international treaties, such as the IWC.

3.2 *Evolutionary Interpretation in Relation to the ICRW (Pre- and Post-Judgment)*

Bowman is his seminal article on whaling made some very valid observations relating to the possible evolutionary interpretation of the revisions of the object and purpose of the ICRW and also of the contents of the Convention itself. It is worthwhile to give an overview of his views on this subject, which are especially interesting in the light of the Judgment in the *Whaling* case. In his view, the most obvious expression in the Preamble which can be subject to an evolutionary interpretation is the phrase 'the great natural resources represented by the whale stocks'. Bowman analyses the word 'resources' in light of definitions which were adopted elsewhere, e.g. in the Convention on Biological Diversity and include an extensive definition of biological resources including genetic resources, organisms, or parts thereof or any other biotic components or ecosystems, the operative part of the definition being 'with actual or potential

150 Letsas, note 143, 122.

151 Ibid.

use or value for humanity'. The most fundamental issue here is the question of the 'needs' of humanity, which have a fluid and evolving nature. Therefore the question of the needs of humanity has to be accommodated in an evolving fashion by the IWC and the Parties to the ICRW.[152] The use of resources and the needs of humanity may have changed since the negotiation of the ICRW, as have the values within generations, who may find the aspirations and needs of previous generations unacceptable. The change in the approach to nature and resources from utilitarian to having an intrinsic value, have influenced attitudes to natural resources, including whales. Thus, according to Bowman, it is likely that natural resources have gained a recreational, cultural and educational quality prevailing over the commercial and utilitarian. Therefore, whaling, as it stands at present, cannot, according to Bowman, be restricted to activities which, in general terms, can be defined as 'fishing', but has to acquire a larger dimension, including such things as the whale watching, which should be included into the regulatory powers of the IWC. However, as Bowman concedes, there is one phrase in the Preamble which does seem to support the view that the ICRW's objectives and purposes have not radically evolved from 1946, namely, the statement: 'increases in the size of whale stocks will permit increase in the numbers of whales which may be captured'.

Bowman, suggests, that close scrutiny of the phrase indicates that the increased capture of whales in not an objective in a strict sense, even less so the sole or principal aim of international regulation; but it was presented in the guise of an opportunity generated by the ultimate objective, i.e. the protection of whales from overharvesting in order to achieve the optimum level of stocks. Bowman, however, agrees that the non-lethal means of the exploitation of whales do not exclude *per se* the more traditional means, which are still part and parcel of the ICRW's objectives and purposes.[153]

The same author discusses other points which, in his view, evidence the evolutionary character of the IWRC. These are, for example, the evolving notion of conservation, which is indicated from the switch in the IWC from the 'blue whale unit' approach to the species-by-species and stock-by-stock assessment of the Revised Management Procedure, which although still falling short of the modern ecosystem approach to conservation, still amounts to progress within the IWC. From the point of view of conservation, Bowman also singles out the following expressions included in the Preamble to the Whaling Convention: 'proper and effective conservation' and 'the orderly development of the whaling industry', which lend themselves to evolutionary

152 Bowman, note 9, 358–361; 449–545.

153 Idem, 467.

interpretation—i.e. a modern ecosystem-based approach, which takes into consideration the impact of the whale population on conservation programmes of other marine species.

Far reaching interpreting of the ICRW's object and purpose in terms of the evolutionary interpretation is not, however, fully embraced by all authors and experts. Professor Burke, for example, was of the view that while the Convention's object and purpose is not meant to remain entirely static, but must evolve, as it has, since 1946, it does not warrant the view that large cetaceans should be completely protected from artificial mortality. Such a view would have to be based on the premise that the 'whaling industry' only includes non-consumptive uses such a whale watching. According to Burke:

> the difficulty with these and similar interpretations is that they turn the terms of the ICRW completely upside down and defeat major purpose of the original agreement. Where the fundamental goal of the initial treaty was to conserve whales in order to permit a sustainable harvest, the purpose would now be to protect whales against any harvest. Optimum utilization, reasonably understood by the parties to mean actual harvest at a conservative level, would not mean not any harvest at all.[154]

Bowman asserts that the Convention itself gives a ground for assuming that it was designed to allow for some evolution. This is indicated by the wording of its Article IV(1), which empowers the IWC to:

(a) encourage, recommend, or if necessary, organize studies and investigations relating to whales and whaling;
(b) collect and analyze statistical information concerning the current condition and trend of the whale stocks and the effects of whaling activities thereon;
(c) study, appraise, and disseminate information concerning methods of maintaining and increasing the populations of whale stocks.

Article IV (2) authorises it to publish 'such reports as it deems appropriate, as well as statistical, scientific, and other pertinent information relating to whales and whaling'. Article V(1) gives the IWC power to 'make recommendations to all or any Contracting Governments which relate to whales and whaling and to the

---

154 William T. Burke, 'Memorandum of Opinion on the Legality of the Designation of the Southern Ocean Sanctuary by the IWC', *Ocean Development and International Law* 27 (1996): 315, 323.

objectives and purposes of this Convention'. Under Article III (4), it may establish various committees to perform authorised functions. Such a broad definition of the IWC's functions appears to give ample scope for the re-definition of certain objectives and purposes of the ICRW. However, as Bowman observes, the ultimate redefining of the focus of the IWC towards the management and conservation of whales in a general sense may pose certain difficulties, as it appears that the ICRW is drafted in such a way that it is doubtful whether the IWC is a suitable organ to undertake such a task. The Whaling Convention, as it stands at present, does not provide a suitable legal framework for the holistic management of whales; and likewise it does not appear that the ICRW as it is drafted imposes a general conservation obligation on its Parties. Some doubts may also arise whether the existing amendment procedure to the Schedule would permit such a task. Therefore, according to Bowman, a fundamental change in the functions of the IWC and, generally, a completely new and redefined approach to the objectives and purposes of the ICRW does not appear feasible at present.[155]

As briefly referred to above, the Court did, itself, though not in any extended way, address the issue of the possible evolutionary nature of the ICRW in Paragraph 45 of the Judgment, which stated:

> In contrast to the 1931 and 1937 treaties, the text of the ICRW does not contain substantive provisions regulating the conservation of whale stocks or the management of the whaling industry. These are to be found in the Schedule, which "forms an integral part" of the Convention, as is stated in Article I, paragraph 1, of the latter. The Schedule is subject to amendments, to be adopted by the IWC. This Commission, established under Article III, paragraph 1, of the Convention, is given a significant role in the regulation of whaling. It is "composed of one member from each Contracting Government". The adoption by the Commission of amendments to the Schedule requires a three-fourths majority of votes cast (Art. III, para. 2). An amendment becomes binding on a State party unless it presents an objection, in which case the amendment does not become effective in respect of that State until the objection is withdrawn. The Commission has amended the Schedule many times. The functions conferred on the Commission have made the Convention an evolving instrument.

However, this statement of the Court has to be interpreted jointly with the one made at the end of paragraph 56 of the Judgment, where the Court stated: '[a]mendments to the Schedule and recommendations by the IWC may put an

155 Bowman, note 9, 471.

emphasis on one or the other objective pursued by the Convention, but cannot alter its object and purpose'.

The question is how to interpret these two statements. On the one hand, the Court in strong language made a very important statement that no amendments to the Schedule could change the object and purpose of the ICRW; on the other hand, it made a pronouncement that the functions conferred on the IWC make the Convention 'an evolving instrument'. It may be presumed that the Court referred to the amendments to the Schedule as well to the Recommendations (non-binding) of the IWC. According to Article VI of the ICRW '[t]he Commission may from time to time make recommendations to any or all Contracting Governments on any matters which relate to whales or whaling and to the objectives and purposes of this Convention'.

Therefore, in the view of the author of this chapter, the main statement of the Court is that the object and the purpose of the ICRW remain as they were originally devised by the negotiators in 1946. What is permitted is the change of 'emphasis on one or the other objective pursued by the Convention'. From this it may be inferred that the sustainable use of whales and the orderly development of the whaling industry together with conservation of whale stocks remain the objectives and purposes of the Whaling Convention. Thus within the framework of this dual object and purpose of the Convention, which was set in 1946, the IWC may evolve the focus of the Convention through amendments to the Schedule, but may not re-define and redesign its object and purpose fundamentally, especially in light of the lack of consensus in the adoption of many of the IWC's Resolutions (with many objecting State-Parties).

4 Object and Purpose of the ICRW and the Principle of Systemic Integration (Article 31(3)(c))[156]

This section of the chapter will be structured in the following manner: first a brief consideration will be presented on the general question of the principle of systemic integration (Section 4.1); next systemic integration will be analysed within the pre-Judgment context (Section 4.2); and finally, the treatment of the issue by the Parties and the Court in the Whaling case will be analysed (Section 4.3).

4.1 *General Issues Concerning the Systemic Integration*

The question of systemic integration in relation to treaty interpretation has, like evolutionary interpretation of treaties, been the subject of much analysis in numerous publications; and like evolutionary interpretation remains an enigma.

156 'Any relevant rules of international law applicable in the relations between the parties'.

Likewise the relationship between evolutionary interpretation, systemic integration and subsequent practice is obscure, entangled and confusing.[157]

The debates within the ILC concerning Article 31(3)(c) has already quite clearly evidenced the lack of agreement of its members as to the legal nature of this provision. The main issues were, firstly, the intertemporality of this article (i.e. whether rules should be applied which were in force at the time of the conclusion of the treaty, or whether perhaps an evolutionary interpretation should be followed); and, secondly the issue of what rules had to be taken account of—treaty rules only, or perhaps also customary international law.[158] At present these issues are also unresolved and Article 31(3)(c) is variously approached as a rule of conflict; a tool of the unification of international law in order to combat its fragmentation; an instrument expressing a temporal element in interpretation; and as an expression of the evolutionary aspect of treaty interpretation.

A very illuminating example of such a lack of uniformity in the approaches to certain law of treaties concepts is provided by the *Costa Rica v. Nicaragua* case.[159] The interpretation in that case of the term '*comercio*', the meaning of which, according to the ICJ, had evolved is, on the one hand, treated as a manifestation of the principle of systemic integration,[160] and, on the other, as a

157 The classic publication is of Campbell Mclachlan, 'The Principle of Systemic Integration and Article 31(3)(c) of the Vienna Convention', *ICLQ* 54(2005): 279–320 (cited hereafter as 'Mclachlan'). See also *idem*, 'The Evolution of Treaty Obligations in International Law', in *Treaties and Subsequent Practice*, Nolte, note 118, 69–79; Philippe Sands, 'Treaty, Custom and Cross-fertilization of International Law', *Yale Human Rights and Development Journal* 1 (1998): 85–21; Panos Merkouris, 'Debating the *Ouroboros* of International Law: The Drafting History of Article 31(3)(c)', *ICLR* 9 (2007): 1–31, (cited hereafter as Merkouris, Debating…'); *idem, Article 31(3)(c) VCLT and the Principle of Systemic Integration,* (Brill, Martinus Nijhoff Publishers, 2015); Dirk Pulkowski, *The International Law and Politics of International Conflict* (Oxford: Oxford University Press, 2014), in particular pp. 272–298 (cited hereafter as 'Pulkowski'). Gabriel Orellana Zabalza, *The Principle of Systemic Integration: Towards a Coherent International Legal Order* (LIT, Muster, 2012) (cited hereafter as 'Orellana Zabalza'); Han van Gellecum, 'Environmental Law in the Context of Article 31(30)(c) of the Vienna Convention on the Law of Treaties. Reconciling Treaty Interpretation and Progressive Environmental Norms: the Pulp Mills Case and Beyond', 2011, http://papers.ssrn.com/sol3/papers.cfm?abstract_id=1989468, accessed 4 August 2015 (cited hereafter as 'Gellecum').

158 Merkouris, Debating…, note 158, 22.

159 *Dispute Regarding Navigational and Related Rights* (*Costa Rica v. Nicaragua*), Judgment, 13 July 2009, I.C.J. Reports 2009, p. 21, para. 58, http://www.icj-cij.org/docket/files/133/15321.pdf, accessed 5 August 2015.

160 Gellecum, note 158, p. 18.

judicial example of the application of evolutionary interpretation of treaties.[161] Both approaches may be fully substantiated and justified; but evidence an ongoing problem of the identification of individual characteristics of these concepts. There is a disagreement as to whether Article 31(3)(c) contains a temporal element. For example, Helmersen stated that 'Article 31.3.c is about interpretation in light of other law, while evolutive interpretation is about interpretation in light of some current meaning'.[162] Orellana Zabalza argued that the rule contained in Article 31(3)(c) of the 1969 VCLT was originally intended to reflect the rule of intertemporal law, in the way it was stated by the Court in the *Legal Consequences for States of the Continued Presence of South Africa in Namibia (South West Africa) notwithstanding Security Council Resolution 276*. However, neither element of this rule, namely, that acts should be judged under the law contemporary to their creation, i.e. *contemporanea exposition*, and the idea that rights validly acquired under contemporaneous law may lost if they fail to reflect the development of international law, was transcribed into conventional law. At present, as Orellana Zabalza argued:

> ...even if the origin of the rule established under article 31(3)(c) of the VCLR may be found in the rule of intertemporal law, as it stands today the article expresses a more general principle—the principle of *systemic integration*. This formulation opens international law to the workings of time, politics and change, although it offers little guidance as to how to achieve it.[163]

It may be remembered that the principle of interpretation contained in Article 31(3)(c) was rarely relied upon and almost dormant until the statement of the Court in the *Oil Platforms* case and the work of the Working Group on Fragmentation, which revived it. We can now observe the continued interest in this principle (almost unstoppable) which has elevated its role in the interpretation of treaties from marginal to very significant[164]—perhaps, even, over

161 Arato, Subsequent Practice, note 119, 447.

162 Helmersen, note 127, 147.

163 Orellana Zabalza, note 158, 232–233.

164 See, e.g. some of the latest publications on the EJIL! Talk blog: Kushtrim Istrefi, 'R.M.T. v. The UK: Expanding Article 11 of the ECHR Through Systemic Integration', http://www.ejiltalk.org/r-m-t-v-the-uk-expanding-article-11-of-the-echr-through-systemic-integration/, accessed 6 August 2015 and Panos Merkouris, 'Keep Calm (no, not batman but...) Articles 31–32. A Comment on Itrefi's Recent Post on R.M.T. v. UK', http://www.ejiltalk.org/keep-calm-and-call-no-not-batman-but-articles-31-32-vclt-a-comment-on-istrefis-recent-post-on-r-m-t-v-the-uk/, accessed 5 August 2015.

enthusiastically, considering the lack of any uniform understanding of its interpretative function, either in theory or judicial practice, as was evidenced in the *Oil Platform* case by the rather reluctant, if not overtly negative, reactions to the application of this article.[165]

International courts and tribunals have at times manifested different understandings of Article 31(3)(c), as it was evidenced by the famous WTO case, *EC-Biotech* in which the WTO panel had disagreed (albeit not expressly) on the notion of the 'relevant rules'. The Biotech panel addressed the question of whether the Cartagena Protocol could be the 'relevant rule' in the context of Article 31(3)(c), the issue being that all parties to the Cartagena Protocol must simultaneously be WTO members, to which the Panel answered in the affirmative, a ruling which was criticized. The fundamental question was whether Article 31(3)(c) requires the full participation of all parties in both treaties; or whether the relationships between States may be analysed on an individual basis.[166] This line of reasoning appears to be still prevalent in the WTO, as in the 2011 *European Communities—Measures Affecting Trade in Large Civil Aircraft*, the Appellate Body stated that '[o]ne must exercise caution in drawing from an international agreement to which not all WTO members are party'.[167] The final point that may be raised is that over-reliance on Article 31(3)(c) may lead to the possible neglect of the rule of interpretation based on ordinary meaning of terms.

Taking into account all the unresolved questions concerning Article 31(3)(c) and its application, there are, however, a few features of this interpretative tool, identified by Mclachlan, which, in the view of the author of this chapter,

---

165 See e.g. Judge Burgenthal, who took a very restrictive view on Article 31(3)(c). According to his view, the Court's jurisdiction is limited to the matters which the parties agreed to submit to it, and this impacts on the applicable law in interpreting of the treaty before it. Such a limitation will not allow the ICJ to rely on all other sources of international law, notwithstanding the source, including the UN Charter. Separate Opinion, Judge Burgenthal, Separate Opinion, paras. 22–24, http://www.icj-cij.org/docket/files/90/9729.pdf, accessed 5 August 2015. Judge Higgins was quite sceptical as to the use (or rather misuse) of Article 31(3)(c), and she was of the view that the Court should have relied on the ordinary meaning of the terms of Article XX para. 1(d), an approach which was commended in this case. Judge Higgins Separate Opinion, para. 49, http://www.icj-cij.org/docket/files/90/9721.pdf, accessed 5 August 2015.

166 *European Communities-Measures Affecting the Approved and Marketing of Biotech Products*, Report of the Panel, WT/DS291R, WT/DS/29/R, adopted 21 November 2006, para. 7.68, Pulkowski, note 158, 290–291.

167 *European Communities-Measures Affecting Trade in Large Civil Aircraft*, WT/DS316/AB/R, adopted 18 May 2011, para. 843.

without doubt can be accepted as generally characterizing systemic integration. Mclachlan says:

> (a) It refers to *rules* of 'international law' – thus emphasizing that the reference for interpretation purposes must be to rules of law, and not to broader principles or considerations which may not be firmly established as rules; (b) The formulation refers to rules of international law in general. The words are apt to include all of the sources of international law, including custom, general principles, and, where applicable, other treaties; (c) Those rules must be both *relevant* and '*applicable* in the relations between the parties'. The sub-paragraph does not specify whether, in determining relevance and applicability one must have regard to all parties to the treaty in question, or merely to those in dispute; (d) The sub-paragraph contains no temporal provision. It does not state whether the applicable rules of international law are to be determined as at the date on which the treaty was concluded, or at the date on which the dispute arises.[168]

4.2 *Systemic Integration within the Pre-Judgment Interpretative Context of the ICRW*

The question of systemic integration within the interpretative context of the ICRW was also discussed by Bowman, who adopted the view that Article 31(3)(c) has a temporal effect, thus that the provisions of the ICRW should be analysed according to the general state of law at the time of their application, on the basis of which, in the view of author of this chapter, it in fact relates to the issue of the evolutionary interpretation of the Whaling Convention, as above explained. Bowman, however, also refers to the other international law norms which may be applicable, existing in parallel with the ICRW, within the context of systemic integration. He, however, also recommends caution in relying on such a tool, as for example only a small group of States may be Parties to the other treaty with conflicting or inconsistent obligations.[169]

It may be mentioned at the outset that the relationship between the ICRW and other conservation conventions, or quasi-conservation conventions, such as the Convention on International Trade in Endangered Species of Wild Fauna and Flora (CITES) (which is in fact a trade Convention, but which, through trade, regulates conservation), is governed by Memoranda of Understanding (MOUS) which give priority, in the case of whaling management,

168 Mclachlan, supra note 157, 290–291.

169 Bowman, note 9, 343 and 459-471.

to the IWC, and thus, in fact, erode the working of Article 31(3)(c), as it is the IWC which sets the relevant rules for other States in questions relating to whaling. In such an instance, the ICRW, and the decisions and recommendations of the IWC, may be a 'relevant law' for the parties to the CITES in the matter of whaling, but not *vice versa*.[170] The same applies to the Convention on the Conservation of Migratory Species of Wild Animals (CMS) with which there is also a MOU concluded. Australia, Japan and New Zealand are parties to the CITES.

Of possibly the most fundamental importance for the interpretation of the ICRW within the context of systemic integration is the global 1992 Biodiversity Convention (CBD) which sets out the modern principles of conservation. All parties to the case are parties to the CBD. This Convention recognizes the intrinsic value of biological diversity; approaches biological diversity as a common heritage of mankind; and is based on the precautionary approach (the Preamble). The State Parties are obliged to cooperate with each other directly or through an international organization to ensure conservation and sustainable use. The CBD regulates *in situ* and *ex situ* conservation. It is obvious that it can be argued that the obligations under ICRW may be interpreted within the context of the CBD. However, in this respect, Article 22(1) of the CBD, is of relevance. This article states that there is no intention of affecting the rights and obligations stemming from other treaties, with an exception where their exercise would cause serious damage or threat to biological diversity. Such an expression is very unclear; and, as Bowman asserts, the risk of conflict may not always be easy to assess with certainty. The practical implementation of such a provision, therefore, requires a collaborative approach with CBD organs to remove the possibility of a conflict. Article 22(2) requires the CBD parties to implement the Convention consistently with the rights and obligations of States under the law whenever the marine environment is of concern.

The other convention within the paradigm of Article 31(3)(c) is the 1982 United Nations Law of the Sea Convention (UNCLOS), which is considered to be 'a constitution for the oceans'. It has created a global framework for the exploitation and conservation of marine resources, and therefore has great importance in relation to whale management within the ICRW. Of importance

170 Resolution 2.9 of the CITES established the IWC as a supporting institution, while the leadership position was left to IWC, Third Meeting of the Conference of the Parties, Conf. 3.13 Trade in Whale Products, see also Alexander Gillespie, *The Whaling Diplomacy: Defining Issues in International Environmental Law* (Cheltenham, UK: Edward Elgar, 2005), 336.

are two Articles of the UNCLOS 65[171] and 120.[172] Articles 62 and 129 should also be analysed in conjunction with Articles 61 and 62 of the UNCLOS. Within the Exclusive Economic Zone, coastal states exercise sovereign rights 'for the purpose of exploring and exploiting, conserving and managing' marine living resources, subject to obligations under Articles 61 and 62 to take 'proper conservation and management measures' to prevent over-exploitation.

There is no agreement as to the content of these Articles. There are concerns by some States that the UNCLOS provisions allow coastal States to promulgate very strict laws restricting whaling. Other States stress the possible fragmentation of the regulatory regime, due to the plurality of supervisory bodies which may be endorsed by the terms of Article 65.[173]

Indeed, there is some disagreement what is understood by 'the appropriate international organizations'; and there are some views that it is not necessarily the IWC. The provisions of the UNCLOS with respect to the conservation of marine mammals are very broad and vague, and their use would not add much, as an interpretative tool within the meaning of Article 31(3)(c), in relation to the ICRW. At any rate, if the view is adopted that the 'the appropriate international organization' in relation to whaling is the IWC, then the situation is similar to that relating to, e.g., the CITES and the IWC, when the Commission is the leading body in the management of whales. In such an event the relevant provisions of the UNCLOS should be interpreted in the light of the ICRW and in particular the decisions and recommendations of the IWC.

Among other norms which may have an impact on the interpretation of the ICRW are the norms on cultural diversity, of which the soft and hard instruments sponsored by the United Nations Educational, Scientific and Cultural Organisation (UNESCO) are of particular importance. Just as the above-discussed conventions, such as the CBD, can be used in the interpretative process as providing 'relevant rules of international law' in support of the conservation of whales as the objective and purpose of the ICRW, so the instruments relating to cultural diversity can be useful in providing a wider interpretative context,

---

171 Article 65: 'Nothing in this Part restricts the right of a coastal State or the competence of an international organization, as appropriate, to prohibit, limit or regulate the exploitation of marine mammals more strictly than provided for in this Part. States shall cooperate with a view to the conservation of marine mammals and in the case of cetaceans shall in particular work through the appropriate international organizations for their conservation, management and study'.

172 Article 120: 'Article 65 also applies to the conservation and management of marine mammals in the high seas'.

173 WDC website http://uk.whales.org/issues/in-depth/united-nations-convention-on-law-of-sea-1982, accessed 5 August 2015.

in particular relating to indigenous peoples' whaling and also to the whaling of States with whaling traditions, such as Japan and Norway. The most important soft law instrument which may be mentioned here is the 2001 Declaration on Universal Diversity[174] which affirms that '...respect for the diversity of cultures, tolerance, dialogue and cooperation, in a climate of mutual trust and understanding are among the best guarantees of international peace and security' (Preamble). Article 4 states that 'the defense of cultural diversity is an ethical imperative, inseparable from respect for human dignity'. Culture is defined as 'the set of distinctive spiritual, material, intellectual, and emotional features of society or a social group' and 'encompasses, in addition to art and literature, lifestyles, ways of living together, value systems, traditions, and believes'.

Of lesser importance are two Conventions: 2003 Convention for Safeguarding of the Tangible Cultural Heritage[175] and the 2005 Convention on the Protection of the Diversity of Cultural Expression.[176] The objects and purposes of these Conventions are not really linked closely to the preservation of traditional practices of resource exploitation. The main object of the 2003 Convention is rather preservation of traditional manifestations of the performing arts. The 2005 Convention is more relevant as it states in Article 1 that one of its objectives is 'to protect and promote the diversity of cultural expressions' (1(a)); and another 'to promote respect for the diversity of cultural expressions and raise awareness of its value at the local, national and international levels' (1(e)). At any rate, none of the parties to the dispute are parties to the 2003 Convention and Japan is not a party to the 2005 Convention. Therefore these Conventions are not 'relevant rules of international law applicable in the relations between the parties'.

4.3 *Systemic Integration in Relation to the ICRW in the Pleadings, Judgment and Judges' Opinions in the Whaling Case*

4.3.1 Systemic Integration: The Memorial

The Memorial of Australia made only modest reference to Article 31(3)(c) of the 1969 VCLT in relation to the interpretation of the ICRW. It said that there are number of international agreements in force between Australia and Japan

174 UNESCO Universal Declaration on Cultural Diversity, http://portal.unesco.org/en/ev.php-URL_ID=13179&URL_DO=DO_TOPIC&URL_SECTION=201.html, accessed August 2015.

175 The UNESCO 2003 Convention for Safeguarding of the Tangible Cultural Heritage, entry into force 20 April 2006, http://www.unesco.org.uk/convention_on_safeguarding_of_the_intangible_cultural_heritage_%282003%29, accessed 5 August 2015.

176 The UNESCO 2005 Convention on the Protection of the Diversity of Cultural Expression, entry into force 18 March 2007, http://portal.unesco.org/en/ev.php-URL_ID=31038&URL_DO=DO_TOPIC&URL_SECTION=201.html#ENTRY, accessed 6 August 2015.

which promote biological diversity and are based on the precautionary principle.[177] However, the understanding of the principle of systemic integration in the Memorial was based on its supposedly temporal aspect. Australia based its use of Article 31(3)(c) on the Advisory Opinion in *South West Africa*[178] and the *Gabcikovo-Nagymaros Project*,[179] both of which cases are rather meant to be an expression of the of the concept of evolutionary interpretation of treaties than of the principle of systemic integration.

Australia then proceeded to develop an argument which was based on the temporal aspect—i.e. on the development of international environmental law concerning, in particular, the conservation of marine mammals. Australia stressed two developments which it considered of most significance: the development of a treaty-based regime for the protection of marine mammals and the precautionary approach.[180] Further, Australia moved from the temporal approach to one expressing more the essence of Article 31(3)(c) and relied on the 'Stockholm Principles' which emerged from the 1972 Stockholm Conference on Human Environment and some multilateral environmental agreements such as the CITES, the CMS, the CBD, the UNCLOS and the 1980 Convention on Conservation of Antarctic Living Resources (CCAMLR), as evidence of 'significant developments in the law relating to conservation'.[181] Australia argued that all these instruments recognize the intrinsic value of all living things and that biological diversity is the 'common concern of mankind'. These instruments are all relevant to the conservation of whales.[182] Further, Australia relied on the precautionary principle (of which the classical definition is contained in Principle 15 of the 1992 Rio Declaration of Environment and Development) in the interpretation of both the ICRW as a whole and its Article VIII in particular. To support wide acceptance of the precautionary principle, Australia relied on several cases before various international courts

177 The Memorial, note 32, para. 4.81.

178 The Court said as follows: '[a]n international instrument has to be interpreted and applied within the framework of the entire system prevailing at the time of interpretation'. *Legal Consequences for States of the Continued Presence of South Africa in Namibia (South West) notwithstanding Security Council Resolution 276* (1970), Advisory Opinion, I.C.J. Reports, para. 53.

179 In this case the Court noted that the 1977 Treaty in force between Hungary and Slovakia which was subject 'is not static but, and open to adapt to emerging norms of international law', *Gabcikovo-Nagymaris Project* (*Hungary v. Slovakia*), Judgment, *I.C.J. Reports* 1997, 67–68, para. 112, http://www.icj-cij.org/docket/files/92/7375.pdf, accessed 5 August 2015.

180 The Memorial, note 32, para. 4.82.

181 Idem, para. 4.85.

182 Idem, para, 4.86.

and tribunals and argued that the Parties to the ICRW had agreed to the adoption of a precautionary principle in a broad range of matter.[183]

4.3.2 Systemic Integration: The Counter-Memorial

Japan in its Counter-Memorial relied on systemic integration as evidencing that in other treaties conservation is a means to secure sustainable use.[184] To substantiate this assertion, Japan submits several examples of multilateral environmental agreements, such as the CBD, in which, it argues, conservation is not the single aim but, which, according to its Article 1, has also sustainable use of natural resources as an aim.[185] According to Japan, the notion of 'use' in the CBD encompasses both commercial exploitation and use for scientific research. The concept of sustainable use as linked to conservation was further developed by the decisions of the Conference of the Parties (COP) of the CBD, such as the 2004 Addis Ababa Principles and Guidelines for Sustainable Use of Biodiversity.[186] Japan further introduced the temporal element into what appears to be systemic integration, arguing that conservation policies change over time and that this is reflected in the principle of adaptive management. Japan also relied on the concept of the precautionary approach, application of which is evident in the JARPA II in the monitoring of the status of species. Japan was of the view that the endangered status of certain species does not signify that this status cannot be changed. In support of this view, Japan relied on the UNCLOS and 1995 Fish Stocks Agreement and the CCAMLR. Finally, Japan stated that even CITES adopts sustainable use and the adaptive management approach.[187] Japan concluded that '[a] s these examples show, there is no contradiction between conservation and utilization. The objective of sustainable use of whale resources of the ICRW is in line with modern environmental treaty practice'.[188]

Finally, Japan argued the differentiation between the ICRW and conventions on wildlife preservation (such 1949 Convention on Nature Protection and Wild Life Preservation in the Western Hemisphere and the 1950 Convention international pour la protection des Oiseaux), which were contemporaneous to the ICRW. Japan argued that the language used in these conventions was different from that used in the ICRW, indicating the purpose of all these conventions

183 Idem, paras, 4.89–4.90.

184 The Counter-Memorial, note 40, para. 6.15.

185 Idem, paras. 6.16–622, note 39.

186 COP Decision VII/2. Annex II, at para. 2.

187 The Counter-Memorial, note 40, paras, 6.28–6.29.

188 Idem, para. 6.30.

was not the same. The language of the ICRW as compared to the preservation conventions indicates that its purpose was not preservation but the management of natural resources.[189] Having compared the ICRW to other environmental conventions, Japan concluded that '[t]he objective of sustainable use of whale resources of the ICRW is in line with modern environmental treaty practice'.[190]

4.3.3 Systemic Integration: Oral Pleadings by Australia

The principle of systemic integration was pleaded by Australia within the context of relevant principles of general international law.[191] It was claimed that general international law emphasises conservation. It was said that contemporary international law rests on three pillars: intergenerational equity, the principle of prevention and the precautionary principle. These principles, according to Australia, govern the interpretation and the application of the ICRW; and they make it possible for its object and purpose to be achieved. These are also relevant rules of international law between Japan and Australia.[192]

A number of international environmental conventions was mentioned to support the applicability of these principles and that of due diligence, as well a number of cases which support the view that the precautionary principle has crystallized into international environmental law.[193]

4.3.4 Systemic Integration: Oral Pleadings by Japan

Japan's oral pleadings referred directly to systemic integration and Article 31(3)(c).[194] Japan questioned the possibility of the change of the object and purpose of the ICRW in light of the evolutionary interpretation, a point that was dealt with by the Court (see above). It appears that Japan in its oral pleadings equated evolutionary interpretation with systemic integration, but pointed

189 Idem, para. 6.34.

190 Idem, para. 6.35.

191 Boisson de Chazournes, note 65.

192 Idem, para. 50.

193 Idem, para. 53. She mentioned in particular, *Pulp Mills on the River Uruguay (Argentina v. Uruguay), Judgment I.C.J. Reports* 2010 (1), p. 77 para. 187, http://www.icj-cij.org/docket/files/135/15877.pdf, accessed 6 August 2015; and the Sea Bed Disputes case of the International Tribunal for the Law of the Sea, *Responsibilities and Obligations of States Sponsoring Persons and Activities in the Area (Request for Advisory Opinion submitted to Seabed Disputes Chamber), Advisory Opinion* 1 February 2011, paras. 125–135, https://www.itlos.org/fileadmin/itlos/documents/cases/case_no_17/adv_op_010211.pdf, accessed 6 August 2015.

194 Boyle, note 72, paras.24–53.

out that it would be necessary to identify some basis in existing law and practice to invoke Article 31(3)(c). Boyle went on saying that '[i]n the present case there are no norms and principles of international law to support Australia's evolutionary reading of the Whaling Convention'.[195]

The term 'conservation' was interpreted in the light of international jurisprudence in order to evidence that it does not mean total prohibition of exploitation.[196] The remaining part of the oral pleadings analysed the term 'conservation' in other multilateral environmental agreements, such as the CBD, the 1982 UNCLOS, the 1994 Fish Stocks Agreement, the CCMLAR, the CITES and the FAO law in line with pleadings in the Counter-Memorial. During oral pleadings it was stated that contemporary international law promoted scientific research as an integral component of the duty to conserve and sustainably use natural resources.[197] The analysis of the relevant international law between Japan and Australia indicates that

> there is no contradiction between conservation and sustainable use of natural resources, whether they are whales, or fish or any other exploitable living resource. The Whaling convention's objective of optimum or sustainable use of whale resources is fully in line with modern treaties on conservation. And an interpretation pursuant to Article 31(3)(c) of the Vienna Convention supports Japan's reading of 'conservation' in the Whaling Convention.[198]

4.3.5 Systemic Integration: Judgment and Judges' Opinions

The Judgment is silent on the issue of systemic integration; but very instructive is the Separate Opinion of Judge Cançado Trindade. He expressly refers to systemic integration or, as he called it, 'systemic outlook', in the following terms:[199]

> With the growth in recent decades of international instruments related to conservation, not one single one of them is approached in isolation from the others: not surprisingly, the co-existence of international treaties of

---

195 Idem, paras. 28 & 29.

196 Idem, paras. 20–32, note 72 The cases submitted were the *Fisheries Jurisdiction* cases (*Fisheries Jurisdiction case, United Kingdom v. Iceland; Federal Republic of Germany v. Iceland, Merits* Judgment, I.C.J. Reports, 1974, p. 31, para. 72p; Judgment I.C.J. Reports 1974, p. 200, para. 64) Pulp Mills case and the *Southern Blue Tuna Case* (New Zealand v. Japan; Australia v. Japan), order of 27 August 1999, ITLOS Nos. 3&$, paras. 28(2) & 29(2).

197 Idem, para. 38.

198 Idem, para. 53.

199 Judge Cançado Trindade, note 93, para. 25.

> kind has called for a systemic outlook, which has been pursued in recent years. Reference can here be made, e.g., to the 1973 Convention on International Trade in Endangered Species of Wild Fauna and Flora (CITES Convention), the 1979 Convention on Migratory Species of Wild Animals, the 1980 Convention on the Conservation of Antarctic Marine Living Resources, the 1982 U.N. Convention on the Law of the Sea, the 1992 U.N. Convention on Biological Diversity (CBD Convention).[200]

Judge Cançado Trindade referred to the ICRW as a 'living instrument'. He gives ample examples of the living instrument approach in international case law: in the jurisprudence of the ICJ (*Namibia Advisory Opinion, Gabcikovo-Nagymaros* case); in jurisprudence of the ECtHR (the *Tyrer v. United Kingdom* case; *Loizidou v. Turkey*); and in the Inter-American Court of Human Rights (cases such as *The Right to Information on Consular Assistance in the Framework of the Guarantees of the Due Process of Law*).

This Judge stated as follows:

> In sum, international treaties and conventions are a product of their time, being also *living instruments*. They evolve with time; otherwise, they fall into *desuetude*. The ICRW Convention is no exception to that. Those treaties endowed with supervisory organs of their own (like the ICRW Convention) disclose more aptitude to face changing circumstances.
>
> Moreover, in distinct domains of international law, treaties endowed with a supervisory mechanism of their own have pursued a hermeneutics of their own, facing the corresponding treaties and conventions as *living instruments*. International treaties and conventions are products of their time, and their interpretation and application in time, with a temporal dimension, bears witness that they are indeed living instruments. This happens not only in the present domain of conservation and management of living marine resources, but likewise in other area of international law.[201]

Judge Cançado Trindade also referred to the concept of intergenerational equity, as a concept of general international law, in connection to the ICRW (this concept is included in the Preamble of the ICRW)[202] resources represented by the whale stocks'. and to several other Conventions which are based on this

200 Idem, para. 25.

201 Idem, paras. 33 & 34. (footnotes omitted).

202 'The interest of the nations of the world in safeguarding for future generations the great natural.

concept such as the CBD; the CITES as well the 2001 UNESCO Universal Declaration on Cultural Diversity. For example the CITES includes in its Preamble that wild fauna and flora 'must be protected for this and the generations to come', and adds that 'peoples and States are and should be the best protectors of their own wild fauna and flora;' the CBD states as follows: 'to conserve and sustainably use biological diversity for the benefit of present and future generations'; and the CMS 'holds the resources of the earth for future generations and has an obligation to ensure that this legacy is conserved and, where utilized, is used wisely'.

4.4 *Interim Conclusions on Systemic Integration and the ICRW*

In the *Whaling* case, the concept of systemic integration was relied upon, to some extent, by all Parties to the dispute. It cannot be said, however, that it played a significant role in their pleadings (and the Court in its Judgment did not address this issue).

However, the interesting point in relation to their reliance on systemic integration is how the Parties conceptualized its legal nature. As was observed above, both theory and judicial application in relation to it indicate that there is no single uniform understanding of the legal nature of the principle of systemic integration. The *Whaling* case exemplifies this. As noted above, according to some authors (Helmersen) the understanding of Article 31(3)(c) does not include the intertemporal element. However, this is not the approach adopted by several other international lawyers (including Judge Cançado Trindade), who indeed appear to profess the view that systemic integration is somewhat akin to, if not coterminous with, the evolutionary interpretation of treaties, in which respect they rely on the *Namibia Advisory Opinion*, which is a classic case in respect to the dynamic interpretation of treaties. They also cite many cases of the ECtHR, which develop the evolutionary approach to the interpretation of the ECHR, as examples of systemic integration, which they equate with evolutionary interpretation. This approach tallies with the one adopted by the ILC Study Group on Fragmentation which noted that Article 31(3)(c) has two interlinked aspects: a mechanism to make a connection and harmonization between different legal regimes; and the temporal element which gives a ground for the evolutionary interpretation of treaties, thus avoiding the conflict between treaties and fragmentation.[203] As Arato observes 'the two aspects of the VCLT 31(3)(c) are connected and codependent'.[204]

203 Study Group of the ILC, 430, note 116.

204 Julian Arato, 'Constitutional Transformation in the ECHR: Strasbourg's Expansive Recourse to International Rules of International Law', *Brook.J.Int'LL* 37(2912): 349, 355.

In fact, Arato explicitly analyses 'evolutive interpretation' of the ECHR by the ECtHR as based on Article 31(3)(c), therefore treating these two institutions as coterminous.

Notwithstanding that the 1969 VCLT has no temporal element, it appears that through the interpretation of Article 31(3)(c) it has acquired one. Although, it may be said that Article 31(3)(c) has gained much weight in the interpretation of treaties, its impact on, and input to, the final outcome of the *Whaling* case was insignificant. A proper understanding of the workings of this article, as well as its connection with evolutionary interpretation and subsequent practice, has not been resolved by this case and it is still awaiting resolution (if it is at all feasible).

5 Subsequent Agreement[205] and Subsequent Practice[206] and the ICRW

The issue of subsequent practice arose in the *Whaling* case in relation to the Resolutions of the IWC. This a very complex issue. This part of the chapter will be structured in the following manner: firstly general considerations concerning the subsequent agreement and subsequent practice (Section 5.1); secondly the pleadings of the parties; and the Judgment of the Court, with the views of the Judges (Section 5.2).

5.1 *General Considerations*

It can be said that the 1969 VCLT does not specify what is understood by 'subsequent agreement'—i.e. whether it must be a treaty or a have the same formal status as the instrument which is interpreted. Therefore it appears that it is sufficient to have convincing evidence that parties have reached an agreement.[207] Gardiner explains that the less formal the agreement, the greater the

205 Article 31(3)(a): 'Any agreement relating to the treaty which was made between all the parties in connection with the conclusion of the treaty'.

206 Article 31(3)(b): 'Any subsequent practice in the application of the treaty which establishes the agreement of the parties regarding its interpretation'. See also Provisionally adopted Articles by the ILC, 'Subsequent agreements and subsequent practice in relation to the interpretation of treaties: Texts and titles of draft conclusions 24 May 2013, 1 to 10 provisionally adopted by the ILC Drafting Committee on 27 and 28 May and on 2 and 3 June 2014'; A/CN.4/L.813.

207 Gardiner, note 2, 242–253.

significance of subsequent practice confirming less formal agreement or understanding.[208]

The complex issues relating to subsequent practice are analysed in detail by Gardiner and Nolte.[209] Therefore in this chapter only an outline of the most pertinent features will be highlighted. As has been observed, Article 31(3)(b);

> offers four elements for consideration or notice (1) The meaning of 'subsequent practice'; (2) that it be in the application of the treaty; (3) which parties need to participate in the practice; (4) what it is that establishes interpretative agreement.[210]

There is also a question what the relationship between subsequent practice, subsequent agreement and subsequent conduct. Nolte explains that the term 'subsequent practice' encompasses both means of interpretation under Article 31(3)(a) and 31(3)(b) of the 1969 VCLT.[211]

What constitutes practice is very widely understood, namely, executive, legislative, and judicial acts. An important issue is the extent of the participation by parties in a practice and of their agreement to this practice. Gardiner notes that participation in practice of all parties is not required, but what is required is 'manifested or imputable agreement'[212] by all parties to a treaty. One of the most taxing issues is again the relationship between evolutionary interpretation of treaties and subsequent practice. Gardiner explains that evolutionary interpretation is based on the concept on an extension of existing meanings of a treaty which has already given rise to concordant practice of the parties, which may be illustrated by the interpretative practice of the ECtHR However, the author of this chapter cannot agree with this view, as it is exactly the interpretative practice of the ECtHR which gave rise to concerns regarding evolutive interpretation, changing the provisions of the ECHR, even writing therein rights which were not granted explicitly.[213] As Gardiner states, a clear division between amendment of a treaty and subsequent practice is not always easy to

208 Idem, 250.

209 Idem, 253–288. In particular see also Nolte, note 118, the ILC Reports reproduced in Part V.

210 Idem, 254.

211 Introductory Report of the ILC Study Group On Treaties Over Time, 'Jurisprudence of the International court of Justice and Arbitral Tribunals of Ad Hoc Jurisdiction Relating to Subsequent Agreements and Subsequent Practice', in Part V, Nolte, note 118, 169 (cited hereafter as 'Nolte, Introductory Report'), 174.

212 Gardiner, note2, 267.

213 George Letsas, *Theory of Interpretation of the European Convention on Human Rights*, (Oxford University Press, 2007), 67.

distinguish.[214] In the view of the author of this chapter the evolutive interpretation should be added to this Gordian knot.[215] It may be worth noting that the provisionally adopted conclusions by the ILC Drafting Committee addressed the intertemporal element in Draft Conclusion 3 in the set of 10 Draft Conclusions adopted in 2014 (see below):

> Interpretation of treaty terms as capable of evolving over time: Subsequent agreements and subsequent practice under articles 31 and 32 may assist in determining whether or not the presumed intention of the parties upon the conclusion of the treaty was to give a term used a meaning which is capable of evolving over time.

Nolte in his first report also identified a cross-cutting issue of intertemporal law.[216] He explained that both evolutionary and contemporaneous interpretation can coexist, depending on circumstances, and both are possible considerations in the process of treaty interpretation.[217] Nolte further explained that evolutionary interpretation cannot be said to be an independent method of interpretation but 'rather a result of a proper application of the usual means of interpretation'.[218] He continued that it is therefore appropriate that subsequent conduct plays a significant role in cases in which the courts and tribunals recognized and practice evolutionary interpretation. According to his view, evolutionary interpretation and subsequent practice are mutually complimentary, it must be noted that subsequent practice 'can play both a justificatory and a limiting role for evolutionary interpretation'. The justificatory role demonstrates that an evolved understanding of a treaty can be also based on subsequent practice as an authentic means of interpretation. The limiting role of subsequent conduct is emerging with the expanding recognition of the possibility of an evolutive interpretation, as it was illustrated by the *Cost Rica v. Nicaragua* case.[219] This case, however, also demonstrates the difficulties of the

214 Gardiner, note 2, 275–280.

215 The problematic issues related to changes in a treaty and the lack of an agreement as to their legal character are well illustrated by a discussion by Gerhard Hafner, 'Subsequent Agreements and Practice: Between Interpretation, Informal Modification, and Formal Amendment', in Nolte, note 118, 105–122 and Jose Alvarez 'Limits of Change by Way of Subsequent Agreements and Practice', in Nolte, note 118, 123–131.

216 Nolte, Introductory Report, note 212, 184.

217 Idem, 188.

218 Ibid.

219 *Dispute regarding Navigational and Related Rights* (*Costa Rica v. Nicaragua*), ICJ, 2009. Nolte, Introductory Report, note 212, 188.

application of, and distinguishing between, the two concepts, as evidenced by very illuminating Opinions in this respect of Judges Skotnikov and Guilaume in that case.

Nolte posed the question whether treaties establishing an international organization are a special case. Such treaties contain more open-ended competences and obligations, and create higher mutual expectations for member States to contribute to the functioning of an organisation.[220]

With respect to subsequent practice of international organisations, Article 5 of the 1969 VCLT must be taken into consideration.[221] Article 31(3)(b) does not exclude taking into account an organization's practice. It can also be that it is the organization's practice that establishes, or represents, the agreement of the parties. It does not need to be the parties themselves engaged in subsequent practice. The decisions or resolutions of a plenary organ may possibly reflect the agreement of the parties regarding the interpretation of the organisation's constituent instrument.[222] This is relevant to the decisions and recommendations of the IWC.

5.2 *Subsequent Agreement and Practice in Relation to the ICRW in the Pleadings, Judgment and Judges' Opinions in the* Whaling *Case*

5.2.1 Subsequent Agreement and Practice: The Memorial

Australia emphasized the importance of subsequent practice in this case, and in this respect singled out the decisions of the IWC.[223] Subsequent practice analysed by Australia related in particular to the Resolutions of the IWC concerning special permit whaling within the structure of Article VIII of the ICRW, adopted since the commencement of the whaling moratorium, and in the 1986 Guidelines for the Review of Scientific Permits which 'also informs the essential characteristics of a genuine program for "purposes of scientific research"

---

220 Nolte, Introductory Report, note 212, 195.

221 Article 5, 'Treaties constituting international organizations and treaties adopted within an international organization' which reads: 'The present Convention applies to any treaty which is the constituent instrument of an international organization and to any treaty adopted within an international organization without prejudice to any relevant rules of the organization'.

222 Christopher Peters, 'Subsequent Practice and Established Practice of International Organizations: Two Sides of the Same Coin?', *Goettingen Journal of International Law* 3 (2011), 617, 637. Jose Alvarez, *International Organizations as Law-Makers* (Oxford: Oxford University Press, 2005); Gardiner, note 2, 280–287. And see Article 2(1)(j) of the 1986 Convention on the Law of Treaties between States and International Organisations (not in force).

223 The Memorial, note 32, para. 4.66.

under Article VIII'.[224] Further, Australia submitted several examples of relevant Resolutions. Australia, it may be presumed, relied on those Resolutions in order to evidence a changing interpretation of the IWCR due to subsequent practice of States. It importantly also noted that Resolutions of the IWC should endorse a restrictive approach to the permissible scope of lethal means in scientific research, in particular based on the Resolution 1995–9, which recommends the use of non-lethal methods to collect data.

5.2.2 Subsequent Agreement and Practice: The Counter-Memorial

Japan noted that the scope of Article 31(3)(b) is not precisely defined in the 1969 VCLT, either as to the type of practice concerned or to the authors whose practice may be relevant; or to the effects of such practice on a treaty (confirming or modifying its natural meaning).[225] Further Japan referred to the preparatory work of the ILC on the 1969 VCLT, which confirmed that subsequent practice can have either a modifying or a confirming effect on a treaty.[226] Japan stressed in particular the necessary requirement of the agreement of all Parties to the treaty. This fundamental element of the agreement of all Parties to the ICRW was missing from the Resolutions adopted by the IWC and relied upon by Australia. Another important observation was made by Japan—i.e. the mode of the adoption of these Resolutions. Far from having been adopted by consensus; they were adopted 'through a sequence of bitterly contested votes'.[227] Japan also put in doubt the assertion that the practice of the organ established by a multilateral treaty could qualify as subsequent practice. In the present case, it was not the powers of the IWC which were questioned, but the obligations of the Parties under the ICRW and its Article VIII. Therefore, only the practice of the Parties to the ICRW had any relevance. Therefore these Resolutions can only be relevant if they are reflected in practice of all Parties to the ICRW.[228] Japan, relying on the judicial practice of the WTO, which defined what is understood under subsequent practice ('concordant, common and consistent sequence of…pronouncements'), concluded that the practice of State Parties to the ICRW does not follow the pattern that Australia asserted had emerged in respect of the application of Article VIII of the IWCR.[229]

224 Idem, para. 4.69.
225 The Counter-Memorial, note 40, para. 8.44.
226 Idem, para. 8.45.
227 Idem, para. 8.48.
228 Idem, para. 8.49.
229 Idem, para. 8.5.

5.2.3 Subsequent Agreement and Practice: New Zealand's Written Observations

New Zealand raised the value of the resolutions of international organisations (and the IWC in particular) as an interpretative tool. It argued that in interpreting Article VIII of the ICRW, the ordinary terms on this article have to be considered in light of the Convention, of the Schedule (an integral part of the Convention), and of its object and purpose. Consideration must be also given to practice of the Parties to the ICRW, including decisions and Resolutions adopted by the IWC and its Committees. Such decisions and Resolutions are evidence of subsequent practice and can be treated as supplementary mean of interpretation under Article 32 of the 1969 VCLT. They also shed interpretative light on the meaning of the terms of Article VIII and their proper application. Such decisions and Resolutions do not modify the terms of Article VIII, but rather confirm the interpretation based on their ordinary meaning in their context.[230] New Zealand, similarly to Australia, pleaded for a restrictive interpretation of the permissible scope of lethal means in scientific research.

5.2.4 Subsequent Agreement and Practice: Japan's Written Observations

Japan strongly reiterated its position on the subject of subsequent practice, in particular the necessary requirement of an agreement of all Parties to such a practice, which is missing from the practice of Parties members of the IWC. Japan also rebutted the argument of New Zealand that the Resolutions of the IWC qualify as such practice. Japan again reiterated that the case-law relied upon by New Zealand related, in fact, to the competences of the conventional organs under the founding treaty, not to the obligations of States-parties under the treaty, which is what is relevant in the present case.[231]

5.2.5 Subsequent Agreement and Practice: Oral Pleadings by Australia[232]

Australia pleaded that subsequent practice by the IWC constantly emphasized the importance of conservation.[233] Australia argued that several

230 Written Observations of New Zealand, note 51, para. 13.

231 Written Observation of Japan, note 59, paras. 11–14.

232 Boisson de Chazournes, note 65.

233 Idem, para. 43.

Resolutions of the IWC can be considered as subsequent practice, such as the New Management Procedure, the Revised Management Procedure and the Berlin Initiative. Further, Australia argued that, in relation to special permits, Resolutions and Guidelines adopted by the IWC constitute subsequent practice in the meaning of Article 31(3)(b) of the 1969 VCLT.[234] The IWC is an organ established with collective responsibility, a watch dog for the ICRW, to which the States-parties have to report. It adopted paragraph 30 of the Schedule for the prior review of special permits and issued a subsequent series of instruments—'Guidelines'—for this same purpose. Paragraph 30 is mandatory and must be complied with. Guidelines comprise six instruments; and in all the IWC adopted forty Resolutions setting out its views on special permit whaling.[235]

According to Australia Guidelines are a 'textbook illustration of' subsequent practice, aimed at development of 'common understanding amongst the Parties as to the proper scope of Article VIII'.[236] Successive instruments remained, according to Australia, substantively consistent, including those criticizing the Japanese programmes.[237] They raised the following issues: special permits should not undermine conservation measures; special permits whaling should not assume the characteristics of commercial whaling; research should be conducted with non-lethal methods where possible.[238] Australia stressed that one of the above mentioned Resolution had been adopted by consensus (1986–2—it set out the IWC's criteria to be met when whales were killed for research—), being accepted in practice by all States Parties to the ICRW, as was the most recent 'iteration' of the Guidelines.[239]

Australia also denied that its reliance on subsequent practice was aimed at revision of Article VIII of the ICRW, as had been argued by Japan, but was only to clarify the approach to its proper interpretation. As a further point, Australia also argued that there is no requirement for all Parties to a treaty to have engaged in a practice 'in order for it to qualify as subsequent practice', but only a requirement to have accepted it.[240]

234 Crawford, Public Sitting, 26 June 2013, at 3 pm, Peace Palace The Hague, CR 2013/8 (cited hereafter as 'Crawford').

235 Idem, paras. 21–35.

236 Idem, para. 38.

237 Idem, para. 39.

238 Idem, para. 40.

239 Idem, paras. 42 & 43.

240 Idem, para. 46.

5.2.6 Subsequent Agreement and Practice: Oral Pleadings by Japan[241]

Japan was of the view subsequent practice is of little interpretative value in relation to Article VIII, but relied rather on what it referred to as 'the *relevant* practice', which evidences that Articles VIII is an exception to the other rules of the ICRW; that it confers a discretion on State Parties to issue special permits, and to determine their content, as they think fit, including the number of animals to kill or take.[242]

Further, Japan argued, some of the instruments of a secondary nature relied upon by Australia were no more than non-binding recommendations. The only binding provision relevant to Article VIII was paragraph 30 of the Schedule. But, while this had some effect on granting permits by State-parties to the ICRW, this was without impacting on the discretion of States in this respect. Japan further argued that paragraph 30 has not modified the Whaling Convention.

Subsequent practice also indicated that IWC has no power to amend the ICRW.[243] Recommendations are purely recommendatory in nature. The conditions under which subsequent practice may be deemed to modify the treaty are very strict. In relation to the Draft Articles of the VCLT, the ILC, envisaged such a possibility in Draft Article 38; but the Commission stressed the need for the common consent of State-parties to the a treaty, in order to modify it.[244]

Most importantly, subsequent practice of States cannot derive from an accumulation of non-binding Resolutions adopted in an antagonistic manner, which do not represent commonly shared views mostly adopted by majority voting.[245] Importantly, Japan stated that consensus is far from being the same as acceptance, in relation to the Resolution adopted on special permits on the basis of paragraph 30 of the Schedule and with respect to which Japan expressed a major reservation.[246] The mere existence of Resolutions critical of JARPA or JARPA II does not amount to practice. In order to do so, these Recommendations would have to reflect the unanimous agreement of the Parties and be followed by physical acts of implementation by States, and the concerned States would all have to

241 Pellet, Public Sitting, 3 July 2013, at 3 pm, Peace Palace, The Hague, CT/2013/14, http://www.icj-cij.org/docket/files/148/17442.pdf, accessed 5 August 2015 (cited hereafter as 'Pellet, afternoon').

242 Idem, para. 36.

243 Idem, paras. 43–45.

244 Idem, para. 52.

245 Idem, paras. 57 & 62.

246 Idem, para. 62.

follow the alleged practice, which is not the case here.[247] Evidence shows that there is no modificatory practice and there is no change in Article VIII.[248]

5.2.7 Subsequent Agreement and Practice: The Judgment

The Court made the following statements in relation to the interpretative value of the IWC Recommendations:[249]

> Article VI of the Convention states that '[t]he Commission may from time to time make recommendations to any or all Contracting Governments on any matters which relate to whales or whaling and to the objectives and purposes of this Convention'. These recommendations, which take the form of resolutions, are not binding. However, when they are adopted by consensus or by a unanimous vote, they may be relevant for the interpretation of the Convention or its Schedule (para. 46).

The Court clarified the issue of the value of the resolutions of the IWC. It made the following observations (para. 83):

(i) 'Article VIII expressly contemplates the use of lethal methods, and the Court is of the view that Australia and New Zealand overstate the legal significance of the recommendatory resolutions and Guidelines on which they rely';
(ii) 'Many of IWC resolutions were adopted without the support of all States parties to the Convention and, in particular, without the concurrence of Japan';
(iii) 'Such instruments cannot be regarded as subsequent agreement to an interpretation of Article VIII, nor as subsequent practice establishing an agreement of the parties regarding the interpretation of the treaty within the meaning of Article 31(3)(a)(b)';
(iv) 'The relevant Resolutions and Guidelines that have been approved by consensus call upon States parties to take into account whether research

247 Idem, para. 63.

248 Idem, para. 67.

249 See on this: Julian Arato, 'Subsequent Practice in the Whaling Case, and What the ICJ Implies about Treaty Interpretation in International Organisations', EJIL: Talk!, 31 March 2014,. http://www.ejiltalk.org/subsequent-practice-in-the-whaling-case-and-what-the-icj-implies-about-treaty-interpretation-in-international-organizations/, accessed 5 August 2015 (cited hereafter as 'Arato, Whaling Case'); idem: 'Treaty Interpretation and Constitutional Transformation: Informal Change in International Organizations', *Yale Journal of International Law* 38 (2013): 290.

objectives can practically and scientifically be achieved by using non-lethal research methods, but they do not establish a requirement that lethal methods be used only when other methods are not available';

(v) 'The Court, however, observes that the States parties to the ICRW have a duty to cooperate with the IWC and the Scientific Committee and thus should give due regard to Recommendations calling for an assessment of the feasibility of non-lethal alternatives'.

5.3 *Subsequent Agreement and Practice: Interim Conclusions*

The Judgment of the Court in relation to subsequent agreement and practice is very illuminating, as this question, in particular within the framework involving the working of and international organ, is very complex. Some of the Parties to the whaling dispute had a very broad approach as to what constitutes subsequent practice, disregarding the method of the adoption of a Resolution and the meaning of consensus. The issue of the participation of all States in the adoption of Resolutions, in they are to be considered as subsequent practice, seems to be crucial. According to the Court, the adoption of a Resolution by consensus has not changed its recommendatory nature into a requirement. In general, the Court set a high standard for subsequent practice in relation to the Resolutions of an international organ to be considered in the interpretative process. As Arato summed it up: 'unanimous Resolutions of a supervisory body might be considered subsequent agreements or practice relevant to the interpretation of the underlying convention, Resolutions adopted by a disputed majority will not count under the general rule of interpretation'.[250]

The issue of consensus in relation to subsequent practice was the subject of the Third Report of Special Rapporteur Nolte (the ILC Working Group on Treaties over Time, Subsequent Practice). He examined the role of consensus in subsequent practice in relation to Conferences of the Parties (COPs) in multilateral environmental agreements (MEAs), which he shows as being a very intricate subject.[251] In relation to it, the issue has arisen whether the 'interpretative resolution' by the COP, in order to qualify as subsequent practice, must be adopted by consensus. According to Nolte, it depends on whether such an agreement between the parties to the treaty must be specifically related to the interpretative question at hand; or whether an abstract agreement between

250 Arato, Whaling Case, note 250.

251 Third Report of the ILC Study Group on Treaties over Time, 'Subsequent Agreement and Subsequent Practice of States outside of Judicial or Quasi-judicial Proceedings', in Nolte, note 118, 303, 373.

the parties on the rules of procedure which are applicable to the COP are sufficient. Nolte is of the view that the first option is more desirable. Although subsequent agreements within the meaning of Article 31(3)(a) are not binding, they have certain effects according to the 1969 VCLT, which are only justified if the agreement between the Parties is expressly present.[252]

Arato was of the view that in the *Whaling* case, the Court took an unduly cautious position in relation to the requirements of what constitutes subsequent agreement and practice within an international organ. He relied on two landmark Advisory Opinions of the ICJ: *Certain Expenses*[253] and the *Wall*.[254] Arato argued that what is essential is that in these two cases, the ICJ expressly relied on Resolutions of the General Assembly as a substitute for the subsequent practice of the membership. Therefore the Court treated them as authentic criteria for the interpretation of the U.N. Charter (despite the fact that in both cases several of the key Resolutions were taken by majority vote, with dissenting votes). In *Wall*, the Court relied upon 'practice of the organization' to 'hew dramatically from the Charter's plain text, thereby recognizing what some consider an informal modification of the U.N. Charter'.[255]

Arato submitted three possible explanations for such a divergence in the judicial practice of the ICJ. The first one is that the ICJ adopted in the present case a more transparent and sovereigntist approach to treaty interpretation—i.e. 'rejecting its earlier case-line by refusing to recognize the relevance of any treaty body practice that does not reflect the will of every member state' (the IWC or the UNGA). Another possibility is that the Court would treat the 'advisory jurisprudence as opportunistic, adopting a less voluntaristic approach in the context of the UNGA because it is an organ of the U.N.'[256] A third interpretation draws a critical distinction between the institutions at issue in the advisory opinions and the *Whaling* case. It may be that Court viewed these cases as concerning an organization characterized by 'international legal personality (the U.N). and a treaty body with certain functions bearing no autonomous personality on the international stage (the IWC)'.[257]

252 Idem, 375.

253 *Certain Expenses of the United Nations, Advisory Opinion*, 1962, *I.C.J. Reports*, 131, http://www.icj-cij.org/docket/files/49/5259.pdf, accessed 5 August 2015.

254 *Legal Consequences of the Construction of the a Wall in the Occupied Palestinian Territory, Advisory Opinion*, 2004, *I.C.J. Reports*, 136, http://www.icj-cij.org/docket/files/131/1671.pdf, accessed 5 August 2015.

255 Arato, Whaling Case, note 250.

256 Ibid.

257 Ibid.

However, it may be that there is yet another explanation. In both Advisory Opinions on which Arato relies, subsequent practice was that of an organ of the United Nations; whilst in case of the IWC it was not the subsequent agreement and practice of the IWC itself under the scrutiny of the Court, but that of the Parties, Members of the IWC. Admittedly, the practice in issue related to the IWC's Resolutions, but it concerned only the behavior of States. This is a very significant distinction, which probably influenced the Court in arriving at its very strict and cautious approach regarding the assessment of the relevance of Article 31(3)(a) and (b) regarding the evidentiary value of the Resolutions of the IWC in relation to subsequent agreement and practice by the Member-States of the IWC.

6 The Interpretation of Article VIII of the ICRW

6.1 *Introduction*

The final part of the chapter will be devoted to the issues of the interpretation of Article VIII of the Whaling Convention, which in fact was the heart of the matter of the *Whaling* case, as it contains the rules on scientific whaling.[258] The chapter will be mainly focused on the pleadings of the Parties (written and oral) and the Judgment of the Court and Opinions of the Judges. No general points on interpretation will be made as the general observations made in relation

258 Article VIII: '1. Notwithstanding anything contained in this Convention any Contracting Government may grant to any of its nationals a special permit authorizing that national to kill, take and treat whales for purposes of scientific research subject to such restrictions as to number and subject to such other conditions as the Contracting Government thinks fit, and the killing, taking, and treating of whales in accordance with the provisions of this article shall be exempt from the operation of this Convention. Each Contracting Government shall report at once to the Commission all such authorizations which it has granted. Each Contracting Government may at any time revoke any such special permit which it has granted. 2. Any whales taken under these special permits shall so far as practicable be processed and the proceeds shall be dealt with in accordance with directions issued by the Government by which the permit was granted. 3. Each Contracting Government shall transmit to such body as may be designated by the Commission, in so far as practicable, and at intervals of not more than one year, scientific information available to that Government with respect to whales and whaling, including the results of research conducted pursuant to paragraph 1 of this article and Article IV. 4. Recognizing that continuous collection and analysis of biological data in connection with the operations of factory ships and land stations are indispensable to sound and constructive management of the whale fisheries, the Contracting Governments will take all practicable measures to obtain such data'.

to the object and purpose of the ICRW as a whole (Section 2 above) hold also in connection with the interpretation of its Article VIII, and some analysis of subsequent agreement and practice in relation to Article VIII was dealt also in that section.

6.2 *Interpretation of Article VIII in the Pleadings, Judgment and Judges' Opinions in the Whaling Case*

6.2.1 Interpretation of Article VIII: The Memorial

Australia set out its basic interpretative approach to Article VIII of the ICRW on the basis that this was governed by Articles 31 and 32 of the 1969 VCLT. It stated that the application of these interpretative principles 'makes clear that Japan's ability to issue special permits under Article VIII must be exercised on the basis of the objective criteria that determine whether an activity is, in good faith, "for the purpose of scientific research"'.[259] Australia relied on the ordinary meaning of words and therefore argued that Article VIII has a restricted meaning, namely, that whaling under it must be carried out for 'purposes' of scientific research within the context of the object and purpose of the ICRW and in good faith.[260]

This reading of Article VIII implies that permits may be lawfully issued if two separate but related conditions are fulfilled: the activity that is subject to the permit is, in fact, 'scientific research'; and the activity must be 'for the purposes of scientific research', and not for any other purposes. Article VIII is both concerned with the product of the activity and the aim (the reason) for which this activity will be carried out.[261] Australia importantly noted that scientific research is not defined in the ICRW. However, its meaning and effect can be determined by the reference to Article 31 of the VCLT. In this regard, permission accorded to any State-Party under Article VIII(1) to grant to its national a special permit authorising them to kill, take of treat whales for the purposes of scientific research is not 'self-judging' or subjective. The meaning of Article VIII has to be determined on an objective basis, i.e. in this case it is for the Court to determine the scope of the exception under this article in accordance with Article 31 of the 1969 VCLT.

Australia applied the rule of interpretation in order to determine the meaning of 'for the purposes' and of the words 'scientific research'.[262] Australia first

259 The Memorial, note 32, para. 4.33.
260 Idem, paras. 4.33 & 434.
261 Idem, para. 4.36.
262 Idem, para. 3.39.

established what is the ordinary meaning of 'purpose', which, according to the *The Concise Oxford English Dictionary*, is 'the reason for which something is done or for which something exists'. Australia, in pursuit of this ordinary meaning, made a number of propositions in relation to the interpretation of the words 'for purposes of' as used in the phrase 'for purposes of scientific research':

(i) that the activity be assessed to be genuinely motivated by an intent to conduct scientific research (and not for any other purposes);
(ii) that a proper interpretation indicates that activity that is unlikely to generate scientifically useful information cannot be considered to be carried out 'for purposes of scientific research'; and
(iii) that the words 'for purposes of' require that it will not be sufficient if the conduct of scientific research is an incidental reason for the conduct of a particular whaling operation.[263]

Australia further made another very high-level test to ascertain the 'object and purpose'. It said that in

> determining the 'purpose' for which an activity is carried out it is also essential to have regard to the outcome or product of that activity, especially an activity carried out over a prolonged period of time. In the context of whaling purportedly conducted for 'purposes of scientific research' under Article VIII, this means examining the result of the relevant whaling operations to determine whether those operations were conducted with the genuine motivation of carrying out scientific research.[264]

Australia agreed that there is no definition of 'scientific research' in ICRW, but submitted some definitions from other sources. It stated, however, that 'the notion and process of science is inherently evolutionary',[265] not static and cannot be judged by the past. In order to determine the meaning of scientific research, Australia further analaysed the object and purpose of the ICRW (see above).

The outcome of the interpretation Article VIII by Australia was as follows:[266]

(i) The permission granted to Contracting Governments to issue special permits authorising whaling 'for purposes of scientific research' (Article

263 Idem, paras. 4.39 & 4.40.
264 Idem, para. 4.42.
265 Idem, para. 4.47.
266 Idem, paras. 4.116–4.122.

VIII of the ICRW) is not self-judging, but has to be assessed by in reliance on objective criteria. The determination of whether a whaling operation is 'for purposes of scientific research' is not a question that is left to the discretion of each Contracting Government, but which falls to be determined by the Court in accordance with the established principles of treaty interpretation;

(ii) According to the principles of the interpretation of 1969 VCLT, the granting of such permits must be interpreted in such a manner as not to undermine the effectiveness of the ICRW regime (including the moratorium on commercial whaling), and must not negate the conservation measures which are an integral part of this regime;

(iii) In order to give effect to the text of Article VIII, a special permit whaling program must be conducted for 'purposes' of scientific research and for no other purpose; and

(iv) These principles are also consistent with the obligation to perform the requirements of the ICRW in good faith.

Australia asserted that this interpretation of Article VIII is consistent and derives from its ordinary meaning, having regard to the object and purpose of the ICRW and to the context of the Convention as a whole. It is supported by subsequent practice and the principle of systemic integration (incorporating the precautionary approach). These developments inform the scope of the Article VIII exception and are consistent with the very strictly limited application of the exception, especially where there is uncertainty in relation to certain stocks.[267]

6.2.2 Interpretation of Article VIII: The Counter-Memorial

Japan also relied on Articles 31–32 of the 1969 VCLT in its interpretation of Article VIII of the ICRW, and the natural and ordinary meaning in the context in which they occur and in the light of their object and purpose.[268] The interpretation of Article VIII clearly indicates, according to Japan, that the special permit whaling, and likewise the right to issues such a permit, is entirely outside the framework of the ICRW (a proposition which is supported by the *travaux préparatoires*).[269] In fact, Article VIII(1) indicates that killing and taking of whales is entirely within the discretion of a State-party. As Japan stated,

267 Idem, para. 4.117.

268 The Counter-Memorial, note 40, para. 7.4.

269 Idem, paras. 7.8–7.10.

no other body or State can impose restrictions or conditions upon the exercise by a State Party of its right to authorize special permit whaling.[270]

Japan accepted that, in the context of Article VIII, it is established that continuous collection and analysis of biological data is indispensable, and is even recognized as a matter of law—as the Parties to the Convention are bound by it to take all practicable measures to obtain such data in connection with the operations of factory ships, and as a necessary corollary obligation under Article VIII(4). The parties are also obliged to recognize the propriety of the continuous collection of data. Recognition of the need for scientific research, Japan averred, 'is one of the basic and permanent elements of the ICRW regime'.[271] This 'exegesis' Article VIII, according to Japan, is 'clear and straightforward' and follows from the 'ordinary meaning of the terms of Article VIII in their context and in the light of the object and purpose of the ICRW'.[272]

Japan also, as discussed above, rejected, due to the lack of agreement or practice as set out in Article 31(3)(a) and (b) the assertion that, that there was any relevant subsequent practice within the interpretative context, and itself asserted that the rules of international law, taking into account Article 31(3)(c) (see above), 'plainly support the reading of the IWCR Article VIII explicated... and offer no grounds for reading a different meaning into plan words'.[273]

On that basis, Japan asserted, there was no need to have recourse to supplemental means of interpretation in accordance with Article 32 of the 1969 VCLT. However, Japan, only for the 'sake of completeness' set out 'to demonstrate that [the above] interpretation is wholly borne out by an examination of the *travaux préparatoires* of the ICRW'.[274] In this respect, Japan asserted that management, rather than conservation of whales, was the primary object and purpose of the ICRW, as was evident from the very initial draft; and later changes were not materiality different—I need, the more significant changes were aimed at the strengthening of the whaling industry.[275] Japan singled out four points arising from the *travaux préparatoires*. Firstly, the importance of whaling as an activity, an industry and a source of natural resources were affirmed;[276] secondly, the importance of scientific research and the continuous collection of data was stressed; thirdly it was clearly understood that the IWC would not

270 Idem, para. 7.16.

271 Idem, paras. 7.19 & 7.21.

272 Idem, para. 7.22.

273 Idem, para. 7.23.

274 Idem, para. 7.24.

275 Idem, paras. 7.25 & 7.26.

276 Idem, paras. 7.30–7.42.

in itself conduct scientific research and data collection but that this was a matter left to Contracting Parties; and fourthly, th suggestion of prior consultations of the Governments with the IWC relating to the issuance of special permit was rejected.

Japan also rejected the assumption that the Schedule amendments can be qualified as 'context' for the interpretation of Article VIII of the ICRW. It contended that in order to qualify as a part of the relevant 'context' for the interpretation of a provision, a text must have been 'intended by the Parties to the treaty, or at least, by its author, to fix the meaning of that provision'.[277] However, all the amendments to the Schedule relied upon by Australia relate exclusively to commercial whaling and were never intended to apply to special permit whaling, and thus to change or to qualify, or to throw light on the meaning of Article VIII of the IWRC. Therefore, they cannot restrict the rights enjoyed by the parties to the ICRW under Article VIII.[278]

6.2.3 Interpretation of Article VIII: New Zealand's Written Observations

New Zealand in its Written Observations relied on Articles 31 and 32 of the ICRW in the interpretation of both the ICRW and Article VIII. It also referred to subsequent practice and the supplementary means of the interpretation, in which it included not only the *travaux préparatoires*, but also statements and decisions of the parties or related bodies concerning the treaty's interpretation or application.[279] New Zealand argued an opposite view to that of Japan requiring interpretation of Article VIII within the context of the ICRW and the Schedule—not, as Japan had argued, as a self-standing provision outside the framework of the Convention. Both as evidence of subsequent practice under 31(3)(b), and as supplementary means of the interpretation (Article 32 of the 1969 VCLT), decisions and recommendations of the IWC could, according to New Zealand, play a valuable role in interpreting of the terms of Article VIII and their proper application; but such decisions and resolutions do not modify the terms of Article VIII but rather confirm the interpretation which is based on its ordinary meaning in their context. New Zealand has also placed reliance on other rules of international law applicable to the Parties to the ICRW.[280]

It may also be inferred from the Pleadings of New Zealand that the object and purpose of the ICRW has evolved due to the collective regulation (see

277 Idem, para. 8.55.

278 Idem, para. 8.56.

279 Written Observations of New Zealand, note 51, para. 11.

280 Idem, para. 13.

above); and this also should be taken into consideration in relation to the interpretation of its VIII.

New Zealand also averred that, due to the collective character of the Convention, Article VIII should be subject to a restrictive rather than an expansive interpretation.[281]

6.2.4 Interpretation of Article VIII: Japan's Written Observations

Japan strongly argued against the restrictive interpretation of Article VIII of the ICRW advocated by New Zealand. Japan argued that such an interpretation is contrary to that 'evident in the relevant case-law', as restrictive interpretation had been only applied by the Court with a view of protecting the sovereignty and freedom of action of a State. New Zealand, however, had invoked this principle to obliterate rights expressly preserved under Article VIII of the ICRW. No principle of interpretation could be invoked, Japan asserted, in order to support an interpretation of a treaty text contradicting its explicit terms. The prevailing interpretation must the one closest to the object and purpose of the treaty.[282]

As in relation to the object and purpose of the ICRW as a whole, Japan was of the view that the object and purpose of Article VIII could not be silently rewritten. Notwithstanding the individual intentions of the States Parties when adhering to a convention, no legal principle allows for a change of its object and purpose in a particular direction that is antagonistic to that established by the text of the convention, and to the will of its authors. 'In particular, States that accede to a convention that was negotiated without their participation... [must] ...accept the Convention as it stands: to pretend otherwise it would encourage an unprecedented form of hijacking of existing conventions'.[283]

Japan also noted the absence of any amendments relating to the object and purpose which would be material in this case, and further denied that the system of 'collective regulation' alleged by New Zealand had replaced the object and purpose of the ICRW so as to provide for the proper conservation of whale stocks. Japan argued that the ICRW is not a complete system that regulates all activities with respect to whaling and which entirely displaces all other rights and duties. The States-Parties to this Convention are bound only by its provisions.[284] The ICRW is a treaty which constrained unilateral whaling and is a 'treaty which is superimposed on a pre-existing freedom of whaling'; neither

281 Idem, para. 46.

282 Written Observations of Japan, note 59, para. 18.

283 Idem, para. 22.

284 Idem, para. 27.

had it created or crystallized the right to engage in whaling, nor set out an exhaustive code governing all whaling activities. It just consists of a set of rules regulating certain aspects of whaling activity, in accordance with the provisions of the ICRW. Therefore, stated Japan, the claim of New Zealand that the discretion of States in relation to Article VIII is restricted due to the existence of an alleged 'collective system' and is subject to validation by other States, goes against the plain meaning of Article VIII.[285] Japan further explains that the words 'Notwithstanding anything contained in this Convention' and 'shall be exempt from the operation of this Convention' indicate that the system set out under the ICRW does not apply to whaling for the purposes of scientific research, which Japan argues is the most crucial fact.[286] Article VIII, according to Japan, does not have also a 'collective decision-making', aspect, but only refers to decisions the legal character of which vary depending whether they are amendments or recommendations.[287]

6.2.5 Interpretation of Article VIII: Oral Pleadings by Australia[288]

Australia submitted that Article VIII, according to its ordinary meaning, envisaged two functions. The first of these is simply the granting of the permit itself to a national. But the discretion to grant a permit or not does not extend to a discretion as to the purpose for which it is granted, This is expressly limited to 'the purposes of scientific research'; and whether a permit meets that requirement is not a matter for determination by the granting Government alone; and in fact, according to Australia, Japan had eventually concurred that Article VIII does not establish a 'self-judging right'. The second function arises from the second phrase of the article 'subject to such restrictions as to number and subject to such other conditions as the Contracting Governments think fit'. According to Australia, the words 'as the Contracting Governments think fit' relate only to the number of whales to be taken and 'the other conditions attached to the special permit'. Again, it does not concern the description of a special permit itself, i.e. the phrase 'for purposes of scientific research'. But this second function, Australia asserted, has to be exercised in accordance with the principle of good faith, which means it must be exercised in 'a reasonable way, and in such a manner that the purpose of the Convention can be realized'.[289]

285 Idem, paras. 28–30.

286 Idem, paras. 33 & 34.

287 Idem, para. 36.

288 Crawford, note 235.

289 Crawford, note 235, paras. 16–19. On subsequent practice and Article VIII see above, note 235.

6.2.6 Interpretation of Article VIII: Oral Pleadings by Japan[290]

Japan focused its interpretation of Article VIII on the phrase 'Notwithstanding anything contained in this Convention…', which means, according to Japan, that Article VIII entails a derogation from the other provisions of the Convention, and must be read in itself. Neither the Convention, nor the secondary texts (such as the amendments to the Schedule), restrict a State-Party's right to grant special permits. This is governed by a special regime, which is based on the discretion of the parties to the IWCR.[291] The language of Article VIII is clear also in giving full discretion to the Governments with respect to the number of whales to be taken and to other conditions (as Governments 'think fit'). This is a different formulation as compared to Article V para. 1(e) in relation to commercial whaling (or in relation to the provisions concerning aboriginal whaling), which place things like size of catch in the hands of the IWC.

Japan also pointed to Article VIII (2)—'any whales under these special permits shall as far as practicable be processed and the proceeds shall be dealt with in accordance with directions issued by the Government by which the permit was granted'. This provision indicates that the marketing of thus processed whales is not contrary to the ICRW (as pleaded by Australia), as there is a clear obligation not to waste the meat of whales. This provision is also, according to Japan, based upon the discretion of the responsible States.[292]

Japan further asserted that the obligation to inform (under Article VIII, para. 3) is the only element of 'collective regulation' of scientific research whaling, and it is only in this sense that special permit whaling is an integral part of the system.

According to Japan, there is no need for a teleological interpretation of Article VIII, nor for invoking subsequent practice or having recourse to *travaoux preparatoires*, as suggested by Australia and New Zealand. The text of Article VIII can be interpreted on the basis of the ordinary meaning of its words. Though according to Australia conservation and the recovery of stocks are ends in themselves, Japan asserted that what is the ultimate aim of the Convention is sustainable use through stock recovery.[293]

---

290 Pellet, Public Sitting, 3rd July 2013, 10 am Peace Palace, the Hague, http://www.icj-cij.org/docket/files/148/17474.pdf, accessed 6 August 2015 (cited hereafter as 'Pellet, morning)'.

291 Idem, paras. 4–7.

292 Idem, para. 15.

293 Idem, paras, 28–30.

6.2.7 Interpretation of Article VIII: Oral Pleadings by New Zealand[294]

The starting point for New Zealand's interpretation of Article VIII was that it is an integral part of the ICRW, a part of the fabric of the Convention. It is one of several mechanisms for collecting scientific data relevant for the IWC. It is not as such a 'free standing' or 'self-contained regime'. It is also not an affirmation of the right under the freedom of high seas, which can only be exercised if it's not restricted by other rules of international law. It follows, according to New Zealand that, as an integral part of the Convention, Article VIII must be interpreted and applied in consistency with other provisions of the Convention.[295]

The integral role of Article VIII can be derived from its structure. It has four paragraphs, but only two relate to special permits. The remaining two relate to scientific information itself. The integral nature of Article VIII can also be seen from the obligation that State-Parties have to submit any special permit to the Scientific Committee for review and comments before it is issued, and to notify the IWC after the special permit has been issued and to transmit the results after completion of the programme. Therefore, permits are 'inextricably linked to the role of the Commission and the rest of the Convention'.[296]

New Zealand contended that Japan had presented a 'strained' interpretation of Article VIII 'that highlights snippets of the language of the provision at the expense of the whole'. Japan, according to New Zealand,put together parts of the language from the first sentence of paragraph 1 of the article to construct 'a blanket exemption from the rest of the Convention'. In particular, in this respect, New Zealand referred to the following phrases 'notwithstanding anything contained in this Convention', 'as the Contracting Government thinks fit', and 'shall be exempt from the operation of the Convention'.[297] Those three phrases, according to New Zealand, should be read in their ordinary meaning in their context of paragraph 1 of Article VIII, read in its natural sense. It can then, according to New Zealand, be seen to contain three distinct elements.

The first is the phrase: 'Notwithstanding anything contained in this Convention any Contracting Government may grant to any of its nationals a special permit authorizing that national to kill, take and treat whales for purposes of scientific research', which, according to New Zealand, is not as argued by Professor Pellet, 'an overarching *chapeau* to Article VIII as a whole'. It is clearly linked only to the words that follow it: 'may grant to any of its nationals a special Permit'. This formulation enables State-Parties to issue a special permit for

294 Finlayson, note 82.

295 Idem, paras. 22–25.

296 Idem, para. 27.

297 Idem, para. 28.

the specific purposes of 'scientific research', notwithstanding the other rules of the Convention and constitutes in that sense a limited exception, but no greater exemption from the obligations of the Convention than that.[298]

The next element is the passage of Article VIII which reads 'subject to such restrictions as to number and subject to such other conditions as the Contracting Government thinks fit'. According to New Zealand, it is again here clear that the words 'as the Contracting Government thinks fit' are linked only to this element'. It cannot be said that they create a general exemption allowing the Contracting Government to do 'whatever it thinks fit' under this article.

The third element referred to by New Zealand is the passage which reads: 'the killing, taking and treating of whales in accordance with the provisions of this article shall be exempt from the operation of this Convention'. According to New Zealand, this text does not say that 'special permit whaling is "exempt from the operation of this Convention"', as pleaded by Japan in the Counter-Memorial. On the contrary, only the killing, taking and treating of whales 'in accordance with the provisions' of Article VIII is exempt from the ordinary rules of the Convention. According to New Zealand, '[f]ar from creating a blanket exemption, the words create an obligation on the Contracting Government to act "in accordance with the provisions" of Article VIII when issuing a special permit'.

In fact, according to New Zealand, the formulation of Article VIII is very narrow regarding the obligations of State Parties. Thus, it provides that a State issuing a special permit:

(i) must do so for specified and articulated 'purposes of scientific research';
(ii) must set restrictions on the number of whales to be taken or killed under special permit;
(iii) must issue permits only 'in accordance with the provisions of article VIII, including the provisions of paragraph 30 of the Schedule and the duty of the meaningful cooperation they include'.[299]

We may also note New Zealand's argument, which is closely linked to the principle of reasonableness, to the effect that the Court's jurisprudence also required that 'a power must be exercised *properly*, that is, for the purpose for which it has been given'. The Court has also, according to New Zealand, repeatedly stressed the principle of effectiveness, 'that a power must be exercised in a manner that would not undermine object and purpose of the provisions of

298 Idem, para. 29.

299 Idem, para. 30.

the treaty as a whole'.[300] Finally, on the basis of fundamental principles of interpretation, Article VIII must be applied for the stated purpose 'scientific research' in a 'reasonable way' and in such manner that the purpose of the ICRW can be implemented. According to New Zealand, Article VIII cannot be applied to special permit whaling when such a permit would be circumventing the other provisions of the Convention or could even frustrate its central objective.[301]

6.2.8 Interpretation of Article VIII: The Judgment

The Court summarized its understanding of the main arguments of the Parties to the dispute, stating, firstly that Japan approached Article VIII on the basis that, although it must be interpreted as part of Convention, it amounts, nevertheless, to an exemption from the Convention. Australia, on the other hand argued that Article VIII needs to be read in the context of the other provisions of the Convention, to which it provides only a limited exception. Australia had, the Court said, argued that Article VIII has to be interpreted within the general objectives of the ICRW, one of these being conservation, such as the Moratorium and the Sanctuaries, which are also relevant for whaling for scientific purposes. Article VIII cannot have the effect of undermining the effectiveness of the regulatory régime as a whole.[302] New Zealand, the Court said, observed that the phrase 'notwithstanding anything contained in this Convention' (Article VIII para 1) only provides a limited discretion for State-Parties to issue special permits for the specific articulated purpose of scientific research and is not a blanket exemption for Special Permit whaling from all aspects of the Convention. Further, the Court noted, it was argued by New Zealand that the provision setting out that the taking of whales in accordance with Article VIII is outside the scope of the operation of the Convention would have been redundant if the opening words of the paragraph, 'notwithstanding anything in the Convention', were intended to cover all aspects of Special Permit whaling.[303]

The Court, in providing an initial summary of its reading of Article VIII of the ICRW, made the following important observations concerning its interpretation:

(i) Article VIII is an integral part of the Convention. For that reason it has to be interpreted in light of the object and purpose of the ICRW, also in consideration of other provisions of the Convention, including the Schedule.

300 Idem, para. 35.
301 Idem, para. 36.
302 The Judgment, note 1, paras. 52 & 53.
303 Idem, para. 54.

(ii) However, since Article VIII para. 1 states that the killing, taking, and treating of whales in accordance with the provisions of the article is exempt from the operation of the Convention:

> whaling based on a special permit which meets the conditions of Article VIII is not subject to the obligations under the Schedule concerning the moratorium on the catching of whales for commercial purposes, the prohibition of commercial whaling in the Southern Ocean Sanctuary and the moratorium relating to factory ships.[304]

The Court further stated that the Preamble to the Convention indicates that its object and purpose is the conservation of whale species while permitting their sustainable use and continued by setting out the most important paragraphs of the Preamble relating to this dual object and purposes the Convention. Firstly, it said:

> The first perambular paragraph recognizes "the interest of the nations of the world in safeguarding for future generations the great natural resources represented by the whale stocks". Similarly the second paragraph of the preamble expresses the desire "to protect all species of whales from further over-fishing", and the fifth one emphasizes the need "to give an interval for recovery to certain species now depleted in numbers".

However, the Court also stated that the Preamble also refers to the exploitation of whales, indicating that the Preamble noted:

> in the third paragraph that 'increases in the size of whale stocks will permit increases in the number of whales which may be captured without endangering these natural resources', and further in the fourth paragraph that 'it is in the common interest to achieve the optimum level of whale stocks as rapidly as possible without causing widespread economic and nutritional distress' and in the fifth that 'whaling operations should be confined to those species best able to sustain exploitation'.

The Court went on to draw attention to the final preambular paragraph which also refers to the 'objectives' of the Convention 'which states that the Contracting Parties "decided to conclude a convention to provide for the proper conservation of whale stocks and thus make possible the orderly development of the

---

304 Idem, para. 55.

whaling industry"'. The Court continued by making a very important statement to the effect that '[a]mendments to the Schedule and recommendations by the IWC may put an emphasis on one or the other objective pursued by the Convention, but cannot alter its object and purpose'.[305]

The Court noted that in order to substantiate their arguments, both Australia and New Zealand relied on those parts of the Preamble referring to the conservation of whale stocks; while Japan relied on the parts referring to their sustainable exploitation. Both Australia and New Zealand advocated a restrictive interpretation of Article VIII of the ICRW, so as not to undermine the objectives and purposes of the Convention (as interpreted by them). Japan, on the other hand, argued that that the authority to authorize the taking of whales for purposes of scientific research should be seen in the general context of the freedom to engage in whaling enjoyed by States under customary international law.[306] Finally, in its review of the general approach to interpretation of Article VIII, the Court, having taken into account the Preamble and other relevant provisions of the Convention, held that 'neither a restrictive nor an expansive interpretation of Article VIII is justified'.

Following this conclusion through to the more specific issue of what might comprise 'scientific research', the Court proceeded to note that programmes for purposes of scientific research should support scientific knowledge, but they may be broader than just either conservation or sustainable exploitation of whale stocks, a conclusion which, the Court said, can be derived from Guidelines issued by the IWC for the review of scientific permit proposals by the Scientific Committee. 'In particular' the Court said:

> the Guidelines initially applicable to JARPA II, Annex Y, referred not only to programmes that "contribute information essential for rational management of the stock" or those that are relevant for "conduct[ing] the comprehensive assessment" of the moratorium on commercial whaling, but also to those responding to 'other critically important research needs'.

The Court further noted that the current Guidelines, Annex P, list three broad categories of objective which include as follows:

(i) programmes aimed at 'improv[ing] the conservation and management of whale stocks';

305 Idem, para. 56.

306 Idem, para. 57.

(ii) programmes 'which have as an objective to "improve the conservation and management of other living marine resources or the ecosystem of which the whale stocks are an integral part;"' and
(iii) programmes aimed at '"test[ing] hypotheses not directly related to the management of living marine resources"'.[307]

The Court next turned its attention to the question, directly related to the interpretation of Article VIII, of the issuance of the special permits. As the Court noted, there was also a wide difference in the approaches to this question as between Japan, on the one hand, and Australia and New Zealand, on the other. Japan was of the view that such an issuance is within discretionary powers of the State-Party; whilst both Australia and New Zealand argued that this is 'an objective process, to which the State of nationality has to conform'[308] In this respect, the Court expressed the important conclusion that:

> Article VIII gives discretion to a State party to the ICRW to reject the request for a special permit or to specify the conditions under which a permit will be granted. However, whether the killing, taking and treating of whales pursuant to a requested special permit is for purposes of scientific research cannot be decided on the basis of that State's sole understanding.[309]

The Court, at this point, somewhat baldly—that is without having really earlier pronounced upon the basis on which it adopted this method of 'interpreting' Article VIII, stated that it would turn to: 'the standard that it will apply in reviewing the grant of a special permit authorizing the killing, taking and reading of whales on the basis of Article VIII, paragraph 1, of the Convention'.[310] The Court noted that that all the Parties, though Japan only in the course of its oral pleading, had in various terms accepted that some objective standard was to be applied to assessing the propriety of the issuance of special permits; and it went on to make the most important statement, from the point of the interpretation of Article VIII, in paragraph 67 of the Judgment, where it said:

> When reviewing the grant of a special permit authorizing the killing, taking and treating of whales, the Court will assess, first, whether the programme

307 Idem, para. 58.
308 Idem, para. 60 (also 59).
309 Idem, para. 61.
310 Idem para. 62. And see the chapter of Caroline Foster in this book on the standard of review in this case.

> under which these activities occur involves scientific research. Secondly, the Court will consider if the killing, taking and treating of whales is 'for purposes of' scientific research by examining whether, in the use of lethal methods, the programme's design and implementation are reasonable in relation to achieving its stated objectives. This standard of review is an objective one. Relevant elements of a programme's design and implementation are set forth below (see paragraph 88).

This introduction by the Court of a 'standard of review' raises the question whether this is a matter solely of interpretation of Article VIII, or whether it introduces elements into the interpretative process which are, strictly, outside of the purview of Article VIII. In the view of author of this Chapter, the interpretation by means of standard of review in fact combines elements of the interpretation of Article VIII with other elements, extraneous to Article VIII. The interpretative element which falls within the purview of Article VIII is the interpretation of the phrase 'for the purpose of scientific research', as this is the core of Article VIII. However, there is a host of extraneous interpretative elements that are of fundamental importance to the notion of the standard of review, as explained by the Court, such as reasonableness and objectivity, that probably are not inherently built into the fabric of Article VIII of the ICRW. Therefore, it appears that the interpretation by means of standard of review is not solely based on Article VIII, but also on the other factors outside the realm of this provision.

The last issue regarding the interpretation of Article VIII that arose in the Judgment is the question of interpreting it in the light of Recommendations of the IWC. In this respect, the Court held that: 'Australia and New Zealand overstate the legal significance of the recommendatory resolutions and Guidelines on which they rely. It further held that Recommendations which had not been accepted by all Parties (in particular not by Japan) could be regarded as neither subsequent agreement nor subsequent practice in relation to the interpretation of Article VIII (para. 83 of the Judgment). However, the Court did affirm that "the States parties to the ICRW have a duty to co-operate with the IWC and the Scientific Committee and thus should give due regard to recommendations calling for an assessment of the feasibility of non-lethal alternatives" (paragraph 83 of the Judgment)—an obligation which Japan accepted (paragraph 137 of the Judgment).

In this regard, the question may be posed whether an obligation of co-operation (accepted by one of the Parties) is one of the elements to be taken into consideration in interpreting Article VIII. In the view of the present author, accepting an obligation of giving 'due regard' to recommendations relating to non-use of lethal methods is a very weak obligation, lacking well defined legal content. Due regard means giving due consideration only, the outcome of which does not necessarily mean that the use of lethal methods must, after having given 'due regard' to the non-lethal methods, be modified or restricted. Therefore, in view of the present author, such a weak obligation (even if explicitly or implicitly accepted by one of the Parties) cannot contribute significantly to the interpretation of Article VIII of the ICRW.

6.2.9 Interpretation of Article VIII: Judges' Separate and Dissenting Opinions

6.2.9.1 *Dissenting Opinion of Judge Owada*

Judge Owada in his interpretation of Article VIII relied on the essential character of the Convention, which is, according to him, a self-regulatory regime for the regulation of whaling. The provisions of Article VIII provide one 'important component' of this regime. In Judge Owada's view, it is wrong to characterize the exception of Article VIII as merely being in difference to State sovereignty and the customary freedom of the high seas relating to whaling. This, according to Judge Owada would, in fact, be to diminishing the important prerogative under Article VIII in terms of its function within this regulatory regime, which is the carrying out of scientific research and collecting of data in the fulfillment of the objectives and purposes of the ICRW. For that reason, a Contracting Party entrusted with this task will be assessed by the regulatory organ within the ICRW, namely, the IWC (or its Scientific Committee). But there is no power bestowed on an organ of the ICRW which can limit the power of the issuing State, which is only restricted by the provision that special permits shall be issued 'for purposes of scientific research subject to such restrictions as to number and subject to such other conditions as the Contracting Government thinks fit'. Therefore the issuance of such permits is fully discretionary.

Judge Owada commented on the allegations of the Applicant relating to alleged bad faith on part of Respondent, i.e. that scientific research was a guise for other activities. He was of the view that:

> [t]he function of the Court in this respect is to see to it that the State in question is pursuing its activities in good faith and in accordance with the requirements of the regulatory régime for the purposes of scientific research that is conducive to scientific outcomes which would help promote the object and purpose of the Convention.

Judge Owada was of the view that the good faith of the issuing Party in this respect had to be presumed, and continued:

> Allegations made by the Applicant that the activities were designed and implemented for purposes other than scientific research under the cover of scientific research thus cannot be presumed, and will have to be established by hard conclusive evidence that could point to the existence of bad faith attributable to the State in question.[311]

Further Judge Owada did not agree with the distinction made by the Court between 'scientific research' and for 'the purpose of scientific research' (Judgment, paragraphs 70–86), which in the view of the present author relates rather more to the substance of a conducted activity than the interpretation of Article VIII. According to Judge Owada, this 'distinction is so artificial that it loses any sense of reality when applied to a concrete situation, in particular that the Court is not in a position to determine what is "for the purpose of scientific research" with a reasonable objectivity'.[312]

6.2.9.2 *Separate Opinion of Judge Cançado Trindade*

Judge Cançado Trindade, was of the view that Article VIII(1) of the ICRW Convention

> is not to be interpreted broadly, so as to go against the object and purpose of the normative framework of the Convention as a whole. Article VIII(1) appears as an exception to the normative framework of the ICRW Convention, to be thus interpreted restrictively.

The sole purpose of this article is, according to Judge Cançado Trindade, scientific research. States Parties to a multilateral environmental agreement, such as the ICRW, with a regulatory mechanism, should not have a full discretion in relation to granting permits, but should substantiate that the permit research is for the purpose of scientific research. Such unlimited discretion, according to Judge Cançado Trindade, is not in accordance with the object and purpose of the ICRW. A State issuing such permits should also take into consideration the Resolutions and Guidelines on scientific research and abide by the principle of prevention and the precautionary approach.

This Judge said, importantly, that in his view

> Article VIII, part and parcel of the ICRW Convention as a whole, is to be interpreted taking into account its object and purpose. This discards any

311 Judge Owada, note 86, para. 22.

312 Idem, paras. 23–25.

> pretence of devising in it a so-called 'self-contained' regime or system, which would go unduly against the ICRW Convention's object and purpose.[313]

In conclusion, Judge Cançado Trindade, said that a State-Party has no unlimited right to decide what is 'scientific research' and whether a given whaling programme 'for purposes of scientific research'. Further, the development of the interpretation of the IWCR and its application is moving away, according to Judge Cançado Trindade, from unilateralism and towards multilateral conservation of living marine resources.

6.2.9.3 *Separate Opinion of Judge Xue*

According to Judge Xue, the ICRW set up a special category of whaling, pursuant to which, a contracting party may issue special permits to its nationals to kill, take, and treat whales for purposes of scientific research ('scientific whaling'), in a manner which a State-Party 'thinks fit'. The killing of whales under Article VIII is exempt from the limitations regarding commercial whaling, and depends on the discretionary power of a Party, the scope of this discretion being a subject of the dispute.[314] The lack of mandatory regulatory powers of the IWC in respect of scientific whaling, and the expression 'thinks fit', leads Japan to argue that there is an extensive right of States to grant special permits. However, Judge Xue disagrees with such an interpretation as the right derives from the regulatory power of the ICRW therefore cannot be unlimited.

Judge Xue proceeded to give three reasons for this. Firstly, in granting special permits for killing, taking and treating whales for scientific purposes, a State must avoid any adverse effect on the stocks with a view to maintaining sustainable utilization and conservation of the resources, so as not to undermine the object and purpose of the Convention. Secondly, Party granting a special permit must consider commercial whaling limitations which indicate the manageable level of the stocks. There is an intrinsic link between commercial whaling and scientific whaling, particularly when scientific whaling is purportedly to be carried out on a large scale and on a continuous basis.[315] Thirdly, discretion under Article VIII, para. 1, 'as a corollary, also means a duty on every authorizing party to exercise the power properly and reasonably by virtue of

313 Judge Cançado Trindade, note 193, paras. 22–23.

314 Separate Opinion of Judge Xue, paras. 2–5, http://www.icj-cij.org/docket/files/148/18152.pdf, accessed 6 August 2015.

315 Idem, para. 8 and 7.

the principle of good faith under the law of treaties'. Therefore, it cannot be said that Article VIII granted to Parties a self-defined right to the Parties.[316]

In a part of her Separate Opinion in which she was critical of the way in which the Court arrived at, or at least explained, its decision, rather than of the decision itself, Judge Xue drew attention to the passage in the Judgment where the Court said: 'whether the killing, taking and treating of whales pursuant to a requested special permit is for purposes of scientific research cannot depend simply on that State's perception' (para. 61 of the Judgment), She noted that this statement was 'left unexplained' by the Court and, while, on the one hand, considering that it could be taken as self-evident, on the other hand considered that:

> Japan's claim that according to the rules of treaty interpretation the meaning of Article VIII must be given first and foremost by its express terms, and so long as the relevant treaty provision on the issuance of special permits is not revised to that effect, it is up to each authorizing party to determine the granting of special permits, is a relevant issue.

And one which, according to Judge Xue, should have been determined by the Court[317]

In refuting Japan's claim to complete autonomy in respect to Special Permig whaling, Judge Xue pointed to the changes in a number of aspects of the ICRW regime which had had a: 'creeping effect on the way in which scientific research may be conducted, particularly with respect to methodology and scale of sample size'. As a result of these developments, she said, it is now 'hard to claim that scientific whaling is totally detached, free-standing, from the operation of the Convention', so that '[t]he assessment of [a decision to grant a Special Permit] cannot simply rely on the perception of the authorizing party, but must be conducted on an objective basis. The authorizing party should justify its decision with scientific evidence and sound reasoning'.[318]

6.3 *Interpretation of Article VIII: Interim Conclusions*

The interpretation of Article VIII of the ICRW by the Parties and in the Court's Judgment, especially, gives raise to many interesting issues, some of them perhaps not fully resolved. The very complex relationship between the Convention itself and Article VIII makes the interpretation of this Article very difficult.

316 Idem, para. 9.

317 Idem, para. 11.

318 Idem, para. 13.

On the one hand, the ICRW sets a regulatory system which covers in its entirety commercial whaling and aboriginal (subsistence whaling), while, on the other hand, there is an exception in relation to scientific whaling. The Court in interpreting this article had to deal with the following issues: whether the rights in respect to the issue of special permits under Article VIII para. 1 are self-standing (outside the scope of the Convention), as argued by Japan; or whether they are part and parcel of the Convention and should be interpreted accordingly (as argued by Australia and New Zealand). It also had to interpret whether the issuing of special permits is exclusively within the discretionary power of a State (as argued by Japan) or whether, due to the developments in the Convention regime and the changes in its object and purpose (collective regime), it has become subject to a requirement for objective (collective) assessment.

The Court interpreted Article VIII in a balanced way. It first of all stated that Article VIII is an integral part of the Convention, so it has to be interpreted according to the Convention's object and purpose. Then it said that Article VIII grants a discretionary power to issue special permits by States. But the determination, however, whether the killing, taking and treating of whales pursuant to a requested special permit is for purposes of scientific research cannot be decided on the bass of that State's sole understanding. The Court also, importantly, stated that the object and purpose of the Convention cannot be altered by amendments to the Schedule and recommendations by the IWC, though these may put an emphasis on one or the other objective pursued by the Convention.

7 General Conclusions

It is an undisputed fact that the *Whaling in the Antarctic* case has contributed in a significant manner to the further understanding of the canons of treaty interpretation. The pleadings of the Parties to the dispute, and the Judgment itself, raised several very complex questions: the object and purpose of the treaty; evolutionary interpretation; subsequent agreement and practice; and the principle of systemic integration. Not all issues raised by the Parties were addressed in the Court's Judgment, for instance the principle of systemic integration (extensively relied on by Australia); others were only succinctly dealt with (such as evolutionary interpretation).

The Court's analysis of the object and purpose of the IWCR, and of Article VIII, was very extensive and undoubtedly, from the point of view of the law of treaties, the most interesting and important part of the Judgment. Article VIII is an integral part of the Whaling Convention, notwithstanding that it involves an exemption from the general provisions of the ICRW in relation to the killing or taking of whales The object and purpose of the Convention and that of Article VIII, also, are inexorably linked together and cannot be treated in isolation from each other.. However, the Court stated that the issuing of special permits is within the discretion of a State Party, though the assessment of whether a special permit is for the purpose of scientific whaling depends on an objective appraisal, by a regulatory body.

The Court also made a very significant observation regarding the possibility of the alteration of the object and purpose of a treaty (the ICRW) through the secondary acts an international organisation and the amendments to the Schedule. The Court was quite firm in stating that the object and purpose of the Convention could not be changed in such a way, but that 'the IWC may put an emphasis on one or the other objective pursued by the Convention'. This is a very complex issue, in the view of present author. What is understood by 'putting an emphasis'? How is this different from alteration of the object and purpose of the treaty? And how is the regulatory body to decide what objectives have had a new 'emphasis' put on them.?

In paragraph 45 of the Judgement, the Court stated on the other hand that '[t]he functions conferred on the Commission have made the Convention an evolving instrument', in relation to the amendments to the Schedule. This statement can lead to the conclusion that although the object and purpose of the ICRW remains unchanged, the Convention itself is evolving. It would have been very instructive if the Court had shed more light on this important statement.

In the pleadings of the Parties, there were a great number of references to other international treaties and to principles of international environmental law, which it can be said were aimed at application of the principle of systemic integration. The Parties to the dispute, however, have not made a significant contribution to the further understanding of the difference (if there is a difference) between the principle of systemic integration and evolutionary interpretation—one of the most taxing problems of the law of treaties.

Finally, the author of the present chapter is of the view that the Court contributed to further crystallisation of the concept of subsequent agreement and practice. The Court was very cautious, for a reason, in defining what is not subsequent practice regarding the resolutions of the IWC.

In conclusion, the Judgment in this case will have a significant effect on the issues of treaty interpretation, in particular in defining what constitutes the object and purpose of a treaty.

CHAPTER 4

The 'Margin of Appreciation' in the Use of Exemptions in International Law: Comparing the ICJ Whaling Judgment and the Case Law of the ECtHR

*Theodore Christakis**

Introduction

In its 31 March 2014 Judgment in the *Whaling Case*, the International Court of Justice (ICJ) mentioned for the first time expressly the concept of 'margin of appreciation',[1] renowned for its widespread use in the case law of the European Court of Human Rights (ECtHR). According to this concept, described by the ECtHR as a 'tool to define relations between the domestic authorities and the Court',[2] State authorities 'are in principle in a better position than the international judge to give an opinion' on the 'necessity' and 'proportionality' of an exemption, derogation or restriction authorized by International Law.[3] As a consequence, international courts 'should grant national authorities an important degree of deference and respect their discretion'[4] on the implementation of exceptions. Thus, without precluding judicial review of a State's action in this field, the doctrine intends to 'limit the scope of this review'[5] and to impose some degree of judicial self-restraint in the assessment of the attitude of national authorities.

* Professor of International Law, Director of the Centre for International Security and European Studies (CESICE), University Grenoble-Alpes, member of the Institut Universitaire de France. This Chapter covers developments till May 2015.

1 ICJ, *Whaling in the Antarctic* (*Australia v. Japan: New Zealand Intervening*), Judgment, 31 March 2014, para. 59 of the Judgment. The Court had previously mentioned the term 'margin of appreciation', but not in relation to exceptions. See ICJ, *Maritime Delimitation in the Black Sea* (*Romania v. Ukraine*), Judgment of 3 February 2009, para. 162, at 113: 'With regard to the role which may be played by the coastal configuration, Ukraine states that there is a broad margin of appreciation as to its scope as a relevant circumstance'.

2 ECtHR, *A and Others v. United Kingdom*, Application No. 3455/05, Judgment of 19 February 2009, 49 *EHRR* (2009) 29 at para. 184.

3 See *infra* 2.3.

4 Yuval Shany, 'Toward a General Margin of Appreciation in International Law?' EJIL, 2006, at 910.

5 Jean-Pierre Cot, 'Margin of Appreciation', Max Planck EPIL, OUP, § 1.

 DOI 10.1163/9789004313828_006

In previous cases the ICJ made reference to some similar terms such as 'choice of means', 'measure of discretion' or 'a very considerable discretion',[6] but without ever mentioning as such, to our knowledge, the concept of 'margin of appreciation'.[7]

It is interesting to note that it was not a European but an Asian State, unfamiliar with the concept of national 'margin of appreciation', that invited the ICJ to 'introduce' this ECtHR-made concept in its case law. As we will see in Part 1 of this paper, Japan used this argument in order to claim a wide discretion in the implementation of the faculty to grant special permits to its nationals under article VIII, paragraph 1, of the Whaling Convention.[8] Japan argued that 'the principle [of margin of appreciation] must be an axiom of international law and relations'.[9] Australia and New Zealand strongly opposed this argument. And the ICJ itself was reluctant to endorse this concept. While mentioning for the first time the concept of 'margin of appreciation', the ICJ did not follow this path, preferring instead the concept of 'standard of review'. In reality, as we will see, this methodological choice didn't really influence the outcome of the case. Indeed, in Part 2 of this paper, we will try to show that even if the ICJ had followed Japan and applied the 'margin of appreciation' doctrine, and the ECtHR's understanding of it in the *Whaling Case*, there would have been few, if any, substantial differences in the final judgment.

1 The Use of the 'Margin of Appreciation' Argument in the Whaling Case

I will first present the use of the 'margin of appreciation' doctrine by Japan in its pleadings, before referring to Australia's and New Zealand's strong opposition to this concept. I will finally turn to the position of the ICJ, refer to previous

6 See *infra*, 1.3.

7 But see *supra* note 1.

8 Article VIII, paragraph 1, of the Convention reads as follows: 'Notwithstanding anything contained in this Convention any Contracting Government may grant to any of its nationals a special permit authorizing that national to kill, take and treat whales for purposes of scientific research subject to such restrictions as to number and subject to such other conditions as the Contracting Government thinks fit, and the killing, taking, and treating of whales in accordance with the provisions of this Article shall be exempt from the operation of this Convention. Each Contracting Government shall report at once to the Commission all such authorizations which it has granted. Each Contracting Government may at any time revoke any such special permit which it has granted'.

9 See *infra* 1.1.

cases and show how the Court avoided this concept in the *Whaling* Judgment, preferring, instead, the more neutral concept of 'standard of review'.

1.1 *The Position of Japan: The 'Margin of Appreciation' as an 'Axiom of International Law'*

During its written and oral pleadings in the ICJ, Japan took a very broad approach to the concept of 'margin of appreciation', presenting it as part of general International Law and as applicable in all international courts and tribunals.

Japan thus argued in its Counter Memorial, that:

> Australia [should] address...the very obvious point that States must have a margin of appreciation in cases such as this. There is extensive jurisprudence on the subject; and the principle must be an axiom of international law and relations, as well as a sensible safeguard against unwarranted accusations of bad faith.[10]

The consequences of the applicability of the margin of appreciation doctrine in relation to Article VIII of the Whaling Convention were progressively explained by Japan.

Japan thus initially argued that 'special permit whaling under Article VIII is entirely outside the scope of the ICRW'. Article VIII, paragraph 1, it contended, was to be regarded as 'free-standing' and would have to be read in isolation from the other provisions of the Convention, not as an exception under the ICRW.[11] As the Court highlights, nonetheless, 'Japan later acknowledged that Article VIII "must...be interpreted and applied consistently with the Convention's other provisions", but emphasized that a consistent reading would consider Article VIII, paragraph 1, as providing an exemption from the Convention'.[12]

For Japan, in cases like this one, there was under this 'axiom' of International Law a very wide margin of appreciation in favour of States implementing similar exemptions. Japan acknowledged that this did not mean complete deference in favour of the State granting the permits.[13] It established nonetheless a presumption. As Japan argued:

---

10 Counter-Memorial of Japan, Vol. 1, 9 March 2012, para. 9.16, at 415.

11 *Ibid.*, para. 7.8, p. 324.

12 Para. 52 of the Judgment.

13 See oral pleading by Professor Vaughan Lowe CR 2013/15, 4 July 2013, p. 19, para. 27. 'That is not to say that the Court cannot question – and in appropriate cases, declare to be in contravention of international law – decisions taken by national authorities. Let me be

> There is a presumption that States are acting in compliance with their international obligations, and accordingly that in this case JARPA II is consistent with Japan's obligations under the ICRW.[14]

According to Japan, 'this presumption is admittedly rebuttable'[15] but this meant that the heavy burden of proof was upon Australia to show that there has clearly been a misuse of the exemption under article VIII.

Another consequence of the 'margin of appreciation' was that the Court had limited powers in relation with the assessment of the use of this exemption by Japan. Indeed, according to Japan, there was no possibility for a '*de novo* review' by the Court. As the ICJ summarized it:

> Japan accepts that the Court may review the determination by a State party to the ICRW that the whaling for which a special permit has been granted is "for purposes of scientific research". In the course of the written and oral proceedings, Japan emphasized that the Court is limited, when exercising its power of review, to ascertaining whether the determination was "arbitrary or capricious", "manifestly unreasonable" or made in bad faith. Japan also stressed that matters of scientific policy cannot be properly appraised by the Court. It added that the role of the Court therefore is "to secure the integrity of the process by which the decision is made, [but] not to review the decision itself".[16]

In another part of its pleadings, Japan clearly stated that the only thing that the ICJ could control was if there was an '*erreur manifeste d'appréciation*'[17] (a 'manifest error of assessment') by Japan when granting the permits.

It is thus clear that, for Japan, the 'margin of appreciation' doctrine should not only place a heavy burden of proof on the claimant, but also substantially limit the scope of review by the ICJ in relation with Japan's decisions under Article VIII.

---

clear. Japan does not argue that there should be complete deference to decisions taken by Contracting Governments'.

14 Counter-Memorial of Japan, Vol. 1, 9 March 2012, para. 9.7, at 411–412.

15 *Ibid.*

16 Para. 65 of the Judgment.

17 Expression of Japan's counsel Professor Alain Pellet in CR 2013/23 (16 July 2013), para. 33, p. 28.

1.2 *The Position of Australia and New Zealand: No Margin of Appreciation in International Law*

Australia and New Zealand strongly rejected Japan's arguments denying any room for a 'margin of appreciation' doctrine in general International Law.

Australia thus submitted that:

> the margin of appreciation is not, as Japan suggests, an "axiom of international law and relations". Its suggestion that this notion is a "sensible safeguard against unwarranted accusations of bad faith" is entirely unfounded. There is no case where the Court has expressly accepted or applied a general "margin of appreciation" test. In Oil Platforms, you explicitly rejected this latitudinarian idea.[18]

Similarly, New Zealand explained that:

> 32. Japan's assertion that the margin of appreciation must be a generally applicable "axiom of international law and relations" is [...] unsupported. Despite its reference to "extensive jurisprudence on the subject", no citation to any decision of this Court is given in support of the existence of such an "axiom".
>
> 33. The reason for that is clear. There is no decision of this Court to support Japan's proposition. Indeed, outside of the specific context of the European Court of Human Rights, there is no widespread acceptance of a separate doctrine of 'margin of appreciation' as a general principle of international law. The one judicial decision cited by Japan – the *Hormones* case – does not even use the term. Its reasoning turns on considerations specific to the provision and Agreement in question in that case.
>
> [...]
>
> 37. Mr. President, that conclusion is a straightforward application of established principles of interpretation as stated by this Court. To interpret Article VIII as providing a special margin of appreciation to a Contracting Government, placing it beyond the ordinary review of the Court, would be inconsistent with this Court's established jurisprudence.[19]

For both these States, accepting the margin of appreciation doctrine in a case similar to this one would be counterproductive. As Australia emphasized:

---

18 James Crawford, CR 2013/8 (26 June 2013), para. 64, at 45. Footnotes omitted.

19 Christopher Finlayson, Attorney-General of New Zealand in CR 2013/17 (8 July 2013), para. 37 at 26.

> Japan's assertion of a wide "margin of appreciation" would undermine the operation of the regulations adopted by the IWC to ensure the effective conservation and management of whale populations.[20]

Australia and New Zealand stressed that the ICJ should avoid by all means 'importing' the doctrine of 'margin of appreciation'; and should 'only rely on its own principles of interpretation and application',[21] especially the principle 'that a provision must be applied in a reasonable manner',[22] the principle 'that a power must also be exercised *properly*, that is, for the purpose for which it has been given',[23] and 'the principle of *effectiveness* – that a power must not be exercised in a manner that would undermine the object and purpose of provisions of the treaty as a whole'.[24] The Court should not recognize a strong presumption in favour of the authorizing State; and the Court's power of review should not be limited to scrutiny for good faith or existence of a 'manifest error of assessment', 'as this would render the multilateral regime for the collective management of a common resource established by the ICRW ineffective'.[25] Both States urged the Court to 'have regard to objective elements in evaluating whether a special permit has been granted for purposes of scientific research'[26] and to fully exercise its powers of review in this case.

1.3 *The Previous Case Law of the ICJ*

Before examining how the ICJ dealt with Japan's arguments in the *Whaling* case, let's examine briefly if the 'margin of appreciation' doctrine has been applied in any previous cases of the ICJ.

As mentioned in the introduction of this Chapter, I am not aware of any previous cases where the ICJ directly applied, as such, the 'margin of appreciation' doctrine.[27] There are, nonetheless, several cases where the ICJ referred to related terms, such as the concept of 'discretion'.

Let's exclude from the outset from our analysis all those cases where the concept of 'discretion' was used by the ICJ in situations not involving the use of 'exceptions', 'exemptions' or 'derogations' to existing rules. Indeed, there are

---

20 James Crawford, CR 2013/8 (26 June 2013), para. 64, at 45.

21 Christopher Finlayson, in CR 2013/17 (8 July 2013), para. 34, at 25.

22 *Ibid.*

23 *Ibid.*, para. 35.

24 *Ibid.*

25 See para. 63 of the Judgment.

26 *Ibid.*

27 But see *supra* note 1.

plenty of cases where what was at stake was the residual liberty of a State to adopt regulatory measures on its territory (or in other spaces such as rivers or maritime zones), or even the 'choice of means' available for a State in order to comply with some international obligations. The Permanent Court of International Justice had already in 1927, in its famous *Lotus* judgment, put emphasis on this principle of residual liberty by saying that 'The rules of law binding upon States [...] emanate from their own free will. [...] Restrictions upon the independence of States cannot therefore be presumed. [...] [International Law leaves States] a wide measure of discretion which is only limited in certain cases by prohibitive rules'.[28] Of course, the *Lotus* dictum must be read correctly today. It should not be considered as implying the 'absolute power' of the will of States and should not lead either to the extremist view that States are bound only to what they have *explicitly* consented to.[29] Instead, the *Lotus* dictum should be read in a negative way: States should not be bound *against* their will, by what they have not *explicitly* or *implicitly* consented to. In those fields, where no international obligations are binding upon States in any way, and that remain under their 'reserved domain',[30] States are free to act; and the principle *in dubio pro libertate* applies. It is thus perfectly clear that States have a perfect 'discretion' in assessing the need for regulation in their territory and for choosing the most appropriate measures in order to deal with a specific problem. It is also clear that several international rules offer States a choice of means (and thus a large measure of discretion) for their implementation – especially when the resulting obligations are obligations of result.

Our query here is not about 'discretion in International Law' in general – and the attitude of the ICJ. Our query is about situations where States *clearly* accepted specific obligations of conduct and try to use 'exceptions' or 'derogations' in order to escape from their international obligations or to introduce limitations to them. Is there a 'wide margin of appreciation' (and a limited

---

28 *The Case of the S.S. 'Lotus'*, Judgment of 7 September 1927, PCIJ Series A No. 10 (1927) 18–19.

29 See my analysis in Theodore Christakis, Marie-Pierre Lanfranchi, *La licéité de l'emploi d'armes nucléaires devant la Cour internationale de Justice* (Economica 1997) 48–125.

30 For the concept of '*domaine réservé*' see the famous dictum of the PCIJ in its *Nationality Decrees* advisory opinion: 'The question whether a certain matter is or is not solely within the jurisdiction of a State is an essentially relative question; it depends upon the development of international relations. Thus, in the present state of international law, questions of nationality are, in the opinion of the Court, in principle within this reserved domain'. PCIJ, Advisory Opinion n° 4, 7 February 1923, at 24.

scope of judicial review) in the implementation of these exceptions? What was the position of the ICJ in this respect before the *Whaling* Judgment?

We think that the analysis of previous ICJ cases shows that the ICJ is reluctant to accept the existence of a 'general' margin of appreciation doctrine in the use of exceptions by States in International Law. Its case law seems to indicate that the Court could only recognise an important discretion in favour of a State if the wording of the relevant treaty provision clearly indicates that such a wide discretion exists in a particular case in favour of the State that invokes the exception.

In its 1986 *Nicaragua* Judgment, the Court thus clearly rejected the idea that Article XXI of the 1956 Friendship, Navigation and Commerce Treaty between Nicaragua and the US, which provided for an exception in order to protect the 'essential security interests' of contracting parties, could provide a wide margin of discretion or limit the Court's power of review. As the Court said: 'Article XXI defines the instances in which the Treaty itself provides for exceptions to the generality of its other provisions, but it by no means removes the interpretation and application of that article from the jurisdiction of the Court as contemplated in Article XXIV'.[31] Furthermore, 'whether a measure is necessary to protect the essential security interests is not [...] purely a question for the subjective judgment of the party'.[32] The Court thus refused to accept wide discretion in favour of the US in the implementation of the 'security exception' and found that this exception afforded 'no defence for the United States in respect of any of the actions [...] under consideration'.[33] The ICJ clearly indicated, nonetheless, that the discretion of a State in the implementation of such a security exception could have been more important if the wording of the treaty provisions contemplating such exceptions was different and clearly gave such discretion to States.[34]

In a similar way, in its 1997 *Gabčikovo-Nagymaros* Judgment, the Court focused on the 'necessity' defense provided by general international law (and what the ILC called the 'circumstances precluding wrongfulness'[35]) and took

31 ICJ, *Military and Paramilitary Activities in and against Nicaragua*, ICJ Rep (1986), para. 222.

32 *Ibid.*, para. 282.

33 *Ibid.*

34 See para. 222.

35 We have written extensively about the fact that it would be better to consider 'state of necessity' as a circumstance precluding or attenuating responsibility. See 'Nécessité n'a pas de Loi? Rapport général sur la nécessité en droit international', in Theodore Christakis, Karine Bannelier (eds), *La nécessité en droit international*, Société française pour le droit international, Paris, Pedone, 2007, at 9–62 and Theodore Christakis, 'Les "circonstances

the position that the discretion of States in invoking such a defense was far from being unlimited and that the scope for judicial review remained very important. According to the Court: 'the state of necessity can only be invoked under certain strictly defined conditions which must be cumulatively satisfied; and the State concerned is not the sole judge of whether those conditions have been met'.[36]

In a consistent attitude to its previous positions, the ICJ rejected, in its 2003 *Oil platforms Judgment,* the US argument according to which an important discretion should be recognized in favour of the US in invoking article XX of the Friendship, Navigation and Commerce Treaty between Iran and the US that provided for a 'security exception' similar to the one in the Nicaragua case. What is very interesting is that, in order to reject the American arguments, the ICJ did not only refer to this specific provision, but, more generally, to the exception of self-defense in International Law. According to the ICJ:

> [T]he United States claims that it considered in good faith that the attacks on the platforms were necessary to protect its essential security interests, and suggests that "A measure of discretion should be afforded to a party's good faith application of measures to protect its essential security interests" [...]. The Court does not however have to decide whether the United States interpretation of Article XX, paragraph 1(d) on this point is correct, since the requirement of international law that measures taken avowedly in self-defence must have been necessary for that purpose is strict and objective, leaving no room for any "measure of discretion".[37]

The ICJ did recognize the existence of 'a very considerable discretion' in favor of France in relation with the decision to refuse mutual assistance in criminal matters in the *Djibouti v. France* 2008 Judgment. Indeed, the Court said that:

> while it is correct, as France claims, that the terms of Article 2 provide a State to which a request for assistance has been made with a very considerable discretion, this exercise of discretion is still subject to the obligation

excluant l'illicéité": une illusion optique?' in *Droit du pouvoir, pouvoir du droit, Mélanges offerts à Jean Salmon*, Bruxelles, Bruylant, 2007, at 201–248.

36 ICJ, *Case Concerning the Gabčikovo-NagymarosProject* (*Hungary/Slovakia*), Judgment of 25 September 1997, para. 51, p. 37.

37 ICJ, *Oil Platforms* (*Islamic Republic of Iran v. United States of America*), Judgment of 6 November 2003, para. 73, p. 39.

> of good faith codified in Article 26 of the 1969 Vienna Convention on the Law of Treaties.[38]

The reason for this wide discretion was, nonetheless, that article 2 (c) of the 1977 Treaty of Friendship and Co-operation between France and Djibouti clearly conferred such an important discretion to the requested State to refuse mutual assistance if it 'considers that execution of the request is likely to prejudice its sovereignty, its security, its *ordre public* or other of its essential interests'. During the proceedings, Djibouti acknowledged that under Article 2 (*c*) the requested State enjoyed such a wide discretion, contending nonetheless that, even in reliance on what it described as a 'self-judging clause', the requested State must act 'reasonably and in good faith'.[39]

1.4 *The* Whaling *Case Judgment*

The 2014 *Whaling* case Judgment was probably an excellent opportunity for the ICJ to follow some scholar's invitations and endorse the 'margin of appreciation doctrine'.[40] Indeed, as we have seen, for the first time, a State openly referred to this doctrine and asked the Court to apply it. The Court, nonetheless, refused to follow this path.

Of course, for the first time the Court made an express reference to the 'margin of appreciation' doctrine in paragraph 59 of the Judgment. But it did so by just referring to what Japan said during the pleadings. The Court did not refer to this doctrine anymore, opting, instead, for the term 'discretion' appearing in all previous judgments. And, faithfully to its case law, the Court explained that, lacking a specific wording to the contrary in article VIII of the ICRW, this discretion should be narrow and the power of judicial review should remain intact. According to the Court:

> Article VIII gives discretion to a State party to the ICRW to reject the request for a special permit or to specify the conditions under which a permit will be granted. However, whether the killing, taking and treating of whales pursuant to a requested special permit is for purposes of scientific research *cannot depend simply on that State's perception*.[41]

---

38 ICJ, *Certain Questions of Mutual Assistance in Criminal Matters* (*Djibouti v. France*), Judgment of 4 June 2008, para. 145.

39 See para. 135 of the Judgment.

40 See for example Yuval Shany (*supra note 4*).

41 Para. 61 of the Judgment. Emphasis added.

When the moment came to explain *how* the Court would assess whether or not there had been a violation of the ICRW in granting these permits, the terminology 'slipped' from the concept of the 'margin of appreciation' (as elaborated by the ECtHR) to the 'Standard of Review' concept used by the WTO Organs! The Court explained that:

> When reviewing the grant of a special permit authorizing the killing, taking and treating of whales, the Court will assess, first, whether the programme under which these activities occur involves scientific research. Secondly, the Court will consider if the killing, taking and treating of whales is "for purposes of" scientific research by examining whether, in the use of lethal methods, the programme's design and implementation are reasonable in relation to achieving its stated objectives. This standard of review is an objective one [...]. In this regard, the Court notes that the dispute before it arises from a decision by a State party to the ICRW to grant special permits under Article VIII of that treaty. Inherent in such a decision is the determination by the State party that the programme's use of lethal methods is for purposes of scientific research. It follows that the Court will look to the authorizing State, which has granted special permits, to explain the objective basis for its determination.[42]

We will not analyse here further the law of evidence and the concept of 'standard of review', in general, or as used by the ICJ in the Whaling Case, in particular. These issues are discussed in other Chapters of this book.[43] We will instead maintain our focus to the 'margin of appreciation' doctrine. We will try to examine if the Judgment (and the outcome of the case) could have been different had the ICJ followed Japan's invitation to use the 'margin of appreciation' doctrine.

2 Applying the ECtHR's Parameters of the 'Margin' in the Whaling Case

In his EPIL entry on 'margin of appreciation', Judge Jean-Pierre Cot rightly explains that, while 'WTO Panels and the Appellate Body do not use the terminology "margin of appreciation", [t]he "standard of review" amounts to the

42 Para. 67 and 68.

43 See the Chapters by Caroline Foster and Shotaro Hamamoto.

same'. Indeed, as the Judge explains, 'the problem is always how much deference should be granted to national decision-makers'.[44]

We could similarly observe that the fact that the ICJ did not endorse the 'margin of appreciation' concept proposed by Japan, preferring instead the concepts of 'discretion' and 'standard of review', did not change substantially the outcome of the *Whaling* case. In this second part of this Chapter we will try, precisely, to identify and apply in the *Whaling Case* the main parameters of the 'margin of appreciation' doctrine as applied by the ECtHR to see if something could have been different.

Before starting these comparisons, a word of caution is necessary. The 'margin of appreciation' is a concept of a tremendous importance in the case law of the ECtHR and probably one of the most debated topics in the academic (and, indeed, not only in the academic) community.[45] Since its 'invention'[46] by the European Judge in 1976 in the famous *Handyside* case (in relation to *restrictions* to human rights) and in the 1978 *Ireland v. UK* case (in relation to *derogations* under article 15 of the ECHR), the margin of appreciation has been used hundreds of times, not only in relation to derogations and restrictions but also, more generally, in order to authorise States to give their own solutions concerning some important societal issues such as the beginning of life, abortion, reproductive rights, same-sex marriage, assisted suicide and euthanasia, amongst others.

The basic idea behind this doctrine is that a human rights treaty like the ECHR does not aspire to strict uniformity among States in the protection of human rights and can, on the contrary, accommodate the great diversity of national traditions and practices in different areas. This doctrine is, with some other devices,[47] part of the principle of subsidiarity which reflects the idea that States have the first responsibility to protect Convention rights domestically.

44 Jean-Pierre Cot (*supra* note 5), para. 45.

45 See among many others the article of President Dean Spielman cited *infra* (note 53) and: Jan Kratochvíl, 'The Inflation of the Margin of Appreciation by the European Court of Human Rights' 29 *Netherlands Quarterly of Human Rights* (2011) 324–357; Yutaka Arai-Takahashi, *The Margin of Appreciation Doctrine and the Principle of Proportionality in the Jurisprudence of the ECHR* (Intersentia 2002); Steven Greer, *The Margin of Appreciation: Interpretation and Discretion under the European Convention on Human Rights* (Council of Europe 2000); H.C. Yourow, *The Margin of Appreciation Doctrine in the Dynamics of European Human Rights Jurisprudence* (The Hague: M. Nijhoff, 1996).

46 For a history of the doctrine see Yutaka Arai-Takahashi (*supra* note 45), at 5ff.

47 Such as exhaustion of local remedies or the principle that, in the first place, it is the national courts that interpret and apply the Convention, under the 'final European supervision' of the Court.

These two principles, of subsidiarity and margin of appreciation, are great contributors to the idea of 'vital space' reclaimed by sovereign States, and it comes as no surprise that some States are extremely attached to them. Indeed, the 2012 Brighton Declaration put a high emphasis on the margin of appreciation doctrine and the principle of subsidiarity;[48] and it concluded that a reference to the principle of subsidiarity and the doctrine of the margin of appreciation as developed in the Court's case law, should be included in the Preamble to the Convention – despite the scepticism of the then President of the Court, Sir Nicolas Bratza, in this respect.[49] As a result, ECHR State parties adopted in June 2013[50] Protocol No. 15 amending the ECHR in order, *inter alia*, to add in the preamble to the Convention such a reference.[51]

That said, the doctrine of the 'margin of appreciation' has also been heavily criticized within the framework of the ECtHR. In his often-quoted partly dissenting opinion appended to the *Z. v. Finland* judgment of 25 February 1997, Judge De Meyer wrote that:

> In the present judgment the Court once again relies on the national authorities' "margin of appreciation". I believe that it is high time for the Court to banish that concept from its reasoning. It has already delayed too long in abandoning this hackneyed phrase and recanting the relativism it implies. It is possible to envisage a margin of appreciation in certain domains.[...]. But where human rights are concerned, there is no room for a margin of appreciation which would enable the States to decide what is acceptable and what is not.[52]

In a similar way, one scholar wrote that 'the domestic margin of appreciation is the most controversial "product" of the European Court of Human Rights'

48 High Level Conference on the Future of the European Court of Human Rights: Brighton Declaration, available at http://hub.coe.int/20120419-brighton-declaration.

49 'Speech by Nicolas Bratza', available at: http://echr.coe.int/Documents/Speech_20120420_Bratza_Brighton_ENG.pdf.

50 As of June 2015, only 17 out of 47 State parties have ratified the convention.

51 'Affirming that the High Contracting Parties, in accordance with the principle of subsidiarity, have the primary responsibility to secure the rights and freedoms defined in this Convention and the Protocols thereto, and that in doing so they enjoy a margin of appreciation, subject to the supervisory jurisdiction of the European Court of Human Rights established by this Convention'.

52 *Z v. Finland*, 25 February 1997, *Reports of Judgments and Decisions* 1997-I.

while others criticised the doctrine as a 'quirk of language' or 'an unfortunate Gallicism'.[53]

It is of course impossible to re-open here the impressive and on-going debate about the forces and weaknesses of the margin of appreciation doctrine within the European system of protection of human rights. Let's just observe that the main criticisms against this doctrine concern not only 'a degree of vagueness, or even a certain incoherence in the Court's reliance on the margin of appreciation',[54] but also the fear that over-reliance on this doctrine and the decisions of national authorities by the Court in the future might annihilate the international protection of human rights.

In order to limit the risks of abuse and despite the difficulties in doing so (which sometimes lead to unpredictable results) the ECtHR has developed a series of criteria and parameters for the application of the doctrine. It is precisely to some of these criteria and parameters that we will refer here.

2.1 *Wide Margin versus Narrow Margin in the Case Law of the ECtHR*

The first thing that should be emphasized is that the applicability of the 'margin of appreciation' doctrine in a specific case, does not necessarily mean a *wide discretion* in favour of a State invoking a derogation or restriction to an international obligation.

The ECtHR constantly explains that 'this margin of appreciation goes hand in hand with a European supervision' embracing both the aim of the measure challenged and its 'necessity' and both the law and the decisions applying it. The scope of the margin of appreciation 'is not identical in each case but will vary according to the context'.[55] Depending on the situation, then, the national margin of appreciation could be very limited[56] and the European supervision particularly rigorous. As Judge Jean-Pierre Cot wrote, there is 'no "one size fits

53 See the scholars and extracts quoted by the current President of the ECtHR Dean Spielman, 'Allowing the Right Margin the European Court of Human Rights and the National Margin of Appreciation Doctrine: Waiver or Subsidiarity of European Review?' in CELS Working Papers Series, February 2012, at 28. Paper available in http://www.cels.law.cam.ac.uk/cels_lunchtime_seminars/Spielmann%20-%20margin%20of%20appreciation%20cover.pdf.

54 *Ibid.*

55 *Buckley v. the United Kingdom*, 25 September 1996, § 74.

56 It goes without saying that our analysis only concerns situations where restrictions or derogations to human rights are possible. When it comes for some 'absolute' rights, for which no derogation and no restriction are possible (such as the prohibition of torture) the margin of appreciation will be virtually non-existent.

all" approach. The flexibility of the margin of appreciation has to be tailored to each case'.[57]

The ECtHR has developed some criteria in order to assess the width of the margin and the intensity of the control. Prominent among them is the nature and importance of the right that the State intends to restrict, as balanced against the importance of the legitimate aim at stake that the State tries to protect.

The Court thus considers that the national margin of appreciation is narrow when a State tries to restrict a particularly important right or, better, a particularly important facet of a right. In the famous *Dudgeon v. the United Kingdom* case (1981), the Court rejected the UK's efforts to justify the prohibition of homosexuality in Northern Ireland saying that 'the present case concerns a most intimate aspect of private life. Accordingly, there must exist particularly serious reasons before interferences on the part of the public authorities can be legitimate for the purposes of paragraph 2 of Article 8'.[58] Similarly, in the famous *Sunday Times v. United Kingdom* Judgment of 1979, the ECtHR explained that the margin of appreciation in relation to restrictions to article 10 was narrow because of the importance of the freedom of expression to a democratic society.

On the other hand, the Court considers that some legitimate aims to protect are of particular importance and justify a wider margin of appreciation. The Court didn't have to wait longer than till 1978 and the *Klass v. Germany* Judgment, for instance, in order to recognize how important it is for democratic societies to combat terrorism and protect national security.

If we apply these criteria in the ICJ *Whaling* case, it is hard to see how the use of the 'margin of appreciation' doctrine could have changed something in favour of Japan. Indeed, while the regulation of whaling and the protection of whales by the ICRW is often presented as a particularly important issue by many State parties (including Australia and New Zealand), it would be very hard to maintain convincingly that giving permits to kill in the name of 'scientific research' is linked to an essential interest of the State. It could thus have been hard for Japan to convince the Court that the nature of the legitimate aim at stake calls for a wide margin of appreciation.

2.2 *Necessity, Proportionality (and Reasonableness)*

The ECtHR always proceeds to a test of necessity and proportionality of measures undertaken in relation to exceptions, both for derogations under article 15 and for restrictions permissible for some human rights even in normal circumstances. Indeed, the Court examines not only whether the limitations to

57 Jean-Pierre Cot (*supra* note 5), para. § 10.

58 Judgment of 22 October 1981, para. 52, p. 17.

human rights taken by national authorities were 'necessary' in order to deal with an exceptional situation or to protect a legitimate aim, but also whether these limitations were proportionate to the legitimate aim pursued. It tries to strike a right and reasonable balance between the two countervailing values and interests. The principles of necessity and, especially, proportionality are thus of paramount importance and are often considered as the 'other side [of the coin] of the margin of appreciation'.[59] If an interference with human rights is arbitrary, unreasonable and disproportionate, the derogation and restriction will not be accepted; and as President Spielman wrote: 'It is for the State to justify the interference. The grounds must be "relevant and sufficient", the need for a restriction must be "established convincingly", any exceptions must be "construed strictly" and the interference must meet "a pressing social need"'.[60]

In the *Whaling case,* the ICJ did not refer specifically to a 'proportionality' test. However, by applying the 'reasonableness' standard, by assessing 'if the killing, taking and treating of whales is "for purposes of" scientific research' and by concluding that 'the evidence does not establish that the programme's design and implementation are reasonable in relation to achieving its stated objectives',[61] the Court seemed to proceed to a kind of 'proportionality' test.

Indeed, one could have the impression that the Court considered that Japan went beyond what is authorized by Article VIII of the ICRW by having under JARPA II target sample sizes which were disproportionate in relation to achieving the programme's objectives. The reaction of Japan after the ruling, and the submission to the Scientific Committee of the 'New Scientific Whale Research Program in the Antarctic Ocean' (NEWREP-A), proposed to commence in the year 2015–16 and including more limited target sample sizes, also seems to be based on the idea that Japan tries to strike a better balance in order to meet the requirements of the proportionality test. In any case, it is clear once again that, had the ICJ adopted the 'margin of appreciation' doctrine, the outcome would have been exactly the same.

2.3 *National Authorities are Better Placed to Assess the 'Local Conditions' (and the Principle of Subsidiarity)*

One of the main ideas underlying the 'margin of appreciation' doctrine is that national authorities are, in principle, better placed than an international court to evaluate, in light of local and specific conditions and traditions, both the

59 Yutaka Arai-Takashi, *The Margin of Appreciation Doctrine and the Principle of Proportionality in the Jurisprudence of the ECHR*, Intersentia, 2002, p. 14.

60 *Op. cit.*, at 22.

61 Para. 227.

existence of a situation requiring a derogation or restriction to human rights, and the necessity and proportionality of specific measures.

As the Court said several times in relation to derogations under article 15:

> it falls to each Contracting State, with its responsibility for "the life of [its] nation", to determine whether that life is threatened by a "public emergency" and, if so, how far it is necessary to go in attempting to overcome the emergency. By reason of their direct and continuous contact with the pressing needs of the moment, the national authorities are in principle better placed than the international judge to decide both on the presence of such an emergency and on the nature and scope of the derogations necessary to avert it. Accordingly, in this matter a wide margin of appreciation should be left to the national authorities.[62]

In a similar way, since 1976 and *Handyside*, the ECtHR highlights in relation to restrictions to human rights that:

> 'The Convention leaves to each Contracting State, in the first place, the task of securing the rights and freedoms it enshrines'. [...]. By reason of their direct and continuous contact with the vital forces of their countries, State authorities are in principle in a better position than the international judge to give an opinion on the exact content of these requirements as well as on the "necessity" of a "restriction" or "penalty" intended to meet them.[63]

It is thus clear that the European human rights apparatus appears as subsidiary to the protection of rights in individual nations. In accordance with the principle of subsidiarity, it is indeed primarily for the Contracting States to decide on the measures necessary to secure Convention rights within their jurisdiction and on the necessary restrictions and derogations. The Court constantly adds, nonetheless, that 'States do not enjoy an unlimited power in this respect' and that 'the national margin of appreciation is thus accompanied by a European supervision'.

Some Judges have expressed their reluctance in relation to this deference by the Court to national authorities. As Judge Ch. Rozakis wrote in his concurring opinion appended to the *Egeland and Hanseid v. Norway* Judgment of 16 April 2009:

---

62 See for example *Brannigan and McBride v. UK* [1993] at para. 43, *Aksoy v. Turkey* [1996] at para. 68 and Tanrikulu and others v Turkey (2005) at 38.

63 Para. 48.

> Indeed, if the concept of the margin of appreciation has any meaning whatsoever in the present-day conditions of the Court's case-law, it should only be applied in cases where, after careful consideration, it establishes that national authorities were *really better placed* than the Court to assess the "local" and specific conditions which existed within a particular domestic order, and, accordingly, had greater knowledge than an international court in deciding how to deal, in the most appropriate manner, with the case before them. Then, and only then, should the Court relinquish its power to examine, in depth, the facts of a case, and limit itself to a simple supervision of the national decisions, without taking the place of national authorities, but simply examining their reasonableness and the absence of arbitrariness.[64]

If we go back now to the *Whaling* case, it is hard to see what could have changed with an eventual applicability of the 'margin of appreciation' doctrine. Indeed, it would have been particularly difficult for Japan to convince that national (Japanese) authorities were really 'better placed' than the ICJ to assess the issuance of permits and, accordingly, had greater knowledge than the International Court in deciding how to deal, in the most appropriate manner, with the case before them. This was not a case about 'local' and specific conditions which existed within a particular domestic order. True, the case did have an important 'cultural dimension' often highlighted by Japan during the proceedings.[65] But on the other hand, the permits issued under Article VIII, affected the IWC regime as a whole and concerned some 'public goods'. There was thus no reason to recognize a 'national margin of appreciation' in favor of one or another particular State party in this regime, or to uphold deference to one State as against another. As Australia highlighted before the Court:

---

64 Empasis added.

65 See for example Payam Akhavan., CR 2013/12 of 2 July 2013, at para. 18, p. 44: 'Japan has complied with the moratorium despite a 2,000 year tradition of subsistence whaling. It has done so although the IWC anti-whaling block has opposed even small-type community-based whaling in Japan's own waters. These coastal communities have been anguished because they can no longer practice their ancestral traditions. The Australian Environment Minister once remarked in this regard that "many cultures and traditions...don't belong in a modern world". IWC resolutions, and even United Nations human rights declarations however, recognize an inextricable link between sustainable "customary resource use" and cultural survival. For the anti-whaling moral crusaders, saving whales that are clearly not endangered outweighs saving foreign cultures and communities'.

> [J]udicial review, notably in relation to resources in the public domain which do not belong even *prima facie* to any individual State, and which are a matter of collective interest, should not be regulated by the Court wholly or primarily on the basis of such fluctuating and subjective notions as bad faith. [...] The 1946 Convention is a collective enterprise. It is not a bilateral treaty between Japan and the rest of the world. Japan does not "own" the whales it catches [...]. The whales of the Southern Ocean are subject to collective regulation on the terms of the Convention.[66]

2.4 *Searching for the 'Common Denominator'* (*Existence of a 'Consensus'*)

The existence or not of a 'European consensus' on a specific matter has become, together with the nature and the importance of the right at stake, one of the most important factors in assessing the scope and the width of the margin of appreciation.

As a matter of principle, the ECtHR considers that when there is consensus on a particular issue among the member countries of the Council of Europe, the margin of appreciation should be very narrow. In the *famous Dudgeon v. UK* Judgment of 1981, for example, the Court found that the vast majority of member States no longer prohibited homosexual relations – which left almost no room for such restrictions to article 8 by the UK.

On the other hand, as the Court said, where 'there is no consensus within the member States of the Council of Europe, either as to the relative importance of the interest at stake or as to the best means of protecting it, particularly where the case raises sensitive moral or ethical issues, the margin will be wider'.[67]

At the universal level, the Human Rights Committee also used the concept of 'international consensus' in order to legitimate some bold decisions. In *Roger Judge v. Canada* in 2003, the Human Rights Committee relied heavily on the 'broadening international consensus in favour of abolition of the death penalty' in order to overturn its own previous jurisprudence; and conclude that, for countries that have abolished the death penalty, there is now an obligation not to expose a person to the real risk of its application by extradition or expulsion.[68]

---

66 James Crawford, CR 2013/10 (10 July 2013), para. 22–23, at 65.

67 *Evans v. the United Kingdom* (GC), Judgment of 10 April 2007, para. 77.

68 UNHRC, *Roger Judge v. Canada*, Communication No. 829/1998, Decision of 5 August 2002 at para. 10.3.

In the *Whaling* case, Japan emphasized that there was a lack of consensus concerning the particularly narrow interpretation of Article VIII of the ICRW advanced by Australia and New Zealand. Japan was successful in convincing the Court that:

> Australia and New Zealand overstate the legal significance of the recommendatory resolutions and Guidelines on which they rely. First, many IWC resolutions were adopted without the support of all States parties to the Convention and, in particular, without the concurrence of Japan. Thus, such instruments cannot be regarded as subsequent agreement to an interpretation of Article VIII, nor as subsequent practice establishing an agreement of the parties regarding the interpretation of the treaty within the meaning of [VCLT 31(3)(a) & (b)].[69]

However, this was not decisive in convincing the Court to recognize a wide discretion for the implementation of exceptions under Article VIII. While the Court refused the idea that Article VIII could be read as authorising lethal methods only when other methods are not available, and while the Court did recognize the possibility for Japan to use such methods, it also emphasized that a programme could not use 'lethal sampling on a larger scale than is reasonable in relation to achieving the programme's stated research objectives'.[70]

Conclusion

The *Whaling* case was probably a 'golden opportunity' for the International Court of Justice to 'import', had it wished to do so, the doctrine of 'margin of appreciation' established within the framework of the European Court of Human Rights. For the first time, the Court was strongly invited to do so by a State, Japan, who presented this doctrine as an 'axiom of International Law'. The ICJ mentioned the doctrine in its Judgment, but refused to endorse it, preferring more 'familiar' language such as the concepts of 'standard of review' and 'reasonableness'.

We can probably understand why the ICJ was reluctant to endorse the doctrine of 'margin of appreciation' which, despite its paramount role (and probably because of it), created a lot of controversies within the system of protection of the European Convention of Human Rights. After all, the doctrine has not

69 Para. 83 of the Judgment.

70 Para. 97 of the Judgment.

yet been endorsed as such either by the WTO Panels and Body (which prefer, as we have seen, the 'standard of review' concept)[71] or by the ITLOS.[72]

In this Chapter we have seen that even if the ICJ had acceded to Japan's invitation to use the 'margin of appreciation' doctrine, the Judgment in the Whaling case would have been the same. If Japan's invitation was based on the assumption that the use of the 'margin of appreciation' doctrine gives a wide discretion to States in the implementation of exceptions to their international obligations and substantially limits the scope of judicial review, this assumption was clearly wrong. As we have seen, within the system of the ECtHR, there is a case-by-case approach on the basis of some criteria and factors, and the width of the 'margin of appreciation' can vary substantially from one case to another. The hypothetical application of these criteria in the *Whaling* case left no room to think that they could have resulted in any change in the intensity of the control of the Court, or could have created a challenge to the basic methodology and conclusions of the Court.

---

71 See *supra*, note 44.

72 See Jean-Pierre Cot, 'The Law of the Sea and the Margin of Appreciation', in Tafsir Malick Ndiaye, R. Wolfrum (eds), Law of the Sea, Environmental Law and Settlement of Disputes – Liber Amicorum Judge Thomas Mensah, Martinus Nijhoff, 2007, at 389–403.

PART 3

Procedural Law Aspects

CHAPTER 5

Unfavourable but Unavoidable Procedures: Procedural Aspects of the *Whaling* Case

Dai Tamada

Introduction

At first sight, the *Whaling in the Antarctic* case[1] seems to have provoked no serious problem of procedure which could have influenced the judgment of the ICJ. The main issues of the case, in fact, related to aspects of substantive law, such as the reasonableness test and the standard of review. It is however undeniable that several procedural issues, especially unfavourable to Japan, arose and these influenced the Court's findings in the case. The problem is that those procedural issues were unavoidable, in the sense that Japan had no means of escaping from them. During the course of the proceedings, Japan expressed its regret at the procedural inequality which appeared in the case:

> [t]he equality of the Parties is also a fundamental principle of international justice. In this context, Japan finds itself in a regrettable position.[2]

In similar vein, Judge Owada pointed out, in his dissenting opinion, that:

> under the somewhat unfortunate procedural circumstances, the Parties were not provided in the proceedings with ample opportunities to develop their respective arguments on the issue of jurisdiction.[3]

Procedural problems, however potential and latent they may have been, must have affected substantially the Court's finding. It is thus important to analyse, first, whether this situation was created intentionally by the litigation strategy of the other States parties to the litigation, especially by the Applicant state. If so, secondly, why Japan could not find any viable solution to avoid such litigation strategy. This paper's conclusion is that Japan did not have any chance, by

1 *Whaling in the Antarctic* (*Australia v. Japan: New Zealand Intervening*), Judgment of 31 March 2014, *I.C.J. Reports 2014*, p.226.

2 Written Observations of Japan on New Zealand's Written Observations, 31 May 2013, para. 2.

3 Dissenting Opinion of Judge Owada, *I.C.J. Reports 2014*, p.301, para. 3.

 | DOI 10.1163/9789004313828_007

any procedural means, to avoid such an unfavourable situation and that this is at least one of the reasons why Japan lost the case.

From the outset, there were differences in the extent of their litigation experience between the parties. Japan has never appeared before the ICJ or the PCIJ before the *Whaling* case, except in the *Wimbledon* case in which it was one of the Applicants.[4] However, Japan's appearance in that case was merely formal.[5] In contrast, Australia had been a Party to four litigations before the ICJ before the *Whaling* case, i.e. in the *Nuclear Tests* case[6] as an Applicant; in the *Phosphate* case[7] as a Respondent; the *East Timor* case,[8] as a Respondent; and in the *Documents and Data* case,[9] as a Respondent. As is clear, the parties to the *Whaling case* did have a difference of experience of litigation before the ICJ. This does not necessarily mean that Japan was legally disadvantaged in comparison to Australia. More substantively speaking, however, there were some particular factors that impacted on the litigation strategy of the Parties to the case.

First, Australia and New Zealand had already experienced and elaborated a cooperative litigation strategy in several cases. In the *Nuclear Tests* cases, the two countries initiated two separate cases because they had different final submissions. In reality, however, these two Applicants cooperated with each other by strengthening their substantive law arguments. A more straightforward strategy was adopted in the *Southern Bluefin Tuna* case,[10] in which these two countries chose to plead as one party in the procedure. Thus, in the *Whaling*

4 As one of the Principal Allied Powers under the terms of the Treaty of Peace of Versailles, Japan jointly submitted an Application instituting the proceeding to the PCIJ. *The S.S. Wimbledon, P.C.I.J. Series A, No. 1*, p. 6.

5 According to the PCIJ, Germany had breached the Treaty of Versailles, but it was to pay, only to France, a fixed sum of money as remedy. Therefore Japan, as one of the applicants, obtained nothing. *The S.S. Wimbledon*, Judgment of 17 August 1923, *P.C.I.J. Series A, No. 1*, p. 33.

6 *Nuclear Tests Case* (Australia v. France), Judgment of 20 December 1974, *I.C.J. Reports 1974*, p. 253.

7 *Certain Phosphate Lands in Nauru* (Nauru v. Australia), Preliminary Objections, Judgment of 26 June 1992, *I.C.J. Reports 1992*, p. 240.

8 *East Timor* (Portugal v. Australia), Judgment of 30 June 1995, *I.C.J. Reports 1995*, p. 90.

9 *Questions relating to the Seizure and Detention of Certain Documents and Data* (Timor-Leste v. Australia). The Application was filed in the Registry of the Court on 17 December 2013 and the case was removed from the General List in 2015.

10 *Southern Bluefin Tuna Cases (New Zealand v. Japan; Australia v. Japan)*, Request for Provisional Measures, Order (27 August 1999) and Award on Jurisdiction and Admissibility (4 August, 2000), rendered by the Arbitral Tribunal constituted under Annex VII of the United Nations Convention on the Law of the Sea.

case, Japan was confronted with two States that had experienced, and had been mutually enriching, cooperative litigation strategy for many years.

Second, Australia and New Zealand had particular experiences of being involved, though not of their own motions, in 'preliminary' procedures. In the *Nuclear Tests* cases, for example, the ICJ, applying the so-called mootness doctrine, found it unnecessary to proceed to the merits of the cases. In the *Phosphate Lands* case and the *East Timor* case, the main topics discussed before the ICJ were 'preliminary' issues.

Third, Australia and New Zealand had had experience of litigation comparable to the *Whaling* case in terms of intervention. Australia made an application to intervene in the case initiated by New Zealand in the *Request for an Examination* case, in which the ICJ declined the request of New Zealand and as a result the request of Australia to intervene.[11]

These differences of litigation experience did not have any substantial and explicit impact on the parties in the *Whaling* case. However, there was some implicit influence of the cases discussed above when it came to Australia's and New Zealand's choice of litigation strategy and in relation to their response to the strategy of the opposite side. This paper focuses in particular on two issues of procedure, analysed by the ICJ in the *Whaling* case: the intervention of New Zealand as a non-Party (1); and the examination and cross-examination of experts (2).

1 Intervention of New Zealand

On 20 November 2012, more than two years after the submission of the Application by Australia,[12] New Zealand submitted a 'Declaration of Intervention' to the ICJ.[13] After the exchange of written observations among Japan, Australia and New Zealand,[14] and without any objection by Japan to the admissibility of the

11 *Request for an Examination of the Situation in accordance with paragraph 63 of the Court's Judgment of 20 December 1974 in the* Nuclear Tests (New Zealand v. France) *case*, Order of 22 September 1995, *I.C.J. Reports 1995*, p. 306, para. 67.

12 Application instituting proceedings was filed in the Registry of the Court on 31 May 2010.

13 Declaration of Intervention Pursuant to Article 63 of the Statute of the Court by the Government of New Zealand, filed in the Registry of the Court on 20 November 2012.

14 *Whaling in the Antarctic (Australia v. Japan)*, Declaration of Intervention of New Zealand, Order of 6 February 2013, *I.C.J. Reports 2013*, p.5, para. 5. So far, those written observations have not been made public on the ICJ's website.

New Zealand Declaration of Intervention,[15] the Court admitted this request by its Order of 6 February 2013.[16]

As is well known, New Zealand had been taking a very similar, if not identical, anti-whaling stance and policy to those of Australia and, thus, the decision to intervene was hardly surprising. From the legal point of view, however, several questions may be asked with regard to this intervention, i.e. why New Zealand did not choose to appear before the Court as an independent party, as it had in the *Nuclear Tests* case? Why New Zealand chose not to be a co-party with Australia, as it had in the *Southern Bluefin Tuna* case? Why did New Zealand rely on Article 63 of the Statute rather than on Article 62? Why did New Zealand intervene only in 2012, two years after the initiation of the case by Australia? Examination of the legal background to New Zealand's intervention will enable us to clarify its intention and to respond to the above questions.

1.1 *Why Intervene under Article 63 of the Statute?*

The Declaration of Intervention, submitted by New Zealand, was based on Article 63 of the Statute of the ICJ and aimed at intervening in the case not as a Party but as a non-Party. In order to evaluate this decision, it is worth reminding ourselves of the basic mechanism of the intervention procedure in the ICJ Statue. First of all, the Statute of the ICJ admits two types of intervention, based, respectively, on Articles 62 and 63:

> *Article 62*
> 1. Should a state consider that it has an interest of a legal nature which may be affected by the decision in the case, it may submit a request to the Court to be permitted to intervene.
> 2. It shall be for the Court to decide upon this request.
>
> *Article 63*
> 1. Whenever the construction of a convention to which states other than those concerned in the case are parties is in question, the Registrar shall notify all such states forthwith.
> 2. Every state so notified has the right to intervene in the proceedings; but if it uses this right, the construction given by the judgment will be equally binding upon it.

15 Order of 6 February 2013, paras. 17 and 19. As we see below, Japan alleged that the New Zealand intervention would produce 'certain serious anomalies' in terms of 'procedural equality'.

16 Order of 6 February 2013, *I.C.J. Reports 2013*, pp.9–10, paras. 19 and 23.

In contrast to Article 62, which requires a legal interest of intervening State, the third Party need not have any interest in the litigation under Article 63 provided it is a Party to a convention the interpretation of which is the subject of a case before the Court. As is pointed out, the rationale of Article 63 is 'to foster uniform interpretation of a convention and thus to promote the harmonious development of international law.'[17]

Before the *Whaling* case, there had been three cases in which declarations of intervention were submitted to the Court under Article 63 of the Statute. Intervention was permitted in two cases and denied in one case.

First, in the *Wimbledon* case, Poland was permitted to intervene. Initially, relying on Article 62 of the Statute, Poland observed that it 'was one of the parties to the Treaty of Peace of Versailles' and that 'the refusal in question constitutes a violation of the rights and material advantages guaranteed to Poland by Article 380 of the Treaty of Versailles.'[18] Thereafter Poland changed its position and relied on Article 63 of the Statute.[19] The PCIJ finally admitted the intervention of Poland as follows:

> [i]t will suffice for the Court to note that in this case the interpretation of certain clauses of the Treaty of Versailles is involved in the suit and that the Polish Republic is one of the States which are parties to this treaty.[20]

In this case, the PCIJ did not consider seriously the Polish request for intervention under Article 62, and, in fact, it did not require any other conditions to be satisfied than those stipulated by Article 63.

Second, in the *Haya de la Torre* case, Cuba's Declaration of Intervention was admitted in the same way. The ICJ stated that:

> every intervention is incidental to the proceedings in a case; it follows that a declaration filed as an intervention only acquires that character, in law, if it actually relates to the subject-matter of the pending proceedings.[21]

---

17 Christine Chinkin, 'Article 63', in Andreas Zimmermann et al. (eds.), *The Statute of the International Court of Justice: A Commentary*, Second Edition (Oxford University Press, 2012), p. 1575; Robert Kolb, *The International Court of Justice* (Hart Publishing, 2013), p. 731.

18 Application by the Polish Government to Intervene, the Hague, May 22nd, 1923.

19 *The S.S. Wimbledon (The Governments of His Britannic Majesty, of the French Republic, of His Majesty the King of Italy and of His Majesty the Emperor of Japan v. The Government of the German Empire; the Government of the Polish Republic intervening)*, Question of Intervention by Poland, Judgment of 28 June 1923, *P.C.I.J. Series A, No. 1*, p. 13.

20 *Ibid.*

21 *Haya de la Torre Case (Colombia/Peru)*, Judgment of 13 June 1951, *I.C.J. Reports 1951*, p. 76.

Applying this requirement to this case, the Court found Cuba's request of intervention admissible only in part.[22] These two cases indicate that there has never been any leading case which provides firm and consistent guidance on the interpretation of Article 63 of the Statute. Furthermore, in these two cases, it was not clarified by the Court whether the intervening States were admitted as intervening States as 'Parties' or 'non-Parties'. This point of law was not analysed in the Judgments and has been left unanswered. Consequently, the *Whaling* case is the first case in which a State intended, and was admitted, to intervene under Article 63 of the Statute as a non-Party.

In the third case, the *Nicaragua* case, the request for intervention by El Salvador was rejected by the Court. The reasons for this decision are not at all clear,[23] but the Court's reasoning appears to be that the request was premature in that it was submitted at the jurisdictional phase while it should have been submitted at the merits phase of the case.

Fourth, several Island States made Applications for Permission to Intervene under Article 62 and Declarations of Interventions under Article 63 in the *Request for an Examination* case between New Zealand and France.[24] In this case, the ICJ rejected the original claim of New Zealand and, as a result, the Requests for Permission to Intervene, incidental to the original claim, were all dismissed as well.[25]

1.2 *Differences between Articles 62 and 63*

Even though the requirements of Article 63 have remained unclear in the Court's findings, there are clearly several differences between Articles 62 and 63 of the Statute, which influenced New Zealand in choosing the latter.

22 *Ibid.*, p. 77. This part is cited by Declaration of Judge Gaja (appended to Order of 6 February 2013).

23 *Military and Paramilitary Activities in and against Nicaragua (Nicaragua v. United States of America)*, Declaration of Intervention of the Republic of El Salvador, Order of 4 October 1984, *I.C.J. Reports 1984*, p. 216, paras. 2–3.

24 In this case, 'Application for Permission to Intervene' under Article 62 of the Statute was submitted by Australia. 'Applications for Permission to Intervene' and 'Declarations of Intervention' under Article 63 of the Statute were submitted by Samoa, Solomon Islands, the Marshall Islands and the Federated States of Micronesia. *Request for an Examination of the Situation in accordance with paragraph 63 of the Court's Judgment of 20 December 1974 in the* Nuclear Tests (New Zealand v. France) *case*, Order of 22 September 1995, *I.C.J. Reports 1995*, p. 292, para. 11.

25 *Ibid.*, p. 306, para. 67.

First, the most important difference is whether a request for intervention is subject to the Court's examination or not. While any request for intervention under Article 62 is examined by the Court, no such examination exists under Article 63, since any State party to the convention in question has 'the right to intervene.'[26] In the *Whaling* case, as well, in accordance with New Zealand's perception,[27] the ICJ stated that Article 63 provides 'an incidental proceeding that constitutes *the exercise of a right*.'[28] There is certainly a limit on the exercise of this right to intervene, i.e., as the Court clarifies in its Order, 'such right to intervene exists only when the declaration [of intervention] concerned falls within the provisions of Article 63.'[29] This means that, so far as a declaration or a request is limited to the construction of a convention, it is admissible. Consequently, in any event, the threshold under Article 63 is extremely low.

Second, New Zealand, as well as Australia, was anxious to start the merits phase before the ICJ without any delay.[30] Had New Zealand chosen intervention under Article 62 of the Statute, there would have been an additional procedure before moving to the merit phase of the case, which could have delayed the proceedings. Furthermore, if the intervention procedure had lasted for a long period of time, Japan would have finished all its whaling under JAPRA II. This would have rendered the Australian claim without any purpose or practical application.[31] Therefore, the two countries deliberately chose the course which would waste the least time in the intervention procedure.

---

26 One exceptional case is the request for intervention by El Salvador in the *Nicaragua* case, which was rejected by the Court. *Military and Paramilitary Activities in and against Nicaragua (Nicaragua v. United States of America)*, Declaration of Intervention of the Republic of El Salvador, Order of 4 October 1984, *I.C.J. Reports 1984*, p. 216, paras. 2–3.

27 Declaration of Intervention Pursuant to Article 63 of the Statute of the Court by the Government of New Zealand, filed in the Registry of the Court on 20 November 2012, para. 7.

28 Order of 6 February 2013, *I.C.J. Reports 2013*, p.5, para. 7.

29 Order of 6 February 2013, *I.C.J. Reports 2013*, pp.5–6, para. 8. *See also* the case cited here, *Haya de la Torre (Colombia/Peru)*, Judgment, *I.C.J. Reports 1951*, paras. 76–77.

30 This intention can be found in the Joint Media release of the two countries' Foreign Ministers of 15 December 2010: 'Australia and New Zealand agree on strategy for whaling legal case'. Available at [http://foreignminister.gov.au/releases/Pages/2010/kr_mr_101215.aspx?ministerid=2].

31 This is an application of the mootness doctrine. Had JAPRA II already ended, there would have been no merit in claiming its termination. Recently, the ICJ clarified, as 'case law', that: 'an essential element for the proper discharge of the Court's judicial function is that

Third, there seems to be no difference between Article 62 and Article 63, in terms of New Zealand being bound by the Judgment. Under Article 62 of the Statute, an intervening party is bound by the Judgment of the Court. In the case of Article 63, as well, 'the construction given by the judgment will be equally binding upon it [the intervening State].' This means substantively that there is no difference as to the legally binding character of the Judgment between these two types of intervention.

Fourth, there is no substantial difference between these two Articles on intervention as regards the procedural rights allowed to the intervening country. Articles 85 and 86 of the Rules of the ICJ provide procedures in the case of intervention, i.e. Article 85 corresponds to Article 62 of the Statute and Article 86 to Article 63.

> *Article 85*
> 3. The intervening State shall be entitled, in the course of the oral proceedings, to submit its observations with respect to the subject-matter of the intervention.
>
> *Article 86*
> 2. [...] The intervening State shall be entitled, in the course of the oral proceedings, to submit its observations with respect to the subject-matter of the intervention.

Thus, from the viewpoint of New Zealand, in relation to its right to submit its observations, in both written and oral form, there is no difference between Articles 85 and 86 of the Rules of the Court and, consequently, between Articles 62 and 63 of the Statute of the ICJ.

Fifth, the fixed time limits for intervention by a State under, respectively, Article 62 (Article 81 of the Rules) and Article 63 (Article 82 of the Rules) are different.

---

its judgments "must have some practical consequence in the sense that [they] can affect existing legal rights or obligations of the parties, thus removing uncertainty from their legal relations"'. *Application of the Interim Accord of 13 September 1995 (The Former Yugoslav Republic of Macedonia v. Greece)*, Judgment of 5 December 2011, *I.C.J. Reports 2011*, p. 662, para. 47. *See also Northern Cameroons (Cameroon v. United Kingdom)*, Preliminary Objections, Judgment, *I.C.J. Reports 1963*, p. 34.

> *Article 81*
> 1. An application for permission to intervene under the terms of Article 62 of the Statute [...] shall be filed as soon as possible, and *not later than the closure of the written proceedings*. In exceptional circumstances, an application submitted at a later stage may however be admitted. (emphasis added)
>
> *Article 82*
> 1. A State which desires to avail itself of the right of intervention conferred upon it by Article 63 of the Statute shall file a declaration to that effect [...]. Such a declaration shall be filed as soon as possible, and *not later than the date fixed for the opening of the oral proceedings*. In exceptional circumstances a declaration submitted at a later stage may however be admitted. (emphasis added)

Although the reasons for the later opportunity for intervention under Article 63 of the Statute are not clear,[32] the use of Article 63 is more favourable for an intervening State. Actually, in the *Whaling* case, New Zealand submitted its declaration of intervention on 20 November 2012, later than the closure of written proceedings on 2 May 2012[33] but before the opening of oral proceedings on 26 June 2013.

Sixth, in contrast to Article 62 of the Statute, which requires a jurisdictional link between the original party and the intervenor, Article 63 has no such requirement.[34] In this context, it is necessary to refer to the New Zealand's optional clause declaration, to which a reservation was attached as follows:

> The Government of New Zealand accept as compulsory [...], on condition of reciprocity, the jurisdiction of the International Court of Justice in

32 *supra* note 17, p. 1588.

33 Press Release No. 2012/18 (18 May 2012), 'Whaling in the Antarctic (Australia v. Japan), Closure of written proceedings'.

34 However, Rosenne points out that 'Article 82 makes no mention of any possible basis of jurisdiction existing between the intervenor and the parties in the case. This question must be regarded as open'. Shabtai Rosenne, *Procedure in the International Court: A Commentary on the 1978 Rules of the International Court of Justice* (Martinus Nijhoff, 1983), pp. 177–178.

> conformity with paragraph 2 of Article 36 of the Court over all disputes other than:
>
> (3) Disputes arising out of, or concerning the jurisdiction or rights claimed or exercised by New Zealand in respect of *the exploration, exploitation, conservation or management of the living resources* in marine areas beyond and adjacent to the territorial sea of New Zealand but *within 200 nautical miles* from the baselines from which the breadth of the territorial sea is measured.[35]

As is shown here, disputes concerning 'living resources' within 200 nautical miles are excluded from the scope of jurisdiction. If New Zealand relied on Article 62 of the Statute, which requires the jurisdictional link, its declaration of intervention would have been adjudged inadmissible because of the lack of a jurisdictional link; and, even if it had been admitted, it must at least have delayed the proceedings.

As seen above, the decision of New Zealand to intervene under Article 63 of the Statute, rather than under Article 62, was fully intentional and actually did enable New Zealand to achieve its objectives in the proceedings of the case, as seen below.

1.3 *Why Intervene as a Non-party?*

The next question is why New Zealand tried to intervene not as a Party but as a non-Party.[36] On this point, it should be observed that, from the outset, there seems to be no possibility under Article 63 to intervene as a 'Party', but only as a 'non-Party.' First, there is no requirement for a jurisdictional link under Article 63, in contract to Article 62. This means that the intervenor under Article 63 is not required to establish a jurisdictional link with the original party. Second, under Article 63, an intervening State does not submit any 'claim' of its own, but is limited to presenting its interpretation of the convention in question.[37]

---

35 Declarations Recognizing the Jurisdiction of the Court as Compulsory by New Zealand (22 September 1977), emphasis added. Available at [http://www.icj-cij.org/jurisdiction/index.php?p1=5&p2=1&p3=3&code=NZ].

36 New Zealand 'emphasises that it does not seek to be a party to the proceedings'. Declaration of Intervention Pursuant to Article 63 of the Statute of the Court by the Government of New Zealand, filed in the Registry of the Court on 20 November 2012, para. 9.

37 This interpretation was supported by Australia in this case, according to which, given the limited reach of an intervention under Article 63, the intervening State cannot be considered a party. Written Observation of Australia of 18 December 2012, which is not yet

As a result, it is reasonable to think that any intervening State under Article 63 must be a 'non-Party'.[38]

1.4 *Litigation Strategy*

The *Nuclear Tests* Cases

In general, Australia and New Zealand have pursued common, or at least similar, policies in a lot of areas such as environment, economy and disarmament. This stance has been reflected in their litigation strategy before the ICJ. The first such experience was the *Nuclear Tests* cases, in which the two countries had the same interests in putting an end to the French nuclear tests conducted around their territories. The two countries decided to sue France before the ICJ separately and submitted different applications against France. The ICJ decided not to join the two cases.[39] Probably, this was because the submission of New Zealand was slightly different from that of Australia. Finally, the ICJ decided not to enter into the merits phase, because of the disappearance of the dispute that the Parties had brought before it; and, as a result, it is difficult to evaluate the litigation strategy of the two applicants in maintaining the two cases separately. It is possible to say, at least, that the separation of two cases did not bring any benefit for the two countries. Even though they submitted different claims against France, the ICJ did not pay any attention to this difference in concluding that their claims had lost their purpose.[40]

The *SBT* Case

It is useful to refer to the previous litigation strategy of Australia and New Zealand in the *Southern Bluefin Tuna* case. At the initiation and the provisional measures phases before the ITLOS, two countries submitted two separate applications and requests. During the provisional measures procedure, however, the

published but cited in Separate Opinion of Judge Cançado Trindade (appended to Order of 6 February 2013), *I.C.J, Report 2013*, p.19, para. 12.

38 *supra* note 17, p.1591. In this regard, Article 63, which admits the binding effect of construction in a Judgment upon an intervening *non-party*, constitutes an exception to Article 59 of the Statute which stipulates that '[t]he decision of the Court has no binding force except *between the parties* [...]' (emphasis added).

39 The ICJ provided no information on this decision and this was criticised by Judges. According to Rosenne, 'after the incident in the Australian case concerning a possible premature disclosure of the results of the 1973 proceedings, it would have been difficult to have joined the two cases'. Shabtai Rosenne, *supra* note 34, p. 110.

40 On this point, the ICJ's conclusions should have been different in two cases.

two countries informed the Tribunal that, 'as parties in the same interest,' they had 'jointly nominated Mr Ivan Shearer' as Judge *ad hoc*[41] and, as a result, the ITLOS rendered *one* order for both Applicants.[42] In the jurisdictional phase, before the arbitral tribunal, Australia and New Zealand strengthened their strategy by joining the two cases into one. During the proceedings, the two countries filed a 'joint Reply on Jurisdiction'[43] and, before the opening of the hearing on jurisdiction, informed the president of their intention 'to be considered as a single party in the proceeding.'[44] As there was no objection from the Respondent (Japan), the president of the tribunal decided that the title of the case would be 'Southern Bluefin Tuna Case – *Australia and New Zealand* v. Japan'[45] and, finally, *one* Award was rendered by the Arbitral Tribunal for the Applicants, the result of which was that the Tribunal declined jurisdiction by 4 votes to 1.[46]

In this case, Australia and New Zealand intentionally chose to integrate two separate cases into one. It seems that this strategy was chosen to strengthen the cooperation between the two countries and to consolidate their common position, rather than to raise their claims on different grounds. Even though this decision seems to have been effective at first sight, the result was not satisfactory for the two countries, i.e. the Tribunal declined jurisdiction. As far as this case is concerned, therefore, the integration strategy did not work effectively.

New Litigation Strategy in the *Whaling* Case

In the *Whaling* case, Australia and New Zealand adopted a new litigation strategy, different from the previous one. On 15 December 2010, the Australian Minister for Foreign Affairs (The Hon Kevin Rudd MP) and the New Zealand Minister for Foreign Affairs (The Hon Murray McCully MP) made public a Joint media Release,[47] for the purpose of making clear that New Zealand would

41 Order of 27 August 1999, para. 11.

42 *Southern Bluefin Tuna Cases* (*New Zealand v. Japan; Australia v. Japan*), Request for Provisional Measures, Order (27 August 1999).

43 *Southern Bluefin Tuna Case* (*Australia and New Zealand v. Japan*), Award on Jurisdiction and Admissibility (4 August 2000), rendered by the Arbitral Tribunal constituted under Annex VII of the United Nations Convention on the Law of the Sea, para. 13.

44 Award of 4 August 2000, para. 11.

45 Award of 4 August 2000, para. 11. (emphasis added).

46 Award of 4 August 2000, para. 72. It is not expressed in the award which Judge was in minority. But, as Judge Keith (New Zealand nationality) appended a separate opinion to the award, he may have made a negative vote.

47 Joint Media Release, 'Australia and New Zealand agree on strategy for whaling legal case'. Available at [http://foreignminister.gov.au/releases/Pages/2010/kr_mr_101215.aspx?ministerid=2].

adopt a common litigation strategy with Australia with a view to 'work together' as a 'strong partner of Australia', by providing 'the valuable support', to persuade Japan to end its whaling. The concrete aspects of this strategy were followings: first, New Zealand, following the Australian Government's preference, would not file a claim as a Party to Australia's legal action, but would intervene in the case. Second, by intervening in the case, New Zealand would be able to make both written and oral submissions to the ICJ. Third, 'because New Zealand has a judge on the ICJ, Sir Kenneth Keith, the joining of the two actions would result in Australia losing its entitlement to appoint a judge for the case.' At the same time, 'New Zealand's decision to intervene will allow the case to proceed without delay.' It is clear that New Zealand was closely consulting with Australia on the litigation strategy and, from the procedural viewpoint, its main purpose was to maintain Judge Keith, a national of New Zealand, on the bench.

Before the *Whaling* case, Australia and New Zealand had been adopting a litigation strategy based on their common interests. As a result, their cases and appointment of Judges *ad hoc* were always joined. Compared with this previous approach, it is clear that the new approach of these two States, adopted in the *Whaling* case, resulted an unfavourable situation for Japan; and, consequently, a favourable position for Australia. In this sense, the change of litigation strategy, based on the two counties' consultation between themselves, may be the most effective of the lessons which they learned from their previous experience in the ICJ and the ITLOS. It should be noted that Japan can still learn a lot from its experience in the *Whaling* case.

1.5 *Appointment of Judge* ad hoc *by Australia*

Japan did not raise any objection to the admissibility of New Zealand's Declaration of Intervention under Article 63 of the Statute and, as a result, the Court admitted the intervention of New Zealand—seemingly without any profound discussion. It seems reasonable, at first sight, that Japan did not raise any objection to the New Zealand's intervention itself, since it has been recognised as an exercise of a 'right.' On the other hand, there remained one procedural issue unresolved, namely, how to object to the appointment of a Judge *ad hoc* by Australia in the particular circumstance of the case, i.e. that a Judge of New Zealand nationality was already on the bench.

Joinder of Proceedings in Cases: Article 47 of the Rules of the Court

With regard to the issue of Judge *ad hoc*, an option open to Japan would have been to request a joinder of cases by the Court. Once proceedings in several cases had been joined, the litigating States are required to appoint a Judge *ad hoc* jointly. Had this rule been applied to the *Whaling* case, where New Zealand had a Judge of its nationality (Judge Keith), Australia would not have been allowed to appoint a Judge *ad hoc* (Judge Charlesworth). Article 47 of the Rules of the Court provides that the ICJ

> may at any time direct that *the proceedings in two or more cases be joined.* It may also direct that the written or oral proceedings, including the calling of witnesses, be in common; or the Court may, without effecting any formal joinder, direct common action in any of these respects. (emphasis added)

In its history, the ICJ exercised this discretion in several cases. In the *South-West Africa* case, for example, the Court stated that:

> considering that the two applicant Governments were *in the same interest* and were therefore, so far as the choice of a Judge *ad hoc* was concerned, to be reckoned as one party only, by Order of 20 May 1961, joined the proceedings in the two cases, and fixed the time limit as requested.[48]

Following this decision of the Court, the two Applicant States jointly chose a single Judge *ad hoc*. Also in the *North Sea Continental Shelf* case, the Court joined two cases, by its Order of 26 April 1968, in accordance with the intentions of all the three parties.[49] In this case, the two Applicants were considered as one party for the purposes of the appointment of a Judge *ad hoc*. More recently, the two cases, *Certain Activities Carried Out by Nicaragua in the Border Area* (*Costa Rica v. Nicaragua*) and *Construction of a Road in Costa Rica along the San Juan River* (*Nicaragua v. Costa Rica*), were joined by the Court. The Court stated, in both cases, that Article 47

> leaves the Court a broad margin of discretion. Where the Court [...] has exercised its power to join proceedings, it has done so in circumstances where joinder was consonant not only with the principle of the sound administration of justice, but also with the need for judicial economy.[50]

48 *South West Africa Cases* (*Ethiopia v. South Africa; Liberia v. South Africa*), Preliminary Objections, Judgment of 21 December 1962, *I.C.J. Reports 1962*, p.321 (emphasis added). In this case, one unified judgment was rendered by the Court.

49 *North Continental Shelf Cases* (*Federal Republic of Germany/Denmark; Federal Republic of Germany/Netherlands*), Order of 26 April 1968, *I.C.J. Reports 1968*, p. 10.

50 *Certain Activities Carried Out by Nicaragua in the Border Area* (*Costa Rica v. Nicaragua*), Joinder of Proceedings, Order of 17 April 2013, *I.C.J. Reports 2013*, p. 170, para. 18; *Construction*

In this case, taking account of the principle of the sound administration of justice and the need for judicial economy, the ICJ had in mind the speedy disposal of the two cases simultaneously.[51] If the Court had taken the same approach in the *Whaling* case, it seems reasonable to suppose that the two Parties' cases would have been joined into one. It must be noticed, however, that Article 47 of the Rules of Court stipulates that 'the proceedings in two or more cases be joined' by the Court. In other words, there must be two or more 'cases' to be joined. In the *Whaling* case, New Zealand was permitted by the Court to intervene merely as a non-Party 'to place its interpretation of the relevant provisions of the Convention [ICRW] before the Court.'[52] As New Zealand had never submitted its own claim or submission to the Court, there was no 'case' of New Zealand and, therefore, no possibility of applying Article 47 of the Rules of the Court. From this viewpoint, the decision of New Zealand to intervene as a 'non-Party' should be evaluated as an appropriate strategy for avoiding its case being joined with that of Australia.

Joint Appointment of Judge *ad hoc*: Article 31(5) of the Statute

Another possibility for Japan to have objected to the appointment of a Judge *ad hoc* by Australia, apart from Article 47 of the Rules of the Court, was for it to have requested the application of Article 31(5) of the Statute,[53] which provides that:

> [s]hould there be several parties *in the same interest*, they shall, for the purpose of the preceding provisions, *be reckoned as one party only* (emphasis added).[54]

of a Road in Costa Rica along the San Juan River (*Nicaragua v. Costa Rica*), Joinder of Proceedings, Order of 17 April 2013, *I.C.J. Reports 2013*, p. 187, para. 12.

51 *Costa Rica v. Nicaragua case*, *I.C.J. Reports 2013*, p. 171, para. 24; *Nicaragua v. Costa Rica case*, *I.C.J. Reports 2013*, p. 188, para. 18.

52 Declaration of Intervention Pursuant to Article 63 of the Statute of the Court by the Government of New Zealand, para. 8.

53 In its Written Observations on New Zealand's Declaration of Intervention, submitted to the Court on 21 December 2012, Japan referred to Article 31(5) of the Statute and Article 36 of the Rules of the Court for the purpose of excluding the possibility of appointing a Judge *ad hoc* by Australia. Although this document has not been made public, this part was cited in Separate Opinion of Judge Cançado Trindade, *I.C.J. Report 2013*, p.19, para. 14 (note 16).

54 Article 36(1) of the Rules of the ICJ makes this provision more precise by stipulating that:
If the Court finds that two or more parties are *in the same interest*, and therefore *are to be reckoned as one party only*, and that there is no Member of the Court of the nationality

Even when the proceedings of several cases have not been joined, the Court may order the parties 'in the same interest' to appoint a single Judge *ad hoc* jointly. In the *South West Africa* case two cases were joined as seen above and the two applicant states (Ethiopia and Liberia), 'acting in concert', chose Sir Louis Mbanefo to sit as Judge *ad hoc*.[55] In the *North Sea Continental Shelf* case two cases were joined and the Court considered Denmark and Netherlands 'to be parties in the same interest since they have set out their submissions in almost identical terms.' Consequently, as far as the choice of a Judge *ad hoc* is concerned, they are 'to be reckoned as one Party only.'[56] These two cases are examples in which the ICJ applied Article 47 of the Rules for joining two cases and ordered the joint appointment of a Judge *ad hoc* by the parties in accordance with Article 31(5) of the Statute.

The following cases are examples in which the ICJ applied Article 31(5) of the Statute without joining several cases. In the *Nuclear Tests* cases, these two cases were initiated by Australia and New Zealand separately and were *not* joined. However, two Applicants chose the same person, Sir Garfield Barwick, to sit as Judge *ad hoc* in respective case.[57] Although the reason for this was not explained by the Court, Parties' act should be understood as a result of application of Article 31(5) of the Statute. The same approach was applied by the Court in the *Fisheries Jurisdiction* cases in which two cases were *not* joined, but two Applicants – the U.K. and Germany – were regarded as one party for appointing jointly one Judge *ad hoc*. In this case, Germany had chosen a person as Judge *ad hoc* and, to this decision, there was no objection by Iceland (respondent). However, the Court, taking into account the proceedings instituted by the U.K. against Iceland and the composition of the Court, which included a judge of UK nationality,[58] decided that 'there was in the present phase, concerning the jurisdiction of the Court, *a common interest* in the sense of Article 31, paragraph 5, of

of any one of those parties upon the Bench, the Court shall fix a time-limit within which they may jointly choose a judge *ad hoc* (emphasis added).

55 *South West Africa Cases* (*Ethiopia v. South Africa; Liberia v. South Africa*), Second Phase, Judgment of 18 July 1966, *I.C.J. Reports 1966*, p. 9.

56 *North Continental Shelf Cases* (*Federal Republic of Germany/Denmark; Federal Republic of Germany/Netherlands*), Order of 26 April 1968, *I.C.J. Reports 1968*, p. 10. This decision of the ICJ was based on the three parties' agreement which had been already expressed in the Protocol. *Ibid.*

57 *Nuclear Tests Case* (*Australia v. France*), Judgment of 20 December 1974, *I.C.J. Repots 1974*, p. 255, para. 3. *Nuclear Tests Case* (*New Zealand v. France*), Judgment of 20 December 1974, *I.C.J. Repots 1974*, p. 458, para. 3.

58 In the *Fisheries Jurisdiction Case* (*United Kingdom of Great Britain and Northern Ireland v. Iceland*), the ICJ made no mention to the issue of Judge *ad hoc* in the Germany's case.

the Statute which justified the refusal of the request of the Federal Republic of Germany for the appointment of a judge *ad hoc*' (emphasis added).[59] In the merit phase, as Iceland decided not to appear before the Court and did not avail itself of the right to appoint Judge *ad hoc*, Germany made clear that 'it did not feel it necessary to insist on the appointment of a judge *ad hoc*'.[60]

From the Court's jurisprudence, it is clear that the Court was not reluctant, until the *North Sea Continental Shelf* case, to join the proceedings of several cases and to order the parties to appoint a single Judge *ad hoc* jointly. After that case, on the contrary, the Court became reluctant to join cases, but favourable to the joint appointment of a Judge *ad hoc*. The ICJ changed its attitude on this point without any clear explanation. Seemingly, however, Australia and New Zealand learned significant lessons from the *Nuclear Tests* cases, in which a joint appointment of a judge *ad hoc* was ordered by the Court.

The core criterion for the application of Article 31(5) of the Statute is the notion of 'in the same interest.' However, in applying this procedure, there is a threshold, i.e. there must be 'parties' in the same interest. Without any party, there is no possibility of applying Article 31(5) of the Statute. In this context, it must be remembered that New Zealand intended to intervene as a 'non-Party' in the *Whaling* case. As a result, simply, there was no other 'Party' than Australia in the case and this led the Court to conclude that:

> since the intervention of New Zealand does not confer upon it *the status of party* to the proceedings, Australia and New Zealand cannot be regarded as being "parties in the same interest" within the meaning of Article 31, paragraph 5, of the Statute.[61]

The decision of New Zealand to intervene as a 'non-Party' was a well-planned strategy to avoid the joint appointment of judge *ad hoc*.

1.6 *Procedural Inequality between Parties*

As seen above, Japan had no means for blocking the intervention by New Zealand and the appointment of a Judge *ad hoc* by Australia. Once New

59 *Fisheries Jurisdiction Case* (*Federal Republic of Germany v. Iceland*), Jurisdiction of the Court, Judgment of 2 February 1973, *I.C.J. Reports 1973*, p. 51, para. 7. One question is remained: why did the ICJ use the term 'common interest' which is not the expression used in Article 31(5) of the Statute?

60 *Fisheries Jurisdiction Case* (*Federal Republic of Germany v. Iceland*), Merits, Judgment of 25 July 1974, *I.C.J. Reports 1974*, p. 178, para. 10.

61 Order of 6 February 2013, *I.C.J. Reports 2013*, para. 21. (emphasis added).

Zealand decided to exercise its 'right' to intervene in the case as a 'non-party' under Article 63 of the Statute:

(i) the Court does not decline this request except in exceptional circumstances, since the intervention under Article 63 of the Statute is 'the exercise of a right'.[62]
(ii) there was no 'case' of New Zealand to be joined to that of Australia.
(iii) there were no 'parties in the same interest' which could be required to appoint a single Judge *ad hoc* jointly.

Consequently, the intervention of New Zealand and the appointment of Judge *ad hoc* by Australia were unavoidable not merely for Japan but also for the Court. In addition, from New Zealand's viewpoint, there was no difference between 'Party' intervention and 'non-Party' intervention, since even in the latter case, the interpretation given by the Court is binding on a 'non-Party' intervenor.[63]

In this sense, the two States developed a well planned, common, litigation strategy with a successful result; and Japan did not have any option other than to accept New Zealand's declaration.[64] The same applies to the ICJ's treatment of the declaration.[65]

While it may be true that acceptance of the above procedures was unavoidable, the question still arises whether there were any adverse consequences for Japan. It is true that the presence of Australia's Judge *ad hoc* did not in the end influence the Judgment.[66] However, some Judges refer to 'concerns' of Japan. According to Judge Cançado Trindade, Japan 'manifest[ed] concern mainly with the procedural equality of the Parties in the proceedings.'[67] Judge Owada

62 *I.C.J. Reports 2013*, p.5, para. 7. According to the Court's understanding, there are some conditions of admissibility applicable to the intervention under Article 63. *See*, *I.C.J. Reports 2013*, pp.5–6, para. 8.

63 Article 63 of the Statute of the Court provides that if a state 'uses this right [of intervention], the construction given by the judgment will be equally binding upon it'. *See*, *I.C.J. Reports 2013*, p.4, para. 2.

64 Judgment of 31 March 2014, *I.C.J. Reports 2014*, p.235, para. 11.

65 Ibid. The Court took the view that it was not necessary to hold hearings on the question of the admissibility of New Zealand's Declaration of Intervention.

66 The main parts of the Judgment's dispositif were adopted under 12–4 votes. Thus, even if Australia's Judge *ad hoc* was not sitting in the bench, the result of the Court would have been the same.

67 Separate Opinion of Judge Cançado Trindade, *I.C.J. Reports 2013*, p.20, para. 17.

refers to 'an important aspect of the principle of equality of the Parties, which forms an essential cornerstone of the fair administration of justice'.[68] Relying on this opinion, Japan stated that '[t]he equality of the Parties is also a fundamental principle of international justice. In this context, Japan finds itself in a regrettable position.'[69]

Even though the appointment of Judge *ad hoc* by Australia, in addition to the sitting of a Judge of New Zealand nationality in the bench, was unavoidable due to the relevant articles of the Statute and the Rules of the Court, the decision of the Court might give rise to some fundamental concerns. Judge Owada emphasised on this point that the intervention by New Zealand would provoke issues concerning the fair administration of justice, including the equality of the Parties in the proceedings before the Court, the examination of which is 'inherent in the judicial function of the Court.'[70] Here, Judge Owada is making clear that the intervention of New Zealand breaches the fundamental principle of equality between the Parties and that the Court should have exercised its inherent power to maintain its judicial function.

It is true that the equality of Parties in proceedings is one of the fundamental principles of procedure. However, the Statute and the Rules of the Court do contain a loophole on this point, i.e. an intervening State is not excluded from appointing a Judge *ad hoc* even if it has an almost identical legal position with another party in a case.

1.7 *Substantial Problems of Intervention*

In this case, several procedural inequalities have occurred implicitly. First, the ICJ permitted New Zealand to submit Written Observations with regard to the merits of the case,[71] the content of which was almost the same as that of Australia's Memorial. As a result, while the ICJ rejected Japan's request for a second round of written pleadings, Australia was substantially allowed to submit a second written pleading through its partner's Written Observations.[72] This unfair situation was mitigated, since Japan was also allowed to submit Written Observations in response to those of New Zealand, and Australia decided not to submit the same kind of observations. It may be said that each side submitted two written rounds of pleadings.

68 Declaration of Judge Owada, *I.C.J. Reports 2013*, pp.12–13, para. 6.

69 Written Observations of Japan, para. 2.

70 Declaration of Judge Owada, *I.C.J. Reports 2013*, pp.11–12, paras. 1 and 4.

71 Written Observations of New Zealand.

72 Separate Opinion of Judge Cançado Trindade, *I.C.J. Reports 2013*, p.20, para. 16.

Second, a more significant issue was that New Zealand's interpretation of the ICRW, shown in its Written Observations, was slightly different from that of Australia. This means that the equal number of written pleadings, 2–2, was not substantively equal, since Japan had to deal with two States, even though one was not a 'Party'.

Third, another procedural inequality arose. Because the intervention of New Zealand became known only just before the initiation of the oral pleadings,[73] Japan was forced to react to two different procedures at the same time.[74] Namely, it had to prepare for the oral pleadings against Australia and for the Written Observations against New Zealand. Japan's request for a second written pleading was denied by the Court without any clear explanation,[75] although Japan was mindful that the oral pleading would be quite important[76] and stressed the need to have adequate time for preparation before the oral proceedings.[77] Thus, it is quite understandable that Japan felt there to be deep inequality in this particular circumstance.

However important the issue of procedural inequality seems to be, the ICJ did not take it into account in its findings; the Court stated that:

> intervention under Article 63 of the Statute is limited to submitting observations on the construction of the convention in question and does not allow the intervenor, which does not become a party to the proceedings, to deal with any other aspect of the case before the Court; and [...] such an intervention *cannot affect the equality* of the parties to the dispute.[78]

Clearly, from the ICJ's viewpoint, intervention as a non-Party and the making of submission on the construction of the convention do not affect the equality

---

73 The timing of the Declaration was intentionally chosen in accordance with Article 82(1) of the Rules. This time-limit is different from Article 81(1) which stipulates the case of intervention under Article 62 of the Statute.

74 Written Observations of Japan, para. 4.

75 The ICJ, without explaining any reason, decided that the filing of a Reply by Australia and a Rejoinder by Japan in this case was not necessary. *See* Press Release of the ICJ, No. 2012/18 of 18 May 2012: 'Whaling in the Antarctic – Closure of Written Proceedings'.

76 Japan requested to the Court that 'adequate time be given to Japan to prepare for these oral proceedings, both in the first and second round'. Order of 6 February 2013, *I.C.J. Reports 2013*, p.8, para. 17.

77 Separate Opinion of Judge Cançado Trindade, *I.C.J. Reports 2013*, p.20, para. 17.

78 *I.C.J. Reports 2013*, p.9, para. 18. (emphasis added).

of the parties. This understanding is nothing other than a formal application of the relevant articles, and does not consider the actual loss of equality between the parties. However, since the ICJ is a court of justice and, especially, is an adjudicatory body based on the concept of sovereign equality, it should have paid more attention to the issue of procedural equality.

In conclusion, it is clear that the procedural inequality between Japan and Australia, stemming from the New Zealand's intervention, was not, or could not have been, expressly discussed before the Court. In reality, it is uncertain to what extent it had any impact on the proceedings and the Court's result. However, it is nonetheless undeniable that Japan was put in an unfavourable situation.

1.8 *Scope of Construction*

First of all, an intervention under Article 63 of the Statute is limited to 'the construction of a convention' in question. This means that, if a request for intervention went beyond the issue of 'construction', then the request could not be admissible under that Article. At the same time, even if a State is entitled to intervene in a case, it is admitted only to submit its observations on the 'construction' of the convention in question; and, consequently, it cannot be admitted to make observations on the application of the convention. As the Court states in its Order, 'the limited object of the intervention [under Article 63] is to allow a third State not party to the proceedings [...] to present to the Court its observations on the construction of that convention.'[79]

1.9 *Interim Conclusions on Intervention*

The litigation strategy taken by Australia and New Zealand in the *Whaling* case in relation to the intervention was undeniably successful. At the same time, it became clear that there was no legal obstacle or hurdle in relation to this strategy, i.e. of intervention as a non-party under Article 63 of the Statute with a view to submitting observations to the Court.

It is also undeniable that this strategy will be adopted by these two countries in the future case concerning Japanese whaling which is expected to be launched in relation to a new program called NEWREP-A,[80] the plan of which

79 Order of 6 February 2013, *I.C.J. Reports 2013*, p.5, para. 7.

80 NEWREP-A (New Scientific Whale Research Program in the Antarctic Ocean). Text is available at [http://www.jfa.maff.go.jp/j/whale/pdf/newrep--a.pdf]. Its 'Outline' is also available at [http://www.jfa.maff.go.jp/j/whale/pdf/outline-of-newrep-a.pdf].

has already been submitted to the Scientific Committee of the IWC in 2014. It is possible in such a hypothetical case that it might be New Zealand that submits an Application instituting the proceeding and Australia that makes a declaration of intervention in the case as a non-Party under Article 63 of the Statute—i.e. there might be an exchange of position between these two countries. The purpose of such a strategy would appear to be, as was in the *Whaling* case, to maintain a Judge of Australian nationality, Mr Crawford,[81] and to enable New Zealand to appoint a Judge *ad hoc*. If for the sake of argument, such a decision were to be taken by these two countries, the legal position would be the same as in the *Whaling* case.

2 Examination and Cross-Examination of Experts

One of the most important characteristics of the *Whaling* case is that it stems from a controversy concerning the scientific evaluation of JAPRA II. As such, this case should be characterised as a 'scientific dispute', even though the Court avoided this. In its history, the ICJ has been confronted with scientific disputes, such as the *Nuclear Weapons* case, the *Gabcikovo-Nagymaros* case and the *Pulp-Mills* case. In those cases, the expert knowledge of scientists was needed to clarify the scientific aspects of the issues, and to assist the Court in making its 'legal' evaluation of the parties' allegations.

2.1 *Procedures in Relation to, and Function of, Experts*

For this purpose, the ICJ has several procedures relating to experts. First, the Court itself may appoint experts under Article 50 of the Statute. Second, the litigating parties may appoint experts under Article 43(5) of the Statute, which provides that '[t]he oral proceedings shall consist of the hearing by the Court of witnesses, experts, agents, counsel, and advocates'. In the *Whaling* case, experts were appointed by both parties and they played an important role in influencing the Court's understanding of the scientific evaluations and, more precisely, of providing the Court with justification in the choice of certain lines of reasoning.

To understand the function of experts, it is worth referring to the difference between 'witness' and 'expert'. The Rules of the Court makes this point clear:

81 On 6 November 2014, the General Assembly and the Security Council of the United Nations elected Mr James Crawford as a new member of the Court, for a term of office of nine years beginning on 6 February 2015. Press Release 2014/32 of 7 November 2014.

> *Article 64*
> Unless on account of special circumstances the Court decides on a different form of words,
> (a) every *witness* shall make the following declaration before giving any evidence:
> "I solemnly declare upon my honour and conscience that I will speak *the truth*, the whole truth and nothing but the truth";
> (b) every *expert* shall make the following declaration before making any statement:
> "I solemnly declare upon my honour and conscience that I will speak the truth, the whole truth and nothing but the truth, and that my statement will be *in accordance with my sincere belief*." (emphasis added)

As is shown here, while the witnesses are expected to 'speak the truth', the experts are called upon not only to speak the truth, but also to make statement on their 'sincere belief'. This means that the experts' function is to present their own evaluation, normally scientific evaluation, on some issues in question.

2.2 *Examination of Experts*

Experts may engage in several phases of proceedings. Experts submit their expert statements to the Court, as a written pleading. Also at the stage of oral pleadings, experts are examined by their own side and cross-examined by the other side. On this point, Article 65 of the Rules of the Court provides that:

> Witnesses and experts shall be examined by the agents, counsel or advocates of the parties under the control of the President. Questions may be put to them by the President and by the judges. Before testifying, witnesses shall remain out of court.

Examination and cross-examination had a significant influence on the findings of the ICJ in the *Whaling* case, as we shall see below. The function of examination is to obtain the knowledge and beliefs of witnesses and experts and enable their veracity and credibility to be challenged and tested.[82] In the

82 Shabtai Rosenne (with the Assistance of Yaël Ronen), *The Law and Practice of the International Court 1920–2005, Fourth Edition, Vol. III, Procedure* (Martinus Nijhoff Publishers, 2006), p. 1308.

Whaling case, the following examinations were conducted during the oral pleadings:

(i) Mr. Mangel (Australia's expert) was examined by Mr. Sands (Australia's counsel) and cross-examined by Mr. Lowe (Japan's counsel).
(ii) Mr. Gales (Australia's expert) was examined by Mr. Gleeson (Australia's counsel) and cross-examined by Mr. Lowe (Japan's counsel).[83]
(iii) Mr. Walløe (Japan's expert) was examined by Mr. Lowe (Japan's counsel) and cross-examined by Mr. Gleeson (Australia's counsel).[84]

2.3 *Japan's Perception of Expert Examination*

Some controversies arose as to the strategy adopted by Japan during the examination and cross-examination of the expert appointed by Japan. Plainly, Japan did adopt a faulty strategy of expert examination and may have had an incorrect understanding of the nature of this procedure. In this regard, the following elements must be taken into consideration.

First, during the examination time allocated to Japan, Mr Lowe (advocate of Japan) requested Mr Walløe (expert of Japan) to read, in 20 minutes, the summary of his expert statement, without posing any questions.[85] In other words, there was no interaction between counsel and expert in the examination procedure. This strategy may have been taken intentionally by Japan on the ground that Japan had already submitted a written statement of the expert and, thus, it would have been a waste of time to repeat the same statement in the oral examination, in the style of a TV court drama.

Second, in contrast to Japan's strategy, Australia's strategy was to make the examination procedure interactive and dramatic. During his examination by Australia's counsel (Mr Sands), Australia's expert (Mr Mangel) observed that 'since I'm speaking extemporaneously to you, I will not generally repeat what I have written [in expert statement] because I haven't tried to memorize anything'.[86] Here it is possible to discern a sharp difference of attitude towards the examination process between Japan and Australia. Namely, while Japan had a static perception of expert examination, Australia had a more dynamic and interactive perception of it, even if the latter's examination can be seen as artificially dramatic.

---

83 Judgment of 31 March 2014, *I.C.J. Reports 2014*, p.237, para. 20.
84 Judgment of 31 March 2014, *I.C.J. Reports 2014*, p.237, para. 21.
85 CR 2013/14, pp. 17–22 (Japan, Mr Walløe).
86 CR 2013/9, p. 40 (Australia, Mr Mangel).

It is difficult to pinpoint direct consequences stemming from this difference of attitude towards the process of expert examination. However, at least, it can be safely said that the Japan's attitude did not make its position favourable or advantageous against the position of Australia. First, as a natural reaction to Japan's examination, the President of the Court regarded this as 'less interactive than usual examination.'[87] Second, instead of allowing the reiteration of Japan's expert's written statement, the advocate of Japan should have brought out in his questions, in support of Japan's position, the fact that JARPA II was positively appreciated in the Scientific Committee. However, in adopting the above mentioned strategy, the advocate did not bring up this point at all.[88]

All the examining counsels in the *Whaling* case were experienced lawyers of the common law tradition. Mr Justin Gleeson (a counsel of Australia), S.C. (Senior Counsel) is Solicitor-General of Australia. Mr Philippe Sands (a counsel of Australia), Q.C. (Queen's Counsel) is Barrister in London. Mr Vaughan Low Q.C. (a counsel of Japan) is a member of the English Bar. Thus, due to the appointment of those people for the expert examination, there had been a shared understanding by both States of adopting, in the present case, the common law perception of the nature of expert examination. Despite this perception, however, Japan seems to have taken an attitude towards expert examination procedure which is differed from the one adopted in the common law countries.

2.4 *Negative Impacts of Expert Opinions on Japan's Case*

It is more important to analyse whether the expert's opinions had an impact on the Judgment or not, and if so, to what extent. With hindsight, it can easily be concluded that the experts' opinions, especially that of Japan, had undeniably an adverse effect on Japan's case. In this context, it is necessary to point out relevant parts of statements by Mr Walløe, who was appointed by Japan, and reactions by the ICJ and individual Judges.

> (i) Mr. Walløe stated in his expert statement that 'Japanese scientists have not always given completely transparent and clear explanations of

87 CR 2013/14, p. 23 (Japan, the President). More directly, Australia's counsel regarded Japan's examination as 'interesting non-interactive presentation' (CR 2013/14, p. 23) and 'unhelpful monologue-in-chief' (CR 2013/19, p. 15).

88 CR 2013/19, p. 29 (Australia, Mr Sands).

how sample sizes were calculated or determined.'[89] Judge Greenwood posed a question with regard to this statement.[90] This part of statement was also cited in the Judgment of the Court.[91]

(ii)In his statement[92] and during the cross-examination,[93] Mr. Walløe stated that 'JARPA II sample sizes had been influenced by funding considerations', although he found this unobjectionable. This statement was cited in the Judgment of the ICJ[94] and also in the Separate Opinion of Judge Xue.[95]

(iii) The Court stated in the Judgment that Mr. Walløe raised concerns about the fin whale component of JARPA II that went beyond the issue of sample size.[96] On this point, Mr. Walløe testified that the fin whale proposal was 'not very well conceived' for two reasons. He stated that random sampling of fin whales within the JARPA II research area was not possible, first, because the main fin whale population was beyond the JARPA II research area—further to the north; and secondly, because the JARPA II vessels could only accommodate the lethal take of smaller fin whales (a point also raised by Australia). The Court recalled that Japan identified random sampling as an element of a programme for purposes of scientific research.[97] This part of the statement was also cited in a Separate Opinion of Judge Greenwood.[98]

(iv) Mr. Walløe made clear that the JARPA II Research Plan lacks transparency in its reasons for selecting particular sample sizes for individual research items. This is a matter on which the experts called by the two Parties agreed, according to the Court.[99]

(v) Mr. Walløe stated that 'a six-year period [of the take of mink whales] is arbitrary'.[100] This part of statement was cited in the Judgment.[101]

89 Statement of Mr Lars Walløe (expert called by Japan), 'Scientific review of issues raised by the Memorial of Australia including its two Appendices' (9 April 2013), p. 8.

90 CR 2013/14, p. 49 (Japan, Judge Greenwood).

91 Judgment of 31 March 2014, *I.C.J. Reports 2014*, p.275, para. 159.

92 Statement of Mr Lars Walløe, p. 10.

93 CR 2013/14, p. 46 (Japan, Mr Walløe).

94 Judgment of 31 March 2014, *I.C.J. Reports 2014*, p.275, para. 159.

95 Separate Opinion of Judge Xue, *I.C.J. Reports 2014*, p.426, para. 20.

96 Judgment of 31 March 2014, *I.C.J. Reports 2014*, p.280, para. 180.

97 CR 2013/14, p. 47 (Japan, Mr Walløe).

98 Separate Opinion of Judge Greenwood, *I.C.J. Reports 2014*, pp.415–416, para. 26.

99 Judgment of 31 March 2014, *I.C.J. Reports 2014*, p.283, para. 188.

100 CR 2013/14, pp. 41–42 (Japan, Mr Walløe).

101 Judgment of 31 March 2014, *I.C.J. Reports 2014*, pp.283–284, para. 192.

(vi) The Judgment the ICJ states that Mr. Walløe 'suggested that co-operation with international research programmes "would be difficult for personal and political reasons", given that the use of lethal methods is contentious among scientists.'[102] Mr. Walløe acknowledged that co-operation with other Japanese research institutions, such as the National Institute for Polar Research, could be improved.[103] This statement of Mr. Walløe was a response to a question posed by Judge Keith.[104]

From the above, it is clear that Mr Walløe, although appointed by the Government of Japan, made a lot of statements unfavourable and even damaging to Japan's case. What is important is that his statements in written and oral pleadings had crucially negative and adverse connotations on the 'scientific' nature of JARPA II. From the outset, Mr Walløe, as a scientist, was adopting the opinion that the take of fin whales and humpback whales by Japan was simply wrong in JARPA II, and not justifiable. This position was nothing other than an ultimate blow for Japan. Thus, it can be said that the decision to appoint Mr Walløe, although his appointment was only one of a number of possible options for Japan, had an adverse impact for Japan's case.

2.5 *The Nature and Function of Experts: Independent Scientist, Counsel or Advocate?*

There is a dilemma as to the nature and function of experts in the proceedings before the ICJ. On the one hand, experts are appointed and hired by the government of the litigating State. This means that experts are expected to support the appointing government by providing supportive statements for that government. On the other hand, experts are obliged, at least legally speaking, to speak the truth in their statements 'in accordance with [their] sincere belief' (Article 47 of the Rules of the Court). In other words, it is required for experts to present their own objective and scientific evaluation or understanding. This dilemma has not been clearly settled in the history of the ICJ's proceedings.

In most cases before *Pulp Mills* [2010], actually, experts had been appointed by a government as its counsel, namely as a member of the litigation delegation. This shows that experts are expected to make statements in favour of the government which appointed them.

102 Judgment of 31 March 2014, *I.C.J. Reports 2014*, pp.291–292, para. 221.

103 CR 2013/14, p. 58 (Japan, Mr Walløe).

104 CR 2013/14, p. 58 (Japan, Judge Keith).

On the other hand, the ICJ indicated an intention to change the status of experts. In the *Pulp Mills* case the Court stated that:

> The Court has given most careful attention to the material submitted to it by the Parties, as will be shown in its consideration of the evidence below with respect to alleged violations of substantive obligations. Regarding those *experts* who appeared before it *as counsel* at the hearings, the Court would have found it more useful *had they been presented by the Parties as expert witnesses under Articles 57 and 64 of the Rules of Court, instead of being included as counsel in their respective delegations*. The Court indeed considers that those persons who provide evidence before the Court based on their scientific or technical knowledge and on their personal experience should testify before the Court as experts, witnesses or in some cases in both capacities, rather than counsel, so that they may be submitted to questioning by the other party as well as by the Court.[105]

In this paragraph, the ICJ makes it clear that the experts should not be included in the delegation of each party, i.e. experts should work independently from the governments which appointed them, in order to provide useful and objective information for the bench's considerations.

2.6 *Unfavourable but Unavoidable Results*

Although Japan's expert made many statements unfavourable to Japan, it should be noted that the role of experts is not to justify all aspects of government's acts. Experts are rather required to present their own scientific opinions, which should be, basically, independent, impartial and non-biased. From this viewpoint, the objectivity is the most valuable aspect of expert statement and should be appreciated by the Judges.[106]

In other words, even when a scientific evaluation of experts does not support one side, it necessarily contains opinions favourable or unfavourable to

105 *Pulp Mills on the River Uruguay* (*Argentina v. Uruguay*), Judgment of 20 April 2010, *I.C.J. Reports 2010*, p. 72, para. 167.

106 Judge Abraham emphasized the objectivity of the expert opinion and expressed appreciation of the statement of Mr Walløe; l'expert M. Walløe 'a demontré son indépendence d'esprit en critiquant, devant la Cour, certains des aspects, cependant mineurs, du programme JARPA II – l'arrêt d'ailleurs en tire argument contre le défendeur à plusieurs reprises. Cela ne rend que plus crédible, à mon avis, l'ensemble de sa déposition'. Opinion dissidente de M. le juge Abraham, *C.I.J. Recueil 2014*, p.335, para. 47.

one of the parties. This means at the same time that, from the Court's viewpoint, there must always some elements on which the Court can rely when it elaborates its judgment; it actually relied on the statements unfavourable to Japan. In this sense, inevitably, any expert opinion can be unfavourable to the State which requested it.

It should be remembered that, in the *Whaling* case, the burden of proof was imposed on Japan, not on Australia. Paragraph 68 of the judgment clearly states that 'the Court will *look to the authorizing State* [*Japan*], which has granted special permits, to explain the objective basis for its determination' (emphasis added). Here the ICJ makes clear its intention to examine the standard of review (the reasonableness test, Judgment, para. 67) on the basis of a burden of proof imposed on Japan. Consequently, Japan was required to prove the reasonableness of JARPA II's design and implementation in relation to achieving its objectives. Ultimately, this position was highly detrimental to Japan, since the Judgment relied on the expert statement of Japan, which contained many observations, unfavourable for supporting the reasonableness test.

Conclusions

As is well known, procedural equality is one thing, but the outcome of the procedure is another. Even if the Parties were not treated equally, this does not necessarily amount to a justification for denying completely the validity of the Judgment.[107] Thus, even if Japan was placed in a slightly unfavourable position in the *Whaling* case, this does not mean that the Judgment itself should be evaluated as unfair. On the other hand, what's more important is to realise that some procedures may be unfavourable, but unavoidable for one of the parties. This was the case for Japan in the *Whaling* case.

First, as to the intervention of New Zealand in the case, there were not many legal issues to be discussed, since it involved no more than an exercise of a 'right' by an intervenor. However, if analysed closely, it becomes clear that there was a well-planned litigation strategy which produced the following merits; (1) maintaining Australia's Judge *ad hoc* in the bench, (2) increasing the burden for Japan, by doubling its works during the oral pleadings, (3) allowing

107 Except in the case of some fundamental defects of procedure. However, the ICJ has never admitted its own defects, such as excess of jurisdiction and serious violation of fundamental rules of procedure.

the appearance of New Zealand before the Court in a position almost same as a 'Party', because it has a Judge *ad hoc* and the interpretation of the Court is binding on it. These advantages were produced by the common litigation strategy of the two countries, which has been revised on the basis of the previous experiences of litigation. This shows clearly that analysis of previous experience and making changes to strategy in the light of it may have a successful impact on future litigation.

Second, as to the examination of experts, this brought no benefit to Japan and Japan's style of examination may be controversial. It is, however, almost impossible to evaluate its effects on the Judgment from a legal point of view, since its impact on Judges' impressions cannot be objectively measured. On the other hand, what was objectively unfavourable for Japan was its expert's opinion; Japan's expert did not hesitate to point out many unfavourable evaluations on JARPA II and, quite predictably, many of his views were used by the Court to justify its reasoning in the Judgment.

These two examples are still to be examined more closely with a view to elaborating the future litigation strategy of Japan, as well as of any other countries which may appear before the Court. For the moment, we should appreciate the *Whaling* case as providing useful lessons for addressing the practical question: how to avoid unfavourable procedural situations, and what may be the consequences if a state fails to avoid them.

CHAPTER 6

Roads Not Taken, Opportunities Missed: Procedural and Jurisdictional Questions Sidestepped in the *Whaling* Judgment

Christian J. Tams[*,1]

Introduction

Landmark cases establish new legal principles that alter the interpretation of existing rules. This, courts typically do by putting forward new interpretations of the law, seeking out new frontiers, or boldly going where no court has gone before. But every now and then courts deliver landmark decisions that are noteworthy for their silences – silences that allow for inferences to be drawn.

The *Whaling* judgment is in many respects a landmark case. In interpreting Article VIII of the International Convention on the Regulation of Whaling,[2] the Court (boldly, some would say, perhaps rather too rashly, others) went where no court had gone before. In requiring the revocation of existing permits,[3] the ICJ certainly pushed the boundaries of the law of remedies.[4] However, merely focussing on what the Court *said* or *did* may not do justice to the *Whaling* judgment. The judgment is also noteworthy for its silences, its omissions, its *lacunae*. Some of these were in fact quite surprising given the public and expert debate preceding the judgment;[5] and even some of those less surprising omissions are worthy of comment.

---

* This chapter draws on research conducted during a fellowship Centre for Global Cooperation Research (Duisburg). I am grateful for the Centre's support.

1 Professor of International Law, University of Glasgow; and currently Senior Fellow, Centre for Global Cooperation Research [christian.tams@glasgow.ac.uk]. The author gratefully acknowledges the Centre's support and thanks Dr James Devaney for comments on an earlier draft.

2 *Whaling in the Antarctic,* Judgment of 31 March 2014, available at www.icj-cij.org. (References in the following are to the '*Whaling case*'; references to individual declarations and opinion appended to the judgment are to the '*Whaling case,* Diss. Op./Sep.Op.' followed by the judge's name.)

3 *Whaling case*, paras. 51–97.

4 *Whaling case*, para. 245.

5 See for instance: Frans A. Nelissen and Steffen Van der Velde, 'Australia Attempts to Harpoon Japanese Whaling Program' Hague Justice Portal Online accessed 16 December 2014; Dan Goodman, 'Whaling in the Southern Ocean: A Reply to Nelissen and van der Velde' Hague Justice Portal Online <http://www.haguejusticeportal.net/index.php?id=11940> accessed 16 December 2014; D.R. Rothwell, 'Australia v. Japan: JARPA II Whaling Case before the

 DOI 10.1163/9789004313828_008

Against this background, the following considerations are addressed the silences in the Court's reasoning. Not all of them can be covered: there are in fact too many, as the Court in its treatment of procedural and jurisdictional matters was as selective as the parties had been in their pleadings. But three silences will be addressed and their implications interrogated:

(i) the Court's rather rushed treatment of questions of jurisdiction;
(ii) the absence from the judgment of language reflecting the public interest or '*erga omnes*' character of the application; as well as, more briefly, comments on; and
(iii) the Court's decision to restrict the written proceedings to one round.

In addressing these issues, the paper will situate the Court's approach in the *Whaling case* in its historical context and will assess whether the judgment has broken new procedural or jurisdictional ground.

1 Optional Clause Jurisdiction and Special Treaty Regimes

The Context

As the judgment does,[6] this short note begins with considerations of jurisdiction. Even in a judgment noteworthy for the large majority carrying the decision, the first paragraph of the *dispositif* stands out. In it, the Court *unanimously* upheld jurisdiction on the basis of the two parties' optional clause declarations.[7] In his dissent, Judge Owada, '[w]ith regard to jurisdiction', expressed 'certain reservations on some aspects of the reasoning of the Judgment'.[8] However, notwithstanding his

International Court of Justice' The Hague Justice Portal <http://www.haguejusticeportal.net/index.php?id=11840> accessed 16 December 2014; Michael Bowman, 'Transcending the Fisheries Paradigm: Towards a Rational Approach to Determining the Future of the International Whaling Commission' (2009) 7 New Zealand Yearbook of International Law 85; Joanna Mossop, 'Australia v Japan: Whaling in the International Court of Justice' (2009) 7 New Zealand Yearbook of International Law 169; Trevor Ryan, 'Sea Shepherd v Greenpeace? Comparing Anti-Whaling Strategies in Japanese Courts' (2009) 7 New Zealand Yearbook of International Law 131; J.J.P. Smith, 'A Double-Edged Harpoon: The Trial of Science in the Antarctic Whaling Case before the International Court of Justice' 28 Ocean Yearbook Online 445.

6 *Whaling case,* paras. 30–41.

7 See *Whaling case,* para. 247(2): 'The Court, [u]nanimously, [f]inds that it has jurisdiction to entertain the Application filed by Australia on 31 May 2010'.

8 *Whaling case,* Diss.Op. Owada, para. 3.

concern that 'the Parties were not provided in the proceedings with ample opportunities to develop their respective arguments on the issue of jurisdiction',[9] he decided not to dissent from the judgment on this aspect. As a result, the jurisdictional issues appear from the judgment to be remarkably straightforward.

What the Court Said

Unsurprisingly, given the unanimous nature of the vote, the Court's treatment of jurisdictional questions in the judgment is brief; but perhaps it is too brief. To be sure, there is some discussion of Japan's reference to paragraph (b) of the reservation to Australia's optional clause declaration exempting from the Court's jurisdiction 'dispute[s] concerning or relating to the delimitation of maritime zones, including the territorial sea, the exclusive economic zone and the continental shelf, or arising out of, concerning, or relating to the exploitation of any disputed area of or adjacent to any such maritime zone pending its delimitation'.[10] This discussion, it is submitted, could be brief as Australia's reservation was indeed intended to cover 'overlapping maritime claims' rather than disputes that materialised within disputed areas.[11]

What the Court Did Not Say

For our present purposes, however, a separate matter is of greater relevance. In its engagement with questions of jurisdiction, the Court accepted that in principle, disputes about the interpretation and application of the Whaling Convention could be litigated on the basis of matching optional clause declarations. Whilst this seems the correct outcome, it is perhaps an outcome which required greater explanation.

Under the system of jurisdiction established under the Court's Statute, the optional clause is designed to enable the Court to deal with an all-encompassing range of disputes: Article 36(2) of the Court's Statute speaks of 'all legal disputes', and the categories of disputes listed in lit. (a)-(d) do not contain any relevant restrictions in this regard.[12] In the terms of Article 36(2), the Australian-Japanese

9 *Ibid*, for *Ibid*. For comment comment see below, Section 3.

10 *Whaling case*, paras. 30–41.

11 See *Whaling case* paras. 34–35, for a brief summary of Australia's argument. The point is explored in CR 2013/11, pp. 41 et seq. (Burmester).

12 For details and background C. Tomuschat, 'Article 36' in A. Zimmermann, C. Tomuschat, K. Oellers-Frahm, C.J. Tams (eds), *The Statute of the International Court of Justice* (OUP 2012, 2nd Edition) 637; Shabtai Rosenne and Yaël Ronen, *The Law and Practice of the International Court, 1920–2005* (4th edn, Martinus Nijhoff 2006), 505 et seq; John G. Merrills, 'The Optional Clause Revisited' 64 British Yearbook of International Law 197; C.H.M. Waldock, 'Decline of the Optional Clause' 32 British Yearbook of International

dispute in the *Whaling case* undoubtedly concerned 'the interpretation of a treaty' in the sense of lit. (a) and of course a 'question of international law' within the meaning of lit. (b).

As the optional clause in principle is all encompassing, reservations have assumed significant importance.[13] While the Court was correct to only address Australia's 'maritime claims reservation' in passing, perhaps there would have been virtue in looking more closely at paragraph (a) of Australia's declaration. Under that provision, Australia exempted from the Court's jurisdiction 'any dispute in regard to which the parties thereto have agreed or shall agree to have recourse to some other method of peaceful settlement'.

As is immediately clear, this type of reservation, which is among the oldest[14] and most popular forms found in declarations,[15] is intended to preclude cases

---

Law 244; Shigeru Oda, 'Reservations in the Declarations of Acceptance of the Optional Clause and the Period of Validity of those Declarations: The Effect of the Shultz Letter' 59 British Yearbook of International Law 1; Malgosia Fitzmaurice, 'Optional Clause, International Court of Justce' Max Planck Encyclopedia of Public International Law <http://opil.ouplaw.com/home/EPIL> accessed 16 December 2014; J. Frowein, 'Reciprocity and Restriction Concerning Different Optional Clauses' in N. Ando, E. McWhinney and R. Wolfrum (eds), *Liber Amicorum Judge Shigaru Oda* (Kluwer 2002); M. Fitzmaurice and M. Vogiatzi, 'Optional Clause Delcarations and the Law of Treaties' in M. Fitzmaurice and O. Elias (eds), *Contemporary Issues in the Law of Treaties* (Eleven 2005).

13 C. Tomuschat, 'Article 36' in A. Zimmermann, C. Tomuschat, K. Oellers-Frahm and C.J. Tams (eds), *The Statute of the International Court of Justice* (OUP 2012, 2nd Edition) 683 at marginal note 83; Shigeru Oda, 'Reservations in the Declarations of Acceptance of the Optional Clause and the Period of Validity of those Declarations: The Effect of the Shultz Letter' 59 British Yearbook of International Law 1; Robert Y. Jennings, 'Recent Cases on "Automatic" Reservations to the Optional Clause' 7 International and Comparative Law Quarterly 349.

14 Historically, the Dutch (PCIJ) optional clause declaration made in 1921 was the first to contain a 'special means reservation'.

15 C. Tomuschat, 'Article 36' in A. Zimmermann, C. Tomuschat, K. Oellers-Frahm and C.J. Tams (eds), *The Statute of the International Court of Justice* (OUP 2012, 2nd Edition) 690 at marginal note 98: 'The most "popular" [reservation] is the clause according to which other mechanisms of dispute settlement as agreed between the parties concerned prevail over the general jurisdiction of the Court'. Reservations similar to that made by Australia have been made by the following States: Austria (1971), Barbados (1980), Belgium (1958), Botswana (1970), Cambodia (1957), Canada (1994), Djibouti (2005), Gambia (1966), Germany (2008), Guinea (1998), Honduras (1960, mód. 1986), Hungary (1992), India (1974), the Ivory Coast (2001), Kenya (1965), Lesotho (2000) Lithuania (2012) Luxembourg (1930), Madagascar (1992), Malawi (1966), Marshall Islands (2013), the Netherlands (1956), New Zealand (1977), Nigeria (1998), Peru (2003), the Philippines (1972), Poland (1996), Portugal (2005), Senegal (1985), Slovakia (2004), Spain (1990), Sudan (1958), Suriname (1987), Swaziland (1969), the United Kingdom (2004); see http://www.icj-cij.org/jurisdiction/?p1=5&p2=1&p3=3.

for which a particular form of dispute settlement has been (or will be) agreed from being brought before the Court on the basis of the optional clause. Put differently, the general mechanism is to be applied in a 'respectful' manner, that is with due respect for the parties' choice of means of dispute resolution – including, in Australia's case, non-binding means.[16] In the words of Yuval Shany, these reservations 'denote certain qualms concerning the suitability of the Court to address all international disputes (e.g. complicated trade issues)' and as such 'the ICJ can be utilized against a reserving state as a last resort only'.[17]

Despite their popularity, 'special means reservations' have not generated much judicial activity: they are common features of optional clause declarations but rarely invoked, interpreted or applied.[18] The Court's silence on the matter in the *Whaling case* is in line with this general trend. Nevertheless, given the parties' history of dispute resolution, it may perhaps come somewhat as a surprise. As is well known, fifteen years prior to the present judgment, Australia (with New Zealand in a slightly different role) had sought a binding decision on the legality of Japan's fishing of southern bluefin tuna.[19] That case – *Southern Bluefin Tuna* – was brought on the basis of Article 287 United Nations Convention on the Law of the Sea (UNCLOS). As is equally well known, the Arbitral Tribunal in that case held that the special dispute settlement mechanism

16 Other reservations differ; they only cover special means resulting in a binding decision.

17 Yuval Shany, The Competing Jurisdiction of International Courts and Tribunals (OUP 1999), 196.

18 C. Tomuschat, 'Article 36' in A. Zimmermann, C. Tomuschat, K. Oellers-Frahm and C.J. Tams (eds), *The Statute of the International Court of Justice* (OUP 2012, 2nd Edition) 690 at marginal note 98.

19 For comment see; Malcolm D. Evans, 'The Southern Bluefin Tuna Dispute: Provisional Thinking on Provisional Measures?' 10 Yearbook of International Environmental Law 7; Malcolm D. Evans and Alan Boyle, 'The Southern Bluefin Tuna Arbitration' 50 International and Comparative Law Quarterly 447; Francisco Orrego Vicuña, 'From the 1893 Bering Sea Fur-Seals Case to the 1999 Southern Bluefin Tuna Cases: A Century of Efforts at Conservation of the Living Resources of the High Seas' 10 Yearbook of International Environmental Law 40; R.R. Churchill, 'ITLOS: The Southern Bluefin Tuna Cases (New Zealand v. Japan; Australia v. Japan): Order For Provisional Measures Of 27 August 1999' 49 International & Comparative Law Quarterly 979; Barbara Kwiatkowska, 'Southern Bluefin Tuna (New Zealand v. Japan; Australia v. Japan), Order on Provisional Measures (ITLOS cases Nos. 3 and 4)' 94 American Journal of International Law 150; Barbara Kwiatkowska, 'The Australia and New Zealand v Japan Southern Bluefin Tuna (jurisdiction and admissibility) award of the first law of the sea convention annex VII arbitral tribunal' 16 The International Journal of Marine and Coastal Law 239; Cesare Romano, 'The Southern Bluefin Tuna Dispute: Hints of a World to Come…Like it or Not' 32 Ocean Development and International Law 313 Natalie Klein, *Dispute Settlement in the UN Convention on the Law of the Sea* (Cambridge University Press 2005).

of the Southern Bluefin Tuna Convention took precedence over binding dispute resolution established under UNCLOS.[20] That decision, controversial as it was, was based on a 'special means' clause that bears at least some resemblance to Australia's special means reservation. The pertinent section of Article 281 UNCLOS permitted recourse to UNCLOS dispute resolution only if any other 'agreement between the parties does not exclude any further procedure'. This is in effect a 'special means exception' of sorts and is one that is not restricted to binding dispute settlement (like Australia's reservation, but unlike Article 282 UNCLOS). In the view of the Arbitral Tribunal, Article 16 of the Convention on the Conservation of Southern Bluefin Tuna (CCSBT), which envisaged a range of dispute settlement modalities not including arbitration, 'exclude[d] any further procedure within the contemplation of article 281(1) [UNCLOS]'.[21]

Of course, there are important differences between the two cases. The *Southern Bluefin case* was based on one compromissory mechanism (namely dispute resolution under Part XV UNCLOS) and assessed that mechanism's relationship to another, more special, compromissory clause (dispute settlement under the CCSBT). In *Southern Bluefin Tuna* the special treaty, the CCSBT, expressly dealt with questions of dispute resolution. And of course the most striking distinction is that in *Southern Bluefin Tuna* the matter was argued at length between the parties.[22]

However, these differences must not be overstated: there are clear links between the two cases. For instance, the impact of the *Southern Bluefin Tuna* litigation on the parties' approach in *Whaling* is plain and, perhaps more importantly, questions of whaling in fact featured in the *Southern Bluefin Tuna case*. In the words of Natalie Klein:

> In arguing in *Southern Bluefin Tuna* that the UNCLOS dispute settlement regime could not prevail over all other dispute settlement clauses in other treaties, Japan specifically referred to the ICRW.[23]

---

20 Award of 4 August 2000, 39 ILM 1359 (2000) (hereinafter '*SBT award*').

21 *SBT award*, para. 59.

22 In the words of the Tribunal, 'The Agents and counsel of Australia, New Zealand and Japan plumbed the depths of these holdings [on the relationship between UNCLOS and the CCSBT] with a profundity that the time pressures of the ITLOS processes [at the interim measures stage] did not permit' (*SBT award*, para. 37).

23 N Klein, 'Whales and Tuna: The Past and Future of Litigation between Australia and Japan' (2009) 22 *Georgetown International Environmental Law Review* 143, text at FN 399. Japan's view on this point is summarised in the *SBT award*, at para. 38(i).

At the oral hearings, counsel for Japan put things rather eloquently: Australia's and New Zealand's construction of the jurisdictional clause would 'drive a coach and horses – or should I say a harpoon – through the system of the Whaling Convention'.[24] In other words, fifteen years ago, Japan seemed to consider whaling disputes to be excluded from binding dispute resolution under UNCLOS and in turn derived from this a more general argument about the relationship between a special and a general mechanism for dispute resolution (in relation to CCSBT and UNCLOS). In the present case, the matter was not argued with the same 'profundity'[25] (in fact, it was not argued at all).[26] However, Japan's 'SBT argument' could very well have been made: if disputes about whaling were not subject to binding dispute resolution under UNCLOS, could they not also be exempted from binding dispute resolution under Article 36(2) of the ICJ Statute? At the time of *Southern Bluefin Tuna* that issue did not arise, as Japan had not yet joined the club of States accepting the optional clause.[27]

Paragraph (a) of Australia's reservation might have provided the requisite link. If Article 16 CCSBT was an 'opt out' from binding dispute resolution, could not the same have been argued for the ICRW's complex and sophisticated regime for resolving disputes about whaling? As Australia in particular noted, the Convention establishes a collective mechanism for regulating questions relating to the interpretation and application of whaling: 'un régime collectif qui définit des normes de comportement, des obligations de résultat et prévoit des institutions qui en assurent la mise en œuvre'.[28]

For years, Australia and Japan had brought their disputes about whaling to the ICW. However, in this case Australia decided to institute ICJ proceedings because its whaling dispute with Japan could not satisfactorily be settled within that 'régime collectif'. Consequently Klein and Stephens are right to suggest – when surveying legal issues that could have potentially arisen in the

24 *Southern Bluefin Tuna*, Oral Hearing, 10 May 2000, at pp. 16–17 transcript (Sir Eilhu Lauterpacht) available at: https://icsid.worldbank.org/ICSID/FrontServlet?requestType=CasesRH&actionVal=OpenPage&PageType=AnnouncementsFrame&FromPage=NewsReleases&pageName=Archive_%20Announcement7.

25 Cf. *SBT award*, para. 37.

26 But this did not absolve the Court: as affirmed on numerous occasions, jurisdiction is to be assessed *ex officio*, see C. Tomuschat, 'Article 36' in A. Zimmermann, C. Tomuschat, K. Oellers-Frahm and C.J. Tams (eds), *The Statute of the International Court of Justice* (OUP 2012, 2nd Edition) 654 at 30; see *Legality of the Use of Force (Serbia and Montenegro v. Belgium) Preliminary Objections*, Judgement, I.C.J. Reports 2004, p. 279, para 36.

27 Declaration of Japan dated 9 July 2007 see http://www.icj-cij.org/jurisdiction/?p1=5&p2=1&p3=3&code=JP.

28 CR 2013/7, p. 52–53, para. 39 (Boisson de Chazournes).

pending litigation prior to the oral hearings – that the 'special means reservations' may have had 'implications for other treaties' (including the ICRW) and could have played a central role in the *Whaling* case.[29]

Implications

In the light of the foregoing, perhaps there would have been room for the Court to analyse whether the ICRW embodied a 'special dispute settlement agreement' as referred to in the reservation to the Australian optional clause. Had the Court addressed the matter, it undoubtedly could have helped clarify an area of considerable uncertainty – namely the relationship between different dispute settlement mechanisms: a problem that not only mattered in the *SBT case*, but that in one way or the other arose in the *Georgian-Russian case* before the ICJ,[30] in debates about WTO disputes litigated before investment tribunals,[31] or most prominently in *MOX Plant*.[32] So what are we to make of the ICJ's decision to remain silent?

The preceding considerations suggest that the Court's silence, by implication, affirms the autonomy of distinct jurisdictional titles and of separate dispute settlement mechanisms. Unlike the Arbitral Tribunal in *Southern Bluefin Tuna* the Court was not easily persuaded that special treaties would render general dispute settlement mechanisms inapplicable. This, it is submitted, is the correct outcome since compulsory jurisdiction (where it exists, whether under matching optional clause declarations or under general frameworks for the binding settlement of disputes, such as Part XV/2 UNCLOS)

29 N. Klein, T. Stephens, 'Whaling in the Antarctic: Protecting Rights in Areas Beyond National Jurisdiction through International Litigation', in Clive H. Schofield, Seokwoo Lee and Moon-Sang Kwon, *The Limits of Maritime Jurisdiction* (Martinus Nijhoff Publishers 2013) 529.

30 *Case Concerning Application of the International Convention on the Elimination of all Forms of Racial Discrimination* (*Georgia v Russian Federation*) (Preliminary Objections) 2011 <http://wwwicj-cijorg/docket/files/140/16398pdf> at para 26 et seq (relationship between the ICJ and the Committee established under the Convention for the Elimination of Racial Discrimination).

31 See *Corn Products International Inc. v. Mexico* (*ICSID Case No. ARB* (*AF*)/*04/1*); *Archer Daniels Midland Company and Tate & Lyle Ingredients Americas Inc. v. Mexico* (*ICSID Case No. ARB* (*AF*)/*04/5*); *Cargill Inc. v. Mexico* (*ICSID Case No. ARB* (*AF*)/*05/2*); (WTO case: *Mexico, Anti-Dumping Investigation of High Fructose Corn Syrup* (*HFCS*) *From the United States*, Panel Report, WT/DS132/R, 28 November 2000, para 5.89 (relationship between the Panel established under the WTO Dispute Settlement Understanding and proceedings under Chapter 19 of the North American Free Trade Agreement)).

32 C-459/03 *Commission v Ireland* (Mox Plant) [2006] ECR I-4635, para 168 et seq; (relationship between UNCLOS and European Community law).

is too important to be given up lightly. Such a significant number of treaties in contemporary international law provide for separate and nuanced dispute settlement regimes that co-existence should be the norm – and consequently special means reservations such as the one contained in paragraph (a) of Australia's optional clause declaration ought to be construed restrictively. The Court's *Whaling* judgment affirms the principle of autonomous bases of jurisdiction,[33] but does so by implication only. To contain the relevant fall-out from the *Southern Bluefin Tuna award* a fuller discussion would have certainly been welcome.

2 Public Interest Standing Accepted as a Matter of Course

The Context

The second of the judgment's silences perhaps speaks more loudly. This silence concerns the absence of a detailed discussion of the Applicant's *locus standi*. This absence is curious as, in many respects, this was an important public interest case in which Australia claimed to be acting in defence of (and was widely credited with having taken up) a cause of public concern.[34] Reflecting on the proceedings in his contribution to the *Simma Festschrift*, James Crawford put the matter very clearly: in his words, Australia 'invokes Japan's responsibility *erga omnes partes* under the Whaling Convention'.[35] As will be shown, this description is accurate as a matter of law, but it probably only reinforced the

33 Yuval Shany, *The Competing Jurisdiction of International Courts and Tribunals* (OUP 1999), 196; For example, Tomuschat's claim that, irrespective of any express provision, optional clause declarations would yield to special arrangements would seem rather controversial: c.f. C. Tomuschat, 'Article 36' in A. Zimmermann, C. Tomuschat, K. Oellers-Frahm and C.J. Tams (eds), *The Statute of the International Court of Justice* (OUP 2012, 2nd Edition) 690 at marginal note 98: 'Such precedence of special arrangements will obtain in any event, irrespective of any reference to them in a declaration under Art. 36, para. 2, depending on the interpretation to be given to such clauses'.

34 Pars *pro toto*, see M Fitzmaurice, 'The International Court of Justice and International Environmental Law', in C.J. Tams, J. Sloan, *The Development of International Law by the International Court of Justice* (OUP 2013) at 373: 'Because of the general and longstanding public interest in the protection of marine mammals, this case is likely to define the public perception of the Court as a protector of environmental concerns'.

35 J. Crawford, 'Responsibilities for Breaches of Communitarian Norms: An Appraisal of Article 48 of the ILC Articles on Responsibility of States for Wrongful Acts', in Fastenrath et al. (eds.), *Festschrift Simma*, at 236.

expectation that questions of standing would be crucial.[36] After all, the *locus standi* of States seeking to enforce public interests through ICJ litigation – and in particular on the basis of the *erga omnes* (*partes*) concept – continues to be been seen as a notoriously difficult matter requiring detailed discussion.[37] The Court itself is at least partly responsible for this state of affairs owing to the fact that its jurisprudence has not been entirely consistent on this issue, generally oscillating between two broadly opposing positions.[38] To elaborate, the Court has at times favoured bilateralism and consensualism in cases such as in the *South West Africa* case – the classic example of the Court's restrictive approach to standing.[39] This case contains the oft-quoted (and just as oft-misunderstood[40]) *dictum* of the Court that although a right to take action in vindication of a public interest:

> ...may be known to certain municipal systems of law, it is not known to international law as it stands at present: nor is the Court able to regard it as imported by the 'general principles of law' referred to in article 38, paragraph 1(c), of its Statute[41]

As a result of the Court's approach to the relevant Mandate provision in this case, the Court created at least 'a presumption against the existence of treaty-based enforcement rights irrespective of individual injury'.[42] Similarly, in *East Timor*, despite recognising the *erga omnes* character of the relevant right, the Court denied Portugal's claim on procedural grounds, stating that any judgment would have concerned the conduct of a third State not present before the

36 See further below.

37 For the author's attempt to 'demystify' the *erga omnes* concept see C.J. Tams, *Enforcing Obligations Erga Omnes in International Law* (CUP, 2005), especially Chapter 4. The subsequent (condensed) discussion draws on that work, as well as on C.J. Tams, 'Individual States as Guardians of Community Interests', in Fastenrath et al. (eds.) *Festschrift Simma*, 379.

38 For comment see e.g. G. Gaja, 'The Protection of General Interests in the International Community' (2013) 364 *Receuil des Cours – Collected Courses at the Hague Academy*; C.J. Tams, *Enforcing Obligations Erga Omnes in International Law* (CUP, 2005); Maurizio Ragazzi, *The Concept of International Obligations Erga Omnes* (OUP, 2000).

39 C.J. Tams, *Enforcing Obligations Erga Omnes in International Law* (CUP, 2005), 63.

40 *Ibid* at 68.

41 ICJ Reports, 1966, 47 (para 88).

42 C.J. Tams, *Enforcing Obligations Erga Omnes in International Law* (CUP, 2005) at 69, see Ian Brownlie, *Principles of Public International Law* (6th Edition, 2003), 452 and Crawford, J. 'Introduction' in *The International Law Commission's Articles on State Responsibility* (Crawford ed., 2002), 24–25.

Court and as such the public interest claims must be denied on jurisdictional ground.[43]

In contrast,[44] at other times the Court has embraced the public interest and recognised the standing of individual States to vindicate it by instituting ICJ proceedings. The Court's *Barcelona Traction* case is considered the *locus classicus*. In this case the Court took a broad approach to standing, stating that 'all States can be held to have a legal interest in…obligations *erga omnes*'.[45] Through its pronouncements in this case the Court seemed to indicate that all States have standing to respond to *erga omnes* breaches by instituting ICJ proceedings.[46]

While not unequivocal, the Court's jurisprudence as a whole (including most recently the *Obligation to Extradite* case)[47] indeed suggests that the *Barcelona Traction dictum* can be taken at face value.[48] A similarly 'liberal' view on public interest enforcement seemed to inform the International Law Commission's (ILC) project on State responsibility (whose Article 48 recognised a broadly formulated right of all beneficiaries of obligations to respond to breaches of obligations 'owed to a group of States…, and…established for the protection of a collective interest of the group');[49] and the recent advisory opinion of the Deep Seabed Chamber of the International Tribunal for the Law of the Sea[50] (ITLOS).

---

43 ICJ Reports, 1995, 90, see paras 8.01–8.17 of Portugal's *replique,* see Tams 166.

44 In fact *Barcelona Traction* has been seen as an attempt to reverse *South West Africa* completely; see further Kamminga, *Inter-State Accountability for Violations of Human Rights* (1992), 153, H. Thirlway, 'The Law and Procedure of the International Court of Justice – Part One', 60 BYIL (1989), 93 and 98, N. Jørgensen, *The Responsibility of States for International Crimes* (2000), 219; J. Dugard 'The South West Africa Cases, Second Phase, 1966', 83 South African Law Journal (1996), 549 and 554, B. Simma, 'From Bilateralism to Community Interest' 250 Recueil des Cours (1994 VI), 295, and the statement of Judge Schwebel in *Nicaragua* that this case had 'decisively displaced' *South West Africa Nicaragua (Provisional Measures)*, ICJ Reports 1984, 190.

45 ICJ Reports, 1970, 32 (para 33).

46 C.J. Tams, *Enforcing Obligations Erga Omnes in International Law* (CUP, 2005), 165.

47 ICJ Reports 2011, 422.

48 See C.J. Tams, *Enforcing Obligations Erga Omnes in International Law* (CUP, 2005), 179–192 (with many references).

49 Article 48(1) of the ILC's Articles on State Responsibility, reproduced in UN Doc A/56/10. As the ILC's commentary makes clear, 'collective interest treaties' are intended 'to foster a common interest, over and above any interests of the States concerned individually' (*ibid.,* at 118: commentary to Article 48).

50 See ITLOS, Deep Seabed Chamber, *Advisory Opinion on Responsibility and Liability for International Seabed Mining* (ITLOS Case No. 17), especially at para. 180.

However, doubts persisted; and so the *Whaling* case seemed to provide the Court with an opportunity to set the record straight. This certainly was the expectation of many commentators. Tim Stephens urged the Court to follow the ITLOS' jurisprudence and to endorse Article 48 of the ILC Articles on State Responsibility.[51] André Nollkaemper's observation seemed to point in the same direction:

> If we accept the protection of whales as a global public good, Australia's claim before the International Court of Justice against Japan may help to produce that good.[52]

Australia's Unusual Approach to Establish Standing

Australia's litigation strategy seemed to reinforce such expectations. In marked contrast to other applicant States seeking to enforce general interests in earlier ICJ proceedings, Australia did not mask its intentions. Its argument on standing – relatively brief though it may have been[53] – exclusively rested on the claim that, as a treaty party, it could enforce ICRW provisions. Responding to a question posed by Judge Bandhari,[54] Australia explicitly renounced any claim to special status:

> L'Australie, comme tous les autres Etats parties à la convention de 1946, a un intérêt commun à ce que l'intégrité du régime découlant de la convention soit maintenue.[55]

This approach is unusual in light of the fact that, in earlier instances of public interest litigation, claimant States had usually opted for a different, 'dualistic', approach that emphasised the right to vindicate general interests but also stressed the claimant State's special position.[56] For example, as applicants in

51 T Stephens, 'Law of the Sea Symposium – Comment', at http://opiniojuris.org/2013/05/27/law-of-the-sea-symposium/: 'it is to be hoped that the Court will refer to and endorse Article 48 of the ASR which provides an important avenue for public interest international environmental claims to be pursued'.

52 A Nollkaemper, 'International Adjudication of Global Public Goods: The Intersection of Substance and Procedure', SHARES Research Paper 9/2012, at 2.

53 CR 2013/18, p. 33, para. 18 et seq.

54 CR 2013/13, p. 73 ('What injury, if any, has Australia suffered as a result of Japan's alleged breaches of the ICRW through JARPA II?').

55 CR 2013/18, p. 33, para. 18 (Boisson de Chazournes).

56 For details on the following see C.J. Tams, *Enforcing Obligations Erga Omnes in International Law* (CUP, 2005), 180 et seq.

the *Nuclear Tests cases,* Australia and New Zealand had emphasised both the special impact of nuclear testing on their territory and the interest of the international community in the banning of such tests.[57] In *East Timor,* Portugal claimed it was defending the *erga omnes* right of self-determination of the people of East Timor, while also relying on its own special status as a former administering power.[58] More recently, in the proceedings against Senegal concerning the *Obligation to Extradite,* Belgium emphasised both its right to defend a general legal interest in seeing the applicable treaty (the Convention against Torture) complied with and the special effects suffered by it as the State who had sought to exercise jurisdiction over Habré and as the State of nationality of three of his victims.[59]

Such a dualistic approach traditionally seemed prudent, and perhaps advisable, as standing to enforce general interests remained controversial, and as the Court remained unwilling to entertain proper public interest claims. For at least four decades the much-maligned *South West Africa* judgment – notably the Court's refusal to admit 'the equivalent of an "*actio popularis*" or right resident in any member of a community to take legal action in vindication of a public interest'[60] – seems to have haunted applicants.

57 See *Nuclear Tests* (Australia v France), ICJ Pleadings, vol 1, 14 and 43; Nuclear Tests (New Zealand v France), ICJ Pleadings, vol 1, 8 and 49; Fitzmaurice, in C.J. Tams, J. Sloan, *The Development of International Law by the International Court of Justice* (OUP 2013) at 358–359.

58 *East Timor,* ICJ Pleadings, paras. 4.12.–4.56. and 8.09.–8.13. of the Portuguese Reply.

59 The 'duality'of Belgium's argument is captured in the ICJ's summary (ICJ Reports 2011, 422, at 448–449, para. 65): "[Belgium] noted in its Application that '[a]s the present jurisdiction of the Belgian courts is based on the complaint filed by a Belgian national of Chadian origin, the Belgian courts intend to exercise passive personal jurisdiction'....In the oral proceedings, Belgium also claimed to be in a 'particular position' since 'it has availed itself of its right under Article 5 to exercise its jurisdiction and to request extradition'. Moreover, Belgium argued that '[u]nder the Convention, every State party, irrespective of the nationality of the victims, is entitled to claim performance of the obligation concerned, and, therefore, can invoke the responsibility resulting from the failure to perform'".

60 ICJ Reports 1966, 47 (para. 88); see further; Heynes, C. and Killander, M. 'South West Africa/Namibia (Advisory Opinion and Judgments)', in Max Planck Encyclopedia of Public International Law (http://opil.ouplaw.com/home/EPIL) John Dugard, *The South West Africa/Namibia Dispute: Documents and Scholarly Writings on the Controversy Between South Africa and the United Nations* (University of California Press 1973); Solomon Slonim, *South West Africa and the United Nations: an International Mandate in Dispute* (Johns Hopkins University Press 1973); Michla Pomerance, 'Case Analysis: The ICJ and

In the *Whaling* litigation, Australia could have equally adopted a dualistic approach to questions of standing. It does, after all, claim jurisdiction over a relevant maritime area in which some of the actual whaling takes place.[61] With regard to the international regime of standing, this could very well have justified an argument emphasising Australia's general legal interest in protecting a common cause alongside its particular interest as a State specially affected by the whaling.[62] However, unlike Portugal in *East Timor* or Belgium in the *Obligation to Extradite case*, Australia claimed to act exclusively in defence of a general legal interest. In the words of its counsel,

> Australia does not claim to be an injured State because of the fact that some of the JARPA II take is from waters over which Australia claims sovereign rights and jurisdiction. ...Every party has the same interest in ensuring compliance by every other party with its obligations under the 1946 Convention. Australia is seeking to uphold its collective interest, an interest it shares with all other parties.[63]

And in fact, this 'collective interest' argument was developed by reference to the Court's (it is submitted: cautious, inconclusive) jurisprudence on *erga omnes* obligations, including that most community-oriented of all pronouncements, the 1951 advisory opinion on *Genocide*:

> En «raison des valeurs qu'ils partagent» et telles qu'exprimées dans la convention de 1946, tous les Etats parties à cette convention ont un intérêt commun à ce que chaque Etat respecte ses obligations en vertu de la

---

South West Africa (Namibia): A Retrospective Legal/Political Assessment' 12 Leiden journal of international law 425.

61 The latter – particularist – approach was at stake in the long-running domestic litigation, before Australian courts, in *Humane Society International v Kyodo Senpaku Kaisha Ltd [2008] FCA 3 (15 January 2008)*.

62 See the discussion in Articles on Responsibility of States for Internationally Wrongful Acts, with commentaries, 2001, reproduced in UN Doc A/56/10, at 118 (commentary to Article 42), see also; Kamen Sachariew, 'State Responsibility for Multilateral Treaty Violations: Identifying the 'Injured State 'and its Legal Status' 35 Netherlands international law review 273; see also Kamen Sachariew, 'State Responsibility for Multilateral Treaty Violations: Identifying the "Injured State" and its Legal Status' 35 Netherlands international Law Review 273; B. Simma, 'Bilateralism and Community Interest in the Law on State Responsibility' in Y. Dinstein and M. Tabori (eds.), International Law in a Time of Perplexity: Essays in Honour of Shabtai Rosenne (1989), 823; D. N. Hutchinson, 'Solidarity and Breaches of Multilateral Treaties' 59 British Yearbook of International Law 151.

63 CR 2013/18, p. 28, para. 19 (Burmester).

convention et du régime en découlant. Pour reprendre les mots de la Cour de céans, « cet intérêt commun implique que les obligations en question s'imposent à tout Etat partie à la [convention de 1946] à l'égard de tous les autres Etats parties. L'ensemble des Etats parties ont un « intérêt juridique » à ce que les droits en cause soient respectés ».

L'Australie fait ainsi valoir son intérêt juridique à ce que les droits découlant de la convention de 1946 soient respectés par le Japon. Pour reprendre encore les termes de la jurisprudence de la Cour, les Etats parties « n'ont pas d'intérêts propres; ils ont seulement tous et chacun, un intérêt commun… » dans le cadre du régime établi dans la convention de 1946.[64]

Japan's and the Court's Reactions

Australia's strategy of not making a particularist case of course had implications. Whilst on the one hand it could be said that such a strategy was somewhat risky, on the other hand it meant Australia did not have to fear having its jurisdiction over maritime spaces scrutinised.[65] In retrospect it seems that the Australian strategy paid off. Japan did not seriously contest Australia's *locus standi* and neither did the Court. Just why Japan opted not to do so is not entirely clear, but it has been suggested that a number of factors played a role in this decision. Such factors included public relations (had Japan raised jurisdictional issues, Australia's memorial would have become available to the public, exposing Australia's criticism of Japan's treatment of whales; which could have reflected badly on Japan in terms of public opinion, 'whilst Japan's arguments would necessarily [have been] limited to the jurisdiction of the Court');[66] and strategic considerations how Japan chose to present its case (including the important issue of the conduct of Japan's whaling programme in 'adjacent'

64 CR 2013/18, pp. 33–34, paras. 19–20 (Boisson de Chazournes) (references omitted). For New Zealand's view see CR 2013/17, p. 33, para. 63 (Finlayson) (ICRW as 'a collective enterprise, in recognition of the shared interest of its parties in the long-term survival of whale stocks').

65 For comment on this point see N. Klein, T. Stephens, 'Whaling in the Antarctic: Protecting Rights in Areas Beyond National Jurisdiction through International Litigation', in: Clive H. Schofield, Seokwoo Lee and Moon-Sang Kwon (eds), *The Limits of Maritime Jurisdiction* (Martinus Nijhoff Publishers 2013) 529.

66 Mr Masafumi Ishii, Director-General, International Legal Affairs Bureau, Ministry of Foreign Affairs, at the Committee on Agriculture, Forestry and Fisheries, House of Councillors, 22 April 2014; translation by Hamamoto, see Sh. Hamamoto 'Procedural Questions in the Whaling Judgment: Admissibility, Intervention and Use of Experts', The Honorable Shigeru Oda Commemorative Lectures, available at http://www.edu.kobe-u.ac.jp/ilaw/en/whaling_docs/2014manuscript_Hamamoto.pdf at 2.

zones with regard to Australia's optional clause declaration.)[67] Given the contested nature of Australia's territorial claims,[68] Japan had argued that its whaling research took place in the appropriate maritime zones, and in areas adjacent to these zones, in an attempt to exclude the case from the Court's jurisdiction on the basis of reciprocity of lit. (b) of Australia's optional clause declaration which excludes:[69]

> any dispute concerning or relating to the delimitation of maritime zones, including the territorial sea, the exclusively economic zone and the continental shelf, or arising out of, concerning, or relating to the exploitation of any disputed area or adjacent to any such maritime zone pending its delimitation.

Consequently, it has been suggested that Japan's decision to focus heavily on the 'adjacency' aspect of the programme could be the reason as to why Japan chose not to challenge Australia's standing.[70]

In addition to political considerations,[71] an undoubtedly important factor was jurisprudential in nature, namely, the timing of the Court's important *Belgium v. Senegal* judgment which was handed down four months after the submission of Japan's Counter-Memorial. In this judgment the Court recognised the validity of claims made regarding obligations *erga omnes partes*. In fact, even before the Court explicitly backed the admissibility of such claims in *Belgium v. Senegal*, the tide had been turning in favour of *erga omnes* invocations of breaches, as evidenced by high profile developments in international law such as Article 48 (1) of the ILC's Articles on State Responsibility.[72] As such, had Japan chosen to contest the admissibility of Australia's claims, it is clear

67 See Counter-Memorial of Japan, 9 March 2012, paras. 1.31–1.33, 1.43–1.45; CR 2013/12, p. par. 18, 22, 32, 33 (Pellet); CR 2013/11, p. 42, para. 10 (Burmester).

68 See Counter-Memorial of Japan, 9 March 2012, paras. 1.31–1.33, 1.43–1.45; CR/2013/12, p.par. 18.

69 For details of such claims and their current status see See Counter-Memorial of Japan, 9 March 2012, paras. 1.31–1.33, 1.43–1.45; CR/2013/12, p.par. 18.

70 See Sh. Hamamoto, 'Procedural Questions in the Whaling Judgment: Admissibility, Intervention and Use of Experts', The Honorable Shigeru Oda Commemorative Lectures, available at http://www.edu.kobe-u.ac.jp/ilaw/en/whaling_docs/2014manuscript_Hamamoto.pdf, at 3.

71 Ibid., at 2.

72 Article 48(1) states that '[a]ny State other than an injured State is entitled to invoke the responsibility of another State in accordance with paragraph 2 if: (a) the obligation breached is owed to a group of States including that State, and is established for the protection of a collective interest of the group…'.

that it would have had a difficult task on its hands, although we can only speculate as to the exact reasons why Japan chose to pursue the strategy it did.

Regardless of the exact motivation behind Japan's decision not to contest the admissibility of Australia's claims, the judgment completely bypasses questions of standing. Having confirmed its jurisdiction, the Court went on to discuss the nature of the Convention, most notably its object and purpose (upon which Australia had relied) – and then assessed Article VIII in depth. Its discussion is focused on the establishment of a breach of the treaty by the Respondent, which almost seems to take place in the abstract: reading the judgment one gets the impression that Australia, as the Applicant, had merely brought a potential breach to the Court's attention. Similarly, when discussing the legal consequence of a breach (once established), the Court at no point attempted to align the remedies ordered with Australia's claims. To elaborate, Australia had sought two remedies from the Court: firstly that the JAPRA II program be declared not to be scientific research for the purposes of the ICRW and consequently a declaration that Japan was in breach of its obligations under that Convention; and secondly an order for the cessation of the internationally wrongful acts committed by Japan.[73] However, in the judgment the Court stated that:

> The Court observes that JARPA II is an ongoing programme. Under these circumstances, measures that go beyond declaratory relief are warranted. The Court therefore will order that Japan shall revoke any extant authorization, permit or licence to kill, take or treat whales in relation to JARPA II, and refrain from granting any further permits under Article VIII, paragraph 1, of the Convention, in pursuance of that programme.[74]

That the Court simply brushed aside the specific remedies sought by Australia in favour of more far-reaching remedies which, in the Court's opinion, were required is significant. Furthermore, its lack of justification for doing so is a significant gap in an otherwise dense and detailed judgment.

Implications

The reasons for the Court's silence on questions of standing are difficult to identify. Without inside knowledge of the Court's deliberative process, two readings seem plausible. Firstly, the silence could be read as representing an unwillingness on the part of the Court to expressly recognise a right of treaty parties to

73 See Memorial of Australia, at 275.

74 *Whaling case*, para. 245.

raise treaty violations.[75] Whilst this is a simplistic explanation (and one not based on any particular evidence), it is not entirely implausible. In other words, it might be that the judgment was already 'sufficiently progressive' and that an express endorsement of some form of standing in the public interest would have affected the majority carrying the decision. Japan's decision not to contest the claimant's *locus standi*, in contrast to its position on jurisdiction that the Court has to establish *proprio motu*, could explain such an argumentative gap.

However this explanation (which would limit the relevance of the *Whaling* judgment for the pursuit of public interests through ICJ litigation) is not the only plausible one. It could equally be that the Court, after decades of equivocation, has now embraced the idea of public interest standing, at least on the basis of multilateral treaties protecting collective interest.[76] If so, then perhaps the *Whaling case* could be seen as a sequel to the 2011 judgment in *Obligation to Extradite*. In that case, too, the Court had surprised commentators by going out of its way to accept the applicant's standing to enforce a multilateral treaty protecting collective interests.

> The States parties to the Convention have a common interest to ensure, in view of their shared values, that acts of torture are prevented and that, if they occur, their authors do not enjoy impunity. The obligations of a State party to conduct a preliminary inquiry into the facts and to submit the case to its competent authorities for prosecution are triggered by the presence of the alleged offender in its territory, regardless of the nationality of the offender or the victims, or of the place where the alleged offences occurred. All the other States parties have a common interest in compliance with these obligations by the State in whose territory the alleged offender is present. That common interest implies that the obligations in question are owed by any State party to all the other States parties to the Convention. All the States parties "have a legal interest" in the protection of the rights involved (Barcelona Traction, Light and Power Company, Limited, Judgment, I.C.J. Reports 1970, p. 32, para. 33). These obligations may be defined as "obligations *erga omnes partes*" in the sense that each State party has an interest in compliance with them in any given case.[77]

Seen in that light, *Belgium v Senegal* and *Whaling*, taken together, could perhaps reflect a new willingness on the part of the Court to handle disputes transcending

75 See for instance; P. Urs, 'Are States Injured by Whaling in the Antarctic?' available at http://opiniojuris.org/2014/08/14/guest-post-states-injured-whaling-antarctic/.

76 *ibid.*

77 ICJ Reports 2012, 422 (para. 68).

reciprocal inter-State relations.[78] This, it is submitted, would be a healthy development, since too much of international law is now regulated in multilateral treaties or customary rules protecting collective interest. Article 48 of the ILC's Articles on State Responsibility showed the ILC willing to come to terms with this new complexity.[79] The Advisory Opinion of the Deep Seabed Chamber showed ITLOS willing to follow suit.[80] And of course, so many multilateral treaties – from the Geneva Conventions to human rights treaties – expressly recognise the right of each and every State party to respond to treaty breaches by other parties that, indeed, the idea of an *actio popularis* 'ne revêt…plus un caractère exceptionnel'.[81] The Court – whose jurisprudence on standing to enforce public interests has lacked direction for too long – should not hesitate to follow so to avoid falling behind.

Which of the two readings is correct is a matter for speculation. However, regardless of which is correct, the fact remains that for the first time in the ICJ's history, an avowed public interest claim was admitted as a matter of course. This is surely a significant result and makes the Court's silence on this issue all the more surprising.

3 One Round of Written Proceedings – Towards More Robust Case-Management?

Background

A third and final silence calls for brief comment: the Court's reasoning leading to the decision to proceed to the oral hearings after only one round of written

78 Leonardo da Vinci is credited with the expression 'nothing strengthens authority so much as silence'; perhaps this aptly describes the second interpretation advanced in the text.

79 See the discussion in Articles on Responsibility of States for Internationally Wrongful Acts, with commentaries, 2001, reproduced in UN Doc A/56/10, at 118 (commentary to Article 48); J. Crawford, 'Responsibilities for Breaches of Communitarian Norms: An Appraisal of Article 48 of the ILC Articles on Responsibility of States for Wrongful Acts', in Fastenrath et al. (eds.), *Festschrift Simma*, at 236.

80 See ITLOS, Deep Seabed Chamber, *Advisory Opinion on Responsibility and Liability for International Seabed Mining* (ITLOS Case No. 17), especially at para. 180; A. Donald, R. Mackgill, C. Payne, 'Advisory Opinion on Responsibility and Liability for International Seabed Mining (ITLOS Case No. 17): International Environmental Law in the Seabed Disputes Chamber' 41 Environmental Policy and Law (2011) 2, 60; T. Poisel, 'Deep Seabed Mining: Implication of Seabed Disputes Chamber's Advisory Opinion', 19 Australian International Law Journal 213.

81 F Voeffray, *L'actio popularis—ou la défense de l'intérêt collectif devant les juridictions internationales* (Presses Universitaires de France, 2004), 383; C.J. Tams, 'Individual States as Guardians of Community Interests', in Fastenrath et al. (eds.) *Festschrift Simma*, 383–384.

proceedings. This decision is recorded in paragraph 6 of the judgment, but in the most succinct terms:

> At [a] meeting [with the ICJ's President], the Agent of Australia stated that his Government did not consider it necessary to organize a second round of written pleadings; the Agent of Japan, for his part, requested a second round of written pleadings. The Court, having regard to Article 45, paragraph 2, of the Rules of Court, decided that a second round of written pleadings was not necessary.

It is submitted that what may seem to be a mere technicality in fact had rather interesting consequences for the presentation of the parties' claims. For our present purposes, it suffices to mention two institutional aspects. First, with only one round of written argument, the proceedings could be completed (by ICJ standards, that is) relatively quickly. Notwithstanding the complexity of the case, the Court's deliberations began just over three years after Australia had instituted proceedings,[82] which is faster than the running time of many other cases decided by the ICJ in recent years, especially those reaching the merits.[83] Second, as the written pleadings produced by the parties were not quite as overwhelming as in other cases, and notably in Japan's case did not cover the material in its entirety,[84] the oral hearings in the *Whaling case* became far more relevant than in other proceedings. In particular, counsel could not simply condense information already presented during the written phase.[85] All of this was perhaps not quite as insignificant as the terse language

82 Australia lodged its application instituting proceedings on 31 May 2010 whilst the Court began to deliberate on 16 July 2013.

83 By way of illustration, the *Maritime Dispute* (Peru *v.* Chile) took six years from application to judgment (2008–2014), the *Territorial and Maritime Dispute* (Nicaragua *v.* Colombia) eleven (2001–2012), the *Diallo case* fourteen (1998–2012); in *Aerial Herbicide Spraying*, five-and-a-half years passed between application and settlement. (For details see http://www.icj-cij.org/docket/index.php?p1=3&p2=3&sort=2&p3=0.) Of the recently decided cases, *Jurisdictional Immunities, Application of the Interim Accord of 13 September 1995* and *Obligation to Prosecute or Extradite* were dealt with in a similarly swift manner, but may not have been as fact-heavy.

84 See *Whaling Case,* Pleadings, Oral Arguments, CR 2013/12, 2 July 2013, para 26.

85 S. Talmon, 'Article 43' in A. Zimmermann and others (eds), *The Statute of the International Court of Justice* (OUP 2012, 2nd Edition) at 96. See Article 60(1) of the Rules of Court which states 'The oral statements made on behalf of each party shall be as succinct as possible, within the limits of what is requisite for the adequate presentation of that party's contentions at the hearing. Accordingly, they shall be directed to the issues that still divide the

of paragraph 6 of the judgment suggests. In fact, it is telling that in his dissent, Judge Owada referred to the 'somewhat unfortunate procedural circumstances, [under which] the Parties were not provided in the proceedings with ample opportunities to develop their respective arguments on the issue of jurisdiction'.[86] Similarly, Judge Greenwood, having noted that 'the Judgment says almost nothing about [the decision to have one round of written proceedings only]', felt the need to explain the Court's approach in more detail.[87]

The Normative Context

In terms of the applicable procedural framework, the Court was no doubt competent to dispense with a second round of written proceedings. Article 45 of the Rules of Court mentions the mandatory written pleadings (memorial, counter-memorial) in its first paragraph and then notes in paragraph 2:

> The Court may authorize or direct that there shall be a Reply by the applicant and a Rejoinder by the respondent if the parties are so agreed, or if the Court decides, *proprio motu* or at the request of one of the parties, that these pleadings are necessary.

As Talmon has observed, there 'is no right of the parties to a further pleading',[88] and further, '[i]f the Court considers itself to be "sufficiently informed…of the contentions of fact and law on which the Parties rely" it will not authorize the filing of further pleadings'.[89] While the letter of the law was complied with, the Court's approach is interesting in that it seems at odds with prior practice. Notwithstanding the language of Article 45 of the Rules, historically, two rounds of written pleadings seem to have been the norm in merits proceedings before the ICJ[90] (at least in those cases brought by unilateral application.)[91] In fact, it is only over the course of the last decade and a half, which saw a relevant

arties, and shall not go over the whole ground covered by their pleadings, or merely repeat the facts and arguments these contain'.

86 Diss. Op. Owada, para. 3.

87 Sep. Op. Greenwood, paras. 32–38.

88 S. Talmon, 'Article 43' in A. Zimmermann and others (eds), *The Statute of the International Court of Justice* (OUP 2012, 2nd Edition), 1106, at marginal note 43.

89 *Fisheries Jurisdiction case* (Spain/Canada), ICJ Reports (1996), pp. 58, 59 and ICJ Reports (1998), pp. 432, 436 (para. 6).

90 In proceedings addressing preliminary objections, one round of written pleadings has been the norm.

91 Special considerations obtain in special agreement cases, where the parties are likely to have addressed the matter in the *compromis*.

increase in the Court's caseload and perhaps the emergence of a 'market' for inter-State litigation, that this practice has come under strain. Successive ICJ Presidents, as well as commentators, have emphasised the need for the Court to modernise its handling of cases, which often meant more robust case-management. Yet to date the Court has treaded softly in this regard. A series of Practice Directions have been adopted, one of which – Practice Direction III – explicitly seeks to curtail (with modest success so far, it is submitted),[92] the trend towards ever-greater detail in written pleadings and documentary annexes (which are particularly problematic).[93] With regard to rounds of pleadings – potentially an effective time-saving device – the Court boldly announced in a press release in 2002 that:

> a single round of written pleadings is to be considered the norm in cases begun by means of an application. A second round of written pleadings will be directed or authorised only where this is necessary in the circumstances of the case.[94]

However, this statement has so far not been translated into practice. In fact, between the bold language of the press release and the *Whaling* judgment, to the best of the writer's knowledge, all cases bar three had two rounds of written pleadings on the merits; and the exceptions (*Avena, Mutual Assistance, Obligation to Prosecute*) were based on an agreement between the parties.[95] More recently, in *Aerial Herbicide Spraying,* Ecuador, as the applicant, had asked for a second round of pleadings, while Colombia, as the Respondent,

92 S. Talmon, 'Article 43' in A. Zimmermann and others (eds), *The Statute of the International Court of Justice* (2 edn, OUP 2012), 1095 at marginal note 14; Anna Riddell and Brendan Plant, *Evidence before the International Court of Justice* (British Institute of International and Comparative Law 2009) 27; Arthur Watts, 'The ICJ's Practice Directions of 30 July 2004' 3 The Law and Practice of International Courts and Tribunals 385; Arthur Watts, 'New Practice Directions of the International Court of Justice' 1 The Law and Practice of International Courts and Tribunals 247.

93 Practice Direction III provides as follows: 'The parties are strongly urged to keep the written pleadings as concise as possible, in a manner compatible with the full presentation of their positions. In view of an excessive tendency towards the proliferation and protraction of annexes to written pleadings, the parties are also urged to append to their pleadings only strictly selected documents'.

94 ICJ Press Release No. 2002/12 of 4 April 2002, Measure No. 1.

95 See ICJ Reports 2004, 6, 12 (para. 5); ICJ Reports 2008, 177, 182 (para. 9); ICJ Reports 2011, 422, 427 (para. 8).

considered this to be unnecessary. The Court opted to 'play it safe' and went along with Ecuador's request[96] (only to see the case settled three years later).[97]

Implications

Viewed against this background, the Court's decision to dispense with a second round of written pleading in the *Whaling case* was certainly unusual. Whether it marks the beginning of a new approach – one that sees the Court enforce Practice Direction III – will have to be tested in practice. In the writer's view, it would be a useful change if the Court began to treat (as it announced in 2002) 'a single round of written pleadings… [as] the norm in cases begun by means of an application'.

Such a change would need to be communicated to the Court's clients, however; and it would have to be implemented in a principled manner. In this respect, perhaps one might have hoped for at least some explanation in the judgment. Fortunately, Judge Greenwood's separate opinion provides some elucidation; and while we do not know whether his approach was shared by the Court, it identifies useful and workable parameters.[98] In Judge Greenwood's view, three considerations are relevant: *First*, and uncontroversially, the Court can always order a second round of pleadings if it feels this is necessary. *Second*, as there is no automaticity, the 'first round pleadings' – Memorial and Counter-Memorial – must set out the parties' case in its entirety. This is probably the main lesson of the *Whaling case:* in future, litigants will consider it risky to leave matters to one side for a future second round of written pleadings. The ICJ's approach suggests that there will not always be a second chance to present them in writing. *Third*, and more controversially according to Judge Greenwood, where the parties are divided on the desirability of a second round, greater weight should attach to the Applicant's view. This on the face of it would seem to interfere with the principle of equality; or it may simply be a way of rationalising why a second round was authorised in *Aerial Spraying*[99] but not in *Whaling*.

However, the argument offered in support would seem plausible: after the presentation of the Counter-Memorial, the parties have exchanged views, and it is for the Applicant to react to the response. In the words of Judge Greenwood, the Counter-Memorial:

---

96 ICJ Reports 2010, 307, 308.

97 See ICJ Press Release No. 2013/20, 17 September 2013, *Aerial Herbicide Spraying* (*Ecuador v. Colombia*), available at: http://www.icj-cij.org/docket/files/138/17526.pdf.

98 Sep.Op. Greenwood, paras. 32–38.

99 Cf. ICJ Reports 2008, 177, 182 (para. 9).

> will usually be the first indication which the Applicant receives of the Respondent's case. It may raise matters which the Applicant has not considered, or evidence which the Applicant needs the opportunity to refute. There is, therefore, a strong case for ordering a second round of written pleadings when the Applicant so requests; not to do so may occasion serious injustice if the Applicant is denied the opportunity to respond to evidence or argument raised by the Respondent in the Counter-Memorial.[100]

All this sounds as if it might be the beginning of a conversation – a conversation about the balance to be struck between the competing interests of efficiency and completeness in the administration of justice. In the writer's view, it is an overdue discussion and one in which the Court's handling of the *Whaling case* is an interesting opening move. For the benefit of future litigants, it may be hoped that the Court will soon move beyond the silence of the *Whaling* judgment and provide express guidance on its understanding of Article 45(2) of the Rules.

Concluding Thoughts

Speculating about the meaning of silences is a curious business: by definition there is so little to go on. However, perhaps this paper has succeeded in establishing that in not only boldly (or rashly) tackling some legal issues, but also in side-stepping others, the Court's *Whaling* judgment sends important messages. With respect to the three 'non-issues' addressed in the foregoing sections, the main impression is that of a Court firmly proceeding along its chosen course, without too much concern for obstacles along the way. This is evidenced by the 'non-fussed' approach to jurisdiction – in which the Court seemed to try to avoid inventing problems where the parties saw none – and, much more surprisingly, to questions of *locus standi*. This can also be seen in the (relatively novel, for the Court) decision to give short shrift to Japan's request for a second round of written pleadings.

The mid and long-term implications of these side steps are difficult to predict due to the fact the Court chose not to explain them. The preceding discussion highlights potential reasons for, and consequences of, these silences. It also brings out that where the Court wishes to send a clear signal (as perhaps with the decisions on written pleadings and on *locus standi*), silences are not

100 Sep.Op Greenwood, para. 36.

necessarily the best way to proceed. In individual cases, it may be useful to side step some of the issues in the interest of judicial economy or so as to not endanger the consensus carrying a majority decision. However, if the Court wishes to shape the conduct of parties on crucial questions, 'silence may be golden when you cannot think of a good answer' (as Muhammad Ali famously observed), yet on jurisdictional and procedural questions of the sort addressed in the preceding sections, the Court's clients ought to be able to expect answers, not silence.

PART 4

Institutional Implications of the Judgment

CHAPTER 7

The *Whaling* Judgment and the Challenges of Dynamic Treaty Regimes

Mika Hayashi

1 Introduction: Treaty Regimes as Challenges

Treaty regimes based on multilateral treaties[1] frequently present a challenge to the traditional techniques of treaty interpretation. These treaty regimes are dynamic, in the sense that they do not remain the same as the treaties originally provided for. Frequently, looking at the terms of the treaties which were set possibly decades before does not reveal the real and current position of the treaty regimes. Their dynamism is greatly increased by the presence of the supervisory organs they themselves establish. These supervisory organs, typically, express their own views, sometimes with, and at other times without, consensual support of the States Parties to the treaty regimes. The great majority of instruments adopted by such supervisory organs are exhortatory in nature, i.e., they are recommendations, and not binding.[2] Nevertheless, they very much affect the behaviour of States Parties to such treaty regimes. Thus, in order to comprehend treaty regimes, it is necessary to take account of the operation of these regimes, including what their supervisory organs say and do.

The challenge these dynamic treaty regimes pose lies in the plain fact that their dynamic aspects are hard to grasp with the tools available under the law of treaties. Admittedly, this is a technical challenge for lawyers. Nevertheless, the challenge is an important one, for those who are committed to the regime are surely interested in knowing how these instruments affect their obligations

1 'Treaty regime' is not a term used in the Vienna Convention on the Law of Treaties. It is used in this chapter as a term to describe '...[t]reaties which establish a collaborative mechanism for States to regulate a particular area of activity', with a focus on a number of characteristics they share: 'Such treaties normally include purposes and principles, but their operational effect is achieved through decisions of organs of members (...). The treaty arrangements are often referred to as régimes, thereby stressing their institutional character and their continuing regulatory function'. Oscar Schachter, *International Law in Theory and Practice* (Martinus Nijhoff, 1991), 75.

2 See a convenient summary of such supervisory organs in various treaties in Jutta Brunnée, "Compliance Control", in Geir Ulfstein (ed.), *Making Treaties Work* (Cambridge University Press, 2007), 373.

 | DOI 10.1163/9789004313828_009

in the treaty regime. This challenge involves the determination of the place of the instruments adopted by the supervisory organs within the law of treaties. In particular, how such instruments should affect the interpretation of the treaty is far from clear. Taking account of the instruments adopted by the supervisory organs is not an orthodox method of interpretation, unless the instruments in question form part of an element of interpretation enshrined in the Vienna Convention on the Law of Treaties (VCLT). Certain instruments may qualify as subsequent practice agreed upon by the parties, that affects the treaty interpretation (Art. 31(3)), but not all of them. In fact, instruments adopted by supervisory organs very often do *not* qualify as such subsequent practice. It is in these cases that the theoretical challenge becomes apparent. On one hand, discarding recommendations and ignoring them can lead to a presentation of the treaty regime in question without any hint of reality. On the other hand, treating such instruments adopted by supervisory organs as if they were part of the treaty, and therefore binding, also distorts the picture, because in form, clearly, they are not.

Treaty regimes may also operate on tacit assumptions. It is not realistic to expect that every single rule can be written down in any legal order. Certain things are always assumed, deduced, or taken for granted. In the context of dynamic treaty regimes, for example, once a state signs and ratifies a treaty that establishes, and largely operates through, its supervisory organs, it may be assumed that the state has agreed to co-operate with the supervisory organs. A tacit assumption is, however, difficult to analyse in the light of the law of treaties. How should it be taken into account in discussing any treaty interpretation? A similar difficulty may arise whenever the analysis has to involve certain practices of parties to the regimes that do not seem to stem from any particular provision of the treaty. Again, if such practices form a subsequent practice as an element of treaty interpretation, it could be analysed in the light of the law of treaties. However, if the practices in question do not seem to relate to any particular words or terms of the treaty provisions, the challenge is the same again: how should one take it into account in discussing the treaty in question?

The *Whaling* judgment by the International Court of Justice (ICJ)[3] is assessed in the light of these challenges. The treaty regime dealt with in the judgment is the regime established by the International Convention for the Regulation of Whaling (ICRW),[4] and its main supervisory organ is the International Whaling Commission (IWC).

3 Whaling in the Antarctic (Australia v. Japan: New Zealand Intervening), Judgment (Mar. 31, 2014) (hereinafter the Judgment).

4 International Convention for the Regulation of Whaling 161 UNTS 72. Judge Hisashi Owada uses the expression 'self-contained regulatory regime' for the Convention. Dissenting Opinion of Judge Owada, para. 19.

The analysis of the judgment in the light of the law of treaties is thoroughly carried out elsewhere in this book.[5] Thus, in this chapter, a different frame of reference is employed to highlight the challenges described above. It is a framework which measures the softness of a rule in three 'dimensions', as explained below. The framework is useful in highlighting the challenges referred to above. In the field of international relations scholarship, a frequently used concept of soft law shows three dimensions of the rule in question: obligation, precision and delegation.[6] The first dimension regarding the degree of obligation refers to the binding nature of the rule in question.[7] The second dimension regarding the degree of precision refers to the clarity of the content of the rule, i.e., how clear or vague the rule in question is.[8] The third dimension of delegation refers to a delegation of authority concerning the interpretation of the rules to a third party.[9] The three dimensions, and their combinations, matter if one is interested in assessing any power that a rule may have on the behaviour of a State. If the degree of all these three ingredients is high, the rule in question is, indeed, very hard law. As the adverb 'very' indicates, the assessment of the softness or hardness of a given rule is not a binary one. A rule can also be a hard rule in one dimension, yet soft in other dimensions.

This chapter analyses how the *Whaling* judgment tackled the challenges of dynamic treaty regimes. As mentioned previously, the traditional techniques of the law of treaties do not readily offer tools to examine and take into account

5 See the discussion in Chapter 3 by Malgosia Fitzmaurice.

6 Kenneth W. Abbott and Duncan Snidal, "Hard and Soft Law in International Governance", *International Organization* 54 (2000): 421–456, 421–422.

7 International law scholarship tends to focus on this dimension as a threshold between hard law and soft law. *E.g.*, Dinah Shelton, "International Law and 'Relative Normativity'" in M. Evans (ed.), *International Law* (Oxford University Press, 4th ed., 2014), 137, 159; Alan E. Boyle and C.M. Chinkin, *The Making of International Law: Foundations of Public International Law* (Oxford University Press, 2007), 212–213.

8 The second dimension described as precision in international relations scholarship is discussed as an aspect of soft law by a number of international lawyers. Shelton (note 7) 159–160. The names used by international lawyers vary. d'Asprement uses the term '*negotium*' as opposed to '*instrumentum*' to discuss the dimension of precision in Jean d'Asprement, "Softness in International Law: A Self-Serving Quest for New Legal Materials", *European Journal of International Law* 19 (2008): 1075–1093, 1082–1087. See also László Blutman, "In the Trap of a Legal Metaphor: International Soft Law", *International and Comparative Law Quarterly* 59 (2010): 605–624.

9 Many bilateral and multilateral treaties have a clause that effectively stipulates the delegation regarding the interpretation of the rules to the International Court of Justice.

the dynamism of treaty regimes. The *Whaling* judgement, in its effort to take account of the operation of this treaty regime and its dynamism, seems to transform certain soft-law aspects of the treaty regime into hard law. The analysis, therefore, focuses on how this transformation occurs, and its implications.

2 The ICRW on the Soft-Law/Hard-Law Scale

In this section, the ICRW is reviewed in relation to the scale of softness/hardness. It will also serve as a convenient recapitulation of the main issues dealt with in the *Whaling* judgment. From the perspective of this chapter, these issues could be understood as issues arising from the softness of certain provisions and instruments in the ICRW.

2.1 *Degree of Obligation in the ICRW*

(i) Hard-Law Aspect

The binding nature of the ICRW is apparent and undisputed. First, its status as a formal treaty automatically means it is legally binding, constituting a hard obligation. Second, while the ICRW sets out further obligations in the form of the Schedule, the Schedule is explicitly treated as an integral part of the Convention (Art. I(1)). Therefore, the Schedule also undisputedly constitutes hard law.

In the context of the *Whaling* judgment, the confirmation of the nature of the Schedule is important because the moratorium on commercial whaling, adopted in 1982, takes the form of a provision of the Schedule.[10] The moratorium stipulates that: 'Notwithstanding the other provisions of paragraph 10, catch limits for the killing for commercial purposes of whales *from all stocks* for the 1986 coastal and the 1985/86 pelagic seasons and thereafter *shall be zero*'.[11]

10 There is a fair amount of literature on this moratorium and its impact. See, e.g., Shirley V. Scott, "Intergovernmental Organizations as Disseminators, Legitimators, and Disguisers of Hegemonic Policy Preferences: The United States, the International Whaling Commission, and the Introduction of a Moratorium on Commercial Whaling", *Leiden Journal of International Law* 21 (2008): 581–600.

11 Schedule 10(e), emphasis added. While commitments in the form of the Schedule are legally binding, that binding force is made conditional on the consent of each State Party; it is those who do not object, or if they do, those who withdraw their initial objections, that are bound by the obligations stipulated in the Schedule (Art. 5(3)) including the moratorium on commercial whaling. Japan, the respondent in the *Whaling* judgment, had in fact withdrawn its initial objection to this moratorium in 1986, long before the

(ii) Soft-Law Aspect: Flexibility

A preliminary diagnosis of the ICRW in terms of its binding nature, including the moratorium on commercial whaling, indicates that it is, indeed, hard law. There are, however, two aspects of the Convention that raise a kind of *caveat*. One is a series of instruments in the Convention that ensure flexibility to its hard obligations. The other is the non-binding instruments that are used by the supervisory organs created by the treaty.

Generally speaking, while clearly establishing binding obligations, and therefore being hard law, treaties can be flexible instruments. Flexibility in a treaty means that the obligations are either not binding in the same way upon all States Parties, or not set as permanent obligations. A well-known and accepted technique to make treaties flexible is to allow reservations to a treaty (VCLT Art. 2(d)). Another accepted technique is the inclusion of the possibility of unilateral denunciations (VCLT Art. 56) or withdrawals. Making temporary obligations that expire or that have to be reviewed after a certain amount of time is also a way to introduce flexibility into a treaty. Apart from the use of such techniques, many treaties also ensure flexibility by setting out explicit exemptions or exceptions to their principal obligations.[12]

The first series of observations in this light is that the ICRW is a remarkably flexible treaty. The Convention contains all the techniques of flexibility mentioned above, to make its obligations flexible. Moreover, these techniques embedded in the Convention have not remained just simply terms of the treaty, but have been resorted to by the States Parties. First, according to the ICRW, any Contracting Government may withdraw from the Convention by giving notice to that effect to the depository government (Art. XI). A well-known case is the withdrawal of Iceland in 1992, but there are other precedents, too.[13] Second, reservations to the ICRW are not prohibited explicitly, and the technique of reservations has also been resorted to in an attempt to attenuate the impact of the obligations under the ICRW.[14] Third, the moratorium on

litigation in the ICJ. See the description in Maria Clara Maffei, "The Protection of Whales in Antarctica" in Francesco Francioni and Tullio Scovazzi (eds.), *International Law for Antarctica* (Kluwer Law International, 1996), 171–224, 179–181.

12 Laurence R. Helfer, "Flexibility in International Agreements" in Jeffrey L. Dunoff and Mark A. Pollack (eds.), *Interdisciplinary Perspectives on International Law and International Relations: The State of the Art* (Cambridge University Press, 2013), 186–189.

13 See the examples described in Malgosia Fitzmaurice, "The International Convention for the Regulation of Whaling and International Whaling Commission – Conservation or Preservation – Can the Gordian Knot Be Cut (or Untangled)?", *Yearbook of Polar Law* 5 (2013): 451–490. Regarding Iceland's withdrawal, see 478–479.

14 For details, see Alexander Gillespie, "Iceland's Reservation at the International Whaling Commission", *European Journal of International Law* 14 (2003): 977–998; Shigeki Sakamoto,

commercial whaling, an obligation that was at the heart of the whaling litigation, was also a flexible obligation for it was set up as a temporary obligation:[15] 'by 1990 at the latest the Commission will undertake a comprehensive assessment of the effects of this decision on whale stocks and consider modification of this provision and the establishment of other catch limits'.[16] Fourth, there is another flexibility arrangement in the ICRW. It is an exemption clause, Article VIII, that places the so-called scientific whaling outside the operation of the Convention.[17] As a consequence, the article allows exceptions to the moratorium on commercial whaling: 'Notwithstanding anything contained in this Convention any Contracting Government may grant to any of its nationals a special permit authorizing that national to kill, take and treat whales for purposes of scientific research subject to such restrictions as to number and subject to such other conditions as the Contracting Government thinks fit, and the killing, taking, and treating of whales in accordance with the provisions of this Article shall be exempt from the operation of this Convention'. In the context of the *Whaling* judgment, this is the most important soft-law aspect in the Convention, for the provision effectively creates an exemption to the moratorium on commercial whaling referred to previously.[18]

(iii) Soft-Law Aspect: Recommendations

The second series of observations regarding the rules in the ICRW is that its supervisory organ established under Article III, the International Whaling Commission (IWC), uses non-binding instruments in overseeing the implementation of the Convention. It may 'make recommendations to any or all Contracting Governments on any matters which relate to whales or whaling and to the objectives and purposes of this Convention' (Art. VI). The term

---

"Determining the Admissibility of Reservations in International Organisations: The Case of the IWC" in Haruyuki Yamate and Shigeru Kozai (eds.), *Legal Structure of the International Community: Its History and Current State* [*Kokusai Shakai no Hōkōzō: Sono Rekishi to Genjō*], Vol. I (Tōshindō, 2003, in Japanese), 153–188.

15 This appears to be the average understanding of the moratorium. See Steinar Andresen, "The International Whaling Commission (IWC): More Failure Than Success?" in Edward L. Miles (ed.), *Environmental Regime Effectiveness: Confronting Theory with Evidence* (MIT Press, 2002), 379–403, 387; Ed Couzens, *Whales and Elephants in International Conservation Law and Politics: A Comparative Study* (Routledge, 2014), 42.

16 This review has not taken place as planned in the moratorium, and is one of the objects of the grievances of Japan.

17 Fitzmaurice (note 13), 466–467.

18 In fact, prior to and during the litigation in the ICJ, Japan had consistently presented its whaling programs as a form of scientific whaling falling under this clause.

'recommendations' obviously means these instruments are not binding, i.e., soft. Thus, soft-law instruments are used to induce compliance with the hard obligations of the Convention.

In the context of the *Whaling* judgment, these soft-law instruments formed an important backdrop to the litigation. Prior to the litigation, the recommendations had been used by the IWC in its attempt to influence the whaling activities of Japan. The Japanese Whale Research Program under Special Permit in the Antarctic (JARPA II), the whaling program which was the object of the *Whaling* case, was first submitted to, and discussed in, the IWC in 2005. It was a programme that involved lethal methods of whale hunt. The IWC recommended 'the Government of Japan to withdraw its JARPA II proposal or to revise it so that any information needed to meet the stated objectives of the proposal is obtained using non-lethal means'.[19] Japan did not comply with this recommendation, and began its activities under JARPA II later in the same year. In a 2007 resolution, the IWC again called upon Japan 'to suspend indefinitely the lethal aspects of JARPA II conducted within the Southern Ocean Whale Sanctuary'.[20] Japan did not heed this recommendation either. In the same resolution, the IWC 'recall[ed] that the Commission has repeatedly requested Contracting Parties to refrain from issuing special permits for research involving the killing of whales within the Southern Ocean Sanctuary, has expressed deep concern at continuing lethal research within the Southern Ocean Sanctuary, and has also recommended that scientific research involving the killing of cetaceans should only be permitted where critically important research needs are addressed'.[21]

2.2 *Degree of Precision in the ICRW*

As discussed in the Introduction section, the second dimension in measuring the degree of softness of rules is the precision of the rules in question. When 'rules unambiguously define the conduct they require, authorize, or proscribe', these rules are deemed precise,[22] and, therefore, hard law. It is more difficult to get away with deviant behaviour where the rules are hard in this sense. According to this criteria, again, a relative precision, and therefore hardness, of

19 IWC Resolution 2005–1.

20 IWC Resolution 2007.

21 *Ibid.*

22 Kenneth W. Abbott, Robert O. Keohane, Andrew Moravcsik, Anne-Marie Slaughter, and Duncan Snidal, "The Concept of Legalization", *International Organization* 54 (2000): 401–419, 401.

provisions in the ICRW appears to be uncontroversial,[23] including the Schedule that sets out the moratorium on commercial whaling.[24]

The part where the precision begins to falter is the exemption clause.[25] As explained previously, this exemption clause places scientific whaling outside the operation of the Convention: Article VIII allows Contracting Governments to 'grant any of its nationals a special permit authorizing that national to kill, take and treat whales *for purposes of scientific research*…'.[26] However, there is no further explanation or definition of whaling 'for purposes of scientific research' in the Convention. What kind of whaling activities fall under this exception despite the moratorium on commercial whaling is open to interpretation.

This is another important soft-law aspect in the context of the *Whaling* case. In fact, the parties to the *Whaling* case had very different views on this issue.[27] As a consequence, the *Whaling* judgment dealt with a whole range of questions arising from the lack of precision in this exemption clause, in particular, the term 'for purposes of scientific research'. Is a testable or defined hypothesis an indispensable component of scientific research?[28] Is peer review of proposals and results an indispensable component of scientific research?[29] Does scientific research have to be the sole goal of the whaling program in order to say that the program is carried out 'for purposes of' scientific research?[30] Can lethal methods have a place in any program of scientific research?[31] Even if yes, how much analysis of the feasibility of non-lethal methods is required in deciding to use lethal methods? Answers to none of these questions were readily apparent in the exemption clause of the ICRW. In other words, the degree of precision in the exemption clause was extremely low, making it a soft rule within a set of relatively hard rules.

2.3 *Degree of Delegation in the ICRW*

The last dimension of soft law concerns the delegation of authority to interpret the rules to external bodies or third parties. In many treaties, delegation is

23 John K. Setear, "Can Legalization Last? Whaling and the Durability of National (Executive) Discretion", *Virginia Journal of International Law* 44 (2004): 711–757.

24 Fitzmaurice (note 13), 458.

25 Setear (note 23), 726–727.

26 Emphasis added by the author.

27 See the telling summary in Chapter 3 by Malgosia Fitzmaurice.

28 Judgment, para. 77.

29 Judgment, para. 84.

30 Judgment, para. 97.

31 Judgment, para. 82.

authorised by a clause allowing one of the parties to take a dispute about them to the International Court of Justice unilaterally. If such a clause is inserted in a treaty, the authority to interpret the treaty rests with the Court, which functions as a dispute settlement mechanism of the treaty.

There is absolutely no controversy about this aspect of the ICRW. The Convention has no provision that refers to an external dispute settlement mechanism, either as a compulsory venue or simply as an option. While the *Whaling* judgment did not dismiss the case on the ground of lack of jurisdiction,[32] reference of disputes to the International Court of Justice is *not* a built-in procedure of dispute settlement for the ICRW, and there is no *general* delegation of the authority to interpret the treaty to the Court.

The ICRW does have its own institution for overseeing the implementation of the Convention: the IWC (Art. III). The IWC is, however, not designed as a dispute settlement mechanism. It is not given any specific competence to settle disputes among the Contracting Governments or interpret the treaty, either. In theory, the IWC could capitalise on its general competence to 'make recommendations to any or all Contracting Governments on any matters which relate to whales or whaling and to the objectives and purposes of this Convention' (Art. VI) in the context of a specific dispute. However, that is different from a delegation of the authority to interpret the treaty. A recommendation by a treaty organ with a supervisory function such as the IWC may explicitly or implicitly contain an interpretation of the Convention. There is, however, no reason to consider such an interpretation as final and authoritative.

In sum, there is no delegation of the authority to any third party to interpret the rules in the ICRW.

3 The *Whaling* Case: Transforming the ICRW's Soft Law into Hard Law

Seen against this background, the *Whaling* case can be described as transformation of soft law into hard law, in all three dimensions of measurement of the softness of rules. This section will examine what kind of means were employed in the *Whaling* case to transform soft law into hard law. That delegation to an external authority occurred in the process of the *Whaling* case is plain. Thus, the focus in exploring the transformation of soft law into hard law will be on

32 The delegation occurred in the *Whaling* case via the acceptance of compulsory jurisdiction of the ICJ under the optional clause, Article 36, of the ICJ Statute by Australia and Japan, respectively.

the other two dimensions that offer measurement of the softness of rules: obligation and precision.

3.1 *Dimension of Obligation: Turning Recommendations into Obligations via 'Duty to Co-operate'*

(i) *Prima Facie* Evaluation of the IWC Recommendations Regarding Lethal Methods

An issue raised by Australia in the *Whaling* case was the use of lethal methods in JARPA II. While not denying a possible place for lethal methods, Australia was of the view that in a research programme under the exemption clause, Article VIII, the use of such methods is allowed 'only when non-lethal methods are not available'.[33] If such an interpretation of Article VIII were correct, any whaling programme and the special permits relating to it that condoned the use of lethal methods despite the availability of non-lethal alternatives would automatically fall outside this exemption clause.

Australia tried to reinforce its position by relying on the IWC resolutions and Guidelines. Some of them were adopted by consensus in the IWC, but not all of them.[34] There are, in fact, resolutions and Guidelines of the IWC that call upon States to take into account whether their research objectives can be achieved by using non-lethal research methods when they evaluate their research programs.[35]

In the judgment, however, the position that IWC resolutions *in general* can form a subsequent agreement or subsequent practice in the context of treaty interpretation (Art. 31, VCLT) is explicitly refuted; the Court observed that many of the resolutions were 'adopted without the support of all States parties to the Convention and, in particular, without the concurrence of Japan', and cannot be part of subsequent agreement or practice that would influence the treaty interpretation.[36] There is no transformation of soft law, i.e. the non-binding resolutions, into hard law in this reasoning.

In terms of substance, moreover, it was pointed out that even the resolutions adopted by consensus 'do not clearly establish a requirement that lethal methods be used only when other methods are not available'.[37] Thus, while the use of lethal methods in whaling is strongly discouraged in many resolutions, it nevertheless does not place a whaling program involving such methods immediately

33 Judgment, para. 78.

34 *Ibid.*

35 *Ibid.*

36 Judgment, para. 83.

37 *Ibid.*

outside the scope of the exemption clause. Again there is no transformation of soft law into hard law in this part of the reasoning of the Court; the exemption clause is unaffected by the soft-law instruments produced by the IWC.

(ii) A Twist: A Duty to Co-operate with the IWC and the Scientific Committee

There is, however, a twist that follows in the judgment. It occurs in the same paragraph in which the previous discussions take place. While the IWC resolutions in general thus remain what they are, i.e., soft law, and do not appear to influence the interpretation of the exemption clause, there is a duty in the ICRW 'to co-operate with the IWC and the Scientific Committee'.[38] This is further explained in the judgment as a duty to 'give due regard to recommendations calling for an assessment of the feasibility of non-lethal alternatives'.[39] The expression employed in this paragraph is that States Parties 'should' give due regard to recommendations, as opposed to 'must' or 'shall'. Nevertheless, given the way this duty is treated in the judgment, as explained below, there is no doubt that the duty to co-operate is treated as an obligation and not merely an exhortation.

Such a duty to co-operate is, of course, a kind of duty that is often found in multilateral environmental agreements. In that sense, the discussion of a duty to co-operate in the context of the ICRW may not look like an unexpected twist. Nevertheless, numerous international environmental agreements contain an explicit reference to the duty to co-operate, whereas the ICRW has no explicit provision to that effect. Therefore, the question as to its origin and its precise content remains.

While the *Whaling* judgment does not provide any explanation of the origin of the duty to co-operate, there are a number of possible answers to the question. It may be that an unwritten duty to co-operate in the ICRW is based on custom observed within this regime. That, however, would have required some examination of the concrete operation of the ICRW that would reveal a uniform and constant internal practice, comparable to State practice that forms customary law. Such an examination is not provided in the *Whaling* judgment. Another possibility is that a duty to co-operate with supervisory organs is generally assumed to exist in any treaty regime that establishes its own supervisory mechanism. The duty does not depend on actual internal practice or custom. The assumption is that a treaty regime cannot operate in the way it is intended to if States do not co-operate with its supervisory organs, and in that sense, the duty to co-operate is a kind of duty that necessarily exists, even if

38 *Ibid.*

39 *Ibid.*

there is no provision to that effect. The duty to co-operate with supervisory organs of a treaty in a treaty regime, in this sense, may be compared to the rule *pacta sunt servanda* in the law of treaties. The regime of the law of treaties would be greatly harmed, if not rendered pointless, if the rule *pacta sunt servanda* were not assumed to exist and operate.[40] Yet another possible provenance of the duty to co-operate in the ICRW could be the consent of States Parties. The *Whaling* judgment, while not explicitly stating that this is the origin of the duty to co-operate, does observe that: 'Japan has accepted that it is under an obligation to give due regard to such recommendations [recommendations that call upon States Parties to consider the use of non-lethal methods to achieve research objectives]'.[41]

(iii) The Effect of the Duty to Co-operate with Supervisory Organs

A more important question, from a practical point of view, regarding the duty to co-operate in the *Whaling* judgment is its precise content and extent, for that would determine the required conduct of any Contracting Government in the ICRW.

The immediate effect of the duty to co-operate, discussed previously, is the weight of IWC resolutions regarding the use of lethal methods in shaping what Japan was supposed to do when it presented the JARPA II Research Plan. According to the judgment, 'the JARPA II Research Plan should have included some analysis of the feasibility of non-lethal methods as a means of reducing the planned scale of lethal sampling', precisely because of the IWC resolutions and Guidelines that call upon States parties to take into account whether research objectives can be achieved using non-lethal methods.[42] According to the judgment, the JARPA II Research Plan did not contain such analysis.

The Court also stated earlier in the judgment that it would 'look to the authorizing State, which has granted special permits [under the exemption clause], to explain the objective basis for' having determined that the use of lethal methods in that particular whaling program is for purposes of scientific research.[43] In this light, the analysis of the Court effectively concludes that Japan had not undertaken serious feasibility study on the possible use of

40 Gerald Fitzmaurice, "Some Problems Regarding the Formal Sources of International Law" in F.M. van Asbeck et al. (eds.), *Symbolae Verzijl : présentées au Professeur J.H.W. Verzijl à l'occasion de son LXX-ième anniversaire* (Martinus Nijhoff, 1958), 153–176.

41 Judgment, para. 137.

42 *Ibid.* This is one of the three reasons given by the Court why the JARPA II Research Plan should have included such analysis.

43 Judgment, para. 68.

non-lethal methods to achieve the research objectives of JARPA II.[44] The resulting appreciation of the conduct of Japan by the Court is that this 'is difficult to reconcile with Japan's obligation to give due regard to IWC resolutions and Guidelines'.[45]

While Japan had undoubtedly accepted the duty to co-operate, as pointed out by the Court,[46] this was not an acceptance of a duty to carry out unconditionally every single act recommended in IWC resolutions and Guidelines. However, once a Contracting Government acknowledges, and consents to, its duty to co-operate with the IWC in general, it is difficult to delineate those recommended acts which it is obliged to carry out, on one hand, and, on the other hand, those which it can, after a consideration in good faith, nevertheless decline to carry out.

In short, in the *Whaling* judgment, the soft-law instruments yielded by the supervisory organ of the ICRW are not what they seemed at first glance. It is submitted that the duty to co-operate with the IWC, and to give due regard to its recommendations, has effectively turned the IWC recommendations regarding the use of lethal methods into a source of obligation for the respondent.

3.2 *Dimension of Precision: Revealing the Actions Required*

The other aspect in which the judgment transforms soft law into hard law is the change in the degree of precision regarding Article VIII.

(i) Evaluation 'in the Light of Article VIII' in Form

As explained above, the degree of precision in the exemption clause in the ICRW is low: It is hard to tell what falls under this clause, Article VIII, and what does not, from the article itself, because it simply speaks of activities 'for purposes of scientific research'. It was indeed a point of contention in the *Whaling* case, with very different views held by the parties to the dispute. Therefore, it was natural that the evaluation of JARPA II in the *Whaling* judgment is carried out 'in light of Article VIII of the Convention' as explicitly declared by the Court.[47]

(ii) A Twist: Test of Reasonableness

There was, however, again a kind of twist that affects the softness of the treaty, this time, of Article VIII. In its specific examination of whether the activities

44 Judgment, paras. 138–141.

45 Judgment, para. 144.

46 Judgment, para. 80.

47 This is indeed the title given to the Section II, Sub-section 3 of the Judgment.

under JARPA II are '*for purposes of* scientific research and thus may be authorized by special permits granted under Article VIII', the judgment restates its task as that of enquiring into 'whether...the design and implementation of JARPA II are reasonable in relation to achieving its stated objectives'.[48] Then, in order to determine the reasonableness of JARPA II, the Court proceeds to examine a series of elements of JARPA II.[49] The Court concludes that the evidence does not establish the design and implementation of JARPA II as something reasonable in relation to achieving its objectives.[50]

According to the judgment, many explanations were missing in showing that JARPA II was a reasonable programme in its design and implementation. Just as an illustration, the Court's findings on the scale of the programme's use of lethal sampling, and the methodology used to select sample sizes, are summarised below.

Regarding the scale of the sampling using lethal methods in JARPA II, the explanation why there has been a significant sample size increase in relation to minke whales compared to JARPA, the predecessor whaling programme, is missing;[51] the justification of lethal sampling of two additional species in JARPA II, fin and humpback whales that were not used in this way in JARPA, is equally missing;[52] the explanation why 'JARPA II is designed to detect one particular rate or degree of change over another' that could have resulted in a lower sample size is missing;[53] an explanation why a 12-year period is chosen as a research period for fin and humpback whales is missing;[54] an explanation why, on the contrary, a six-year research period is chosen for minke whales is missing, and consequently, an explanation of the use of this research period to calculate the minke whale sample size is missing;[55] an explanation how these two different sets of research time frames for the three species can usefully contribute to the research objectives[56] is missing.[57] In sum, the Court concluded, '[I]n the context of Article VIII', the missing explanations do not allow

48 Judgment, para. 127.

49 Seven elements are referred to in paragraph 88 of the Judgment. What this chapter calls the test of reasonableness is examined as 'standard of review' elsewhere. See the analysis in Chapter 1 by Caroline Foster.

50 Judgment, para. 227.

51 Judgment, para. 153.

52 *Ibid.*

53 Judgment, para. 188.

54 Judgment, para. 180.

55 Judgment, para. 193.

56 The research objectives in question are identified in Judgment, paras. 114–115.

57 Judgment, para. 193.

'one to understand why that sample size is reasonable in relation to achieving the programme's objectives',[58] for there is only 'scant analysis and justification for the underlying decisions that generate the overall sample size'.[59]

Moreover, in the same examination of the reasonableness of the design and implementation of JARPA II, the Court goes on to criticise a few other aspects of the programme. It pointed out the inflexible target sample size set by JARPA II, despite the actual take which was significantly lower than the target sample size in many of the years during the research period;[60] the open-ended time frame, where 'a "time frame with intermediary targets" would have been more appropriate';[61] the scientific output of JARPA II that 'appears limited';[62] rather meagre co-operation record between JARPA II and other domestic and international research institutions.[63]

(iii) The Effect of the Test of Reasonableness

From the perspective of the dimension of precision in the scale of soft law/hard law, the test of reasonableness is striking, for this test of reasonableness effectively serves to produce an extraordinary degree of precision of the rule. As described above, the test of reasonableness could be presented as a list of missing actions and explanations by Japan, that would, had Japan taken them or offered them, have increased the chance of proving the reasonableness of the design and implementation of JARPA II.

It is true that neither the Convention nor the Court names these missing actions and explanations as generally required actions under the exemption clause of the Convention. In fact, these missing actions and explanations are specific to the case under review, i.e. JARPA II, and what each party to the dispute said about JARPA II. Nevertheless, at least regarding JARPA II, they are in practice the requirements that should have been fulfilled in order to render JARPA II reasonable; it is the lack of these actions and explanations that leads the Court to conclude JARPA II does not fall under the exemption clause.[64]

It is submitted that the *Whaling* judgment also transformed the soft-law aspect of precision into a hard one. The list of missing actions identified, and criticisms pronounced, by the Court constitutes a list of requirements to be

58 Judgment, para. 195.

59 Judgment, para. 197.

60 Judgment, paras. 209–212.

61 Judgment, para. 216.

62 Judgment, para. 219.

63 Judgment, para. 222. The precise expression in the Judgment is 'some further evidence of co-operation…could have been expected'.

64 See how this connection is directly made by the Court (Judgment, para. 227).

fulfilled by JARPA II in order to be considered as reasonable, and therefore falling within the exemption under Article VIII. As far as JARPA II is concerned, therefore, the required actions for a program for scientific purposes under Article VIII have been revealed by the Judgment, transforming this soft rule into a hard one.

4 Analysis

From the perspective of soft law/hard law, there are two striking instances in the *Whaling* judgment. First is the use of the duty to co-operate with the IWC and the Scientific Committee, even though such a duty is not explicitly provided in the ICRW. Second is the use of the test of reasonableness in determining whether or not JARPA II was reasonable in its design and implementation in the light of its objectives. According to the framework of soft law/hard law, each instance can be seen as a transformation of soft law into hard law.

Regarding the duty to co-operate with supervisory organs in a treaty regime, in general, it is probably not disputed that such a duty exists. There is no point in creating supervisory organs if States are not ready to co-operate with them. Whether the duty is explicitly provided or not, co-operation with supervisory organs is simply necessary for a treaty regime to function effectively. In the case of the ICRW, this duty happens to be an unwritten rule. Accordingly, the duty to co-operate was identified in the *Whaling* judgment without any help of the law of treaties. There is, however, no frame of reference that replaces the law of treaties in the analysis of the judgment. Without a frame of reference, the exact content of a duty to co-operate is difficult to discern. In the *Whaling* judgment, it was the transformation of soft-law instruments into hard law that made up for this difficulty; the recommendations by the IWC effectively provided the concrete content of the co-operation required under the duty to co-operate in this treaty regime. Consequently, though the identification of the duty to co-operate was not framed as an interpretation of Article VIII, it had a decisive impact on the appreciation of the conduct of Japan in the light of Article VIII, because it opened a way to take into account the soft-law instruments adopted by the IWC in this appreciation.

Regarding the test of reasonableness, again the law of treaties was of little help. While the judgment maintained that its findings regarding JARPA II were based on the interpretation of Article VIII, the traditional techniques of treaty interpretation were referred to only nominally and briefly in its analysis. Instead of scrutinising the term 'for purposes of scientific research' itself, the judgment carried out an in-depth analysis of JARPA II and what the parties to the dispute had to say about this whaling programme. In this exercise, a series of missing actions by Japan became apparent. These missing actions could, had Japan taken them, have justified JARPA II as a whaling programme with a reasonable design and implementation in relation to its objectives. In other words, the exercise revealed actions required for a whaling programme to be considered as a legitimate exception under Article VIII. This is also a transformation of the soft rule into a hard rule, this time by providing much more precision to the rule.

Both instances analysed above reflect challenges in relation to a dynamic treaty regime when it is discussed in a court of law. The basis of the treaty regime being a treaty, any court of law would naturally look at the treaty. The challenge lies in the fact that treaty regimes consist of a whole range of soft and hard rules not limited to the letter of the treaties. Certain parts of soft law in treaty regimes cannot adequately be analysed in the light of the law of treaties, for the simple reason that the law of treaties is conceived as an instrument to deal with static treaties.

This view regarding the law of treaties is not in anyway unorthodox, and surely the one shared by the *Whaling* judgment. The judgment made no pretence that elaborate discussion, and subsequent application, of the law of treaties would serve this case. On the contrary, it found ways to take into account certain aspects of a dynamic regime that do not squarely fall within any concept of the law of treaties, and that only the examination of the operation of the treaty regime can reveal. The nominal and brief discussions in the judgment that relate to the law of treaties only underscore the Court's efforts to balance the two demands, namely, to stick to the law of treaties and its orthodox interpretation methods, while at the same time evaluating JARPA II within this dynamic whaling regime.

CHAPTER 8

IWC and the ICJ Judgment

Joji Morishita[1]

1 Introduction

On 31 March 2014 the International Court of Justice (ICJ) issued its Judgment on the whaling case (*Whaling in the Antarctic,* Australia *v.* Japan: New Zealand intervening[2]) after four years of preparation and presentation of arguments by the Parties and the intervening State. The Case was submitted to the ICJ in 2010 by Australia who claimed that the Second Phase of the Japanese Whale Research Program under Special Permit in the Antarctic (JARPA II) was not a program for purposes of scientific research within the meaning of Article VIII of the International Convention for the Regulation of Whaling (ICRW) and that, therefore, Japan had violated the commercial whaling moratorium and other provisions of the ICRW.

The ICJ '[f]inds that the special permits granted by Japan in connection with JARPA II do not fall within the provisions of Article VIII, paragraph 1, of the International Convention for the Regulation of Whaling' (paragraph 247(2), the ICJ Judgment) and '[d]ecides that Japan shall revoke any extant authorization, permit or licence granted in relation to JARPA II, and refrain from granting any further permits in pursuance of that programme'. (paragraph 247(7))

This article examines the ICJ Judgment and provides the author's observations on its implications for future developments at the International Whaling Commission (IWC).

2 Background

The scientific whaling programs (the Second Phase of the Japanese Whale Research Program under Special Permit in the Antarctic (JARPA II) – Monitoring of the Antarctic Ecosystem and Development of New Management Objectives

1 Japan's Commissioner to the International Whaling Commission. The views and interpretations contained in this article are those of the author and do not represent the position of the Government of Japan.

2 http://www.icj-cij.org/docket/index.php?p1=3&p2=1&case=148&code=aj&p3=4.

 | DOI 10.1163/9789004313828_010

for Whale Resources,[3] and the Second Phase of the Japanese Whale Research Program under Special Permit in the North Pacific (JARPN II))[4] authorized by the Government of Japan have been one of the contentious issues at the meetings of the International Whaling Commission and between countries supporting the sustainable use of whales (pro-whaling countries) and anti-whaling countries. Anti-whaling countries oppose Japan's scientific whaling, arguing that the programs are 'commercial whaling in disguise'.

Scientific whaling programs are sanctioned by the ICRW Article VIII as follows:

> Article VIII
> 1. Notwithstanding anything contained in this Convention any Contracting Government may grant to any of its nationals a special permit authorizing that national to kill, take and treat whales for purposes of scientific research subject to such restrictions as to number and subject to such other conditions as the Contracting Government thinks fit, and the killing, taking, and treating of whales in accordance with the provisions of this Article shall be exempt from the operation of this Convention. Each Contracting Government shall report at once to the Commission all such authorizations which it has granted. Each Contracting Government may at any time revoke any such special permit which it has granted.
> 2. Any whales taken under these special permits shall so far as practicable be processed and the proceeds shall be dealt with in accordance with directions issued by the Government by which the permit was granted.
> 3. Each Contracting Government shall transmit to such body as may be designated by the Commission, in so far as practicable, and at intervals of not more than one year, scientific information available to that Government with respect to whales and whaling, including the results of research conducted pursuant to paragraph 1 of this Article and to Article IV.
> 4. Recognizing that continuous collection and analysis of biological data in connection with the operations of factory ships and land stations are indispensable to sound and constructive management of the whale fisheries, the Contracting Governments will take all practicable measures to obtain such data.

3 http://www.icrwhale.org/pdf/SC57O1.pdf.

4 http://www.icrwhale.org/pdf/SC5402.pdf, http://www.icrwhale.org/pdf/SC56O1.pdf.

3 Perception and the Commercial Whaling Moratorium

The dispute surrounding whales and whaling is often based on perceptions rather than on scientific information and international law.[5]

Wide-spread perceptions about whales and whaling are that all whale species are 'special' and endangered, that whaling is permanently prohibited by international law, and that the world opinion is against whaling, to name a few. These perceptions play at least some part in the ICJ case.

The perceptions have been built over the history of the whaling dispute as a result of, among other things, extensive and persistent campaigns by anti-whaling non-governmental organizations (NGOs) and the sympathetic media. Furthermore, the adoption of the 'commercial whaling moratorium' by the IWC in 1982 reinforced the perceptions, because the public assumes the moratorium is a ban on whaling introduced as a result of the depletion of whales.

However, when the instrument that established the moratorium, and its negotiation history, are reviewed, the meaning of the moratorium proves to be very different from the wide-spread perceptions.[6]

Commonly, paragraph 10(e) of the Schedule to the ICRW is referred to as the 'IWC moratorium on commercial whaling' and anti-whaling NGOs, others opposed to whaling and the media frequently call it 'the ban on commercial whaling', implying or stating that it is a permanent prohibition of something evil or even criminal. Examination of the discourse used in presenting and supporting moratorium proposals, including the IWC's verbatim records of its annual meetings, written opening statements and the Chairman's reports of the annual meetings, clearly shows that this is not the case. These documents, together with the language of paragraph 10(e), clearly demonstrate that the moratorium was intended as a temporary conservation and management measure related to uncertainties of scientific information, without categorically denying whaling as a legitimate activity and without passing moral or value judgments. The negotiation history of paragraph 10(e) confirms that the moratorium was simply an ordinary conservation and management measure to temporarily suspend whaling operations while conducting a comprehensive assessment of whale populations.

---

5 Morishita, Joji, Multiple analysis of the whaling issue: Understanding the dispute by a matrix, Marine Policy 30 (2006) 802–808.

6 Morishita, Joji, The Truth about the Commercial Whaling Moratorium, Senri Ethnological Studies 83: 335–351 ©2013, *Anthropological Studies a/Whaling,* Edited by Nobuhiro Kishigami, Hisashi Hamaguchi and James M. Savelle.

This interpretation was also expressed by the FAO's long-time participant in the work of the IWC, who wrote that 'as presented to the IWC, and as the term implies, the moratorium was to be a temporary measure, to be reviewed not later than 1990. It was intended to give scientists time to remove doubts about figures of sustainable yields, numbers of whales and so on'.[7]

3.1 *Plain Meaning/Straight Reading of Schedule Paragraph 10(e)*

The 'Schedule' of the ICRW is the means for adopting legally binding 'regulations with respect to the conservation and utilization of whale resources (Article V.1., ICRW)' and the so-called moratorium on commercial whaling was established by the addition of paragraph 10(e) to the Schedule in 1982.

Paragraph 10(e) of the Schedule reads as follows:

> (e) Notwithstanding the other provisions of paragraph 10, catch limits for the killing for commercial purposes of whales from all stocks for the 1986 coastal and the 1985/86 pelagic seasons and thereafter shall be zero. This provision will be kept under review, based upon the best scientific advice, and by 1990 at the latest the Commission will undertake a comprehensive assessment of the effects of this decision on whale stocks and consider modification of this provision and the establishment of other catch limits.

The plain meaning of this provision is that when a comprehensive scientific assessment on a whale stock is completed and a sustainable catch quota calculated, a non-zero catch limit can be established for the stock. Therefore it is perfectly appropriate to establish a whaling quota under this 'moratorium' language. Such action does not violate the moratorium or 'lift' it. It should be emphasized that the language of paragraph 10(e) does not introduce a permanent ban on commercial whaling nor establish any value judgement to the effect that commercial whaling is wrong or illegal. It is true that anti-whaling interests had tried to establish a permanent ban on whaling in the course of the negotiation of this provision; but the fact that the Schedule language is formulated as a temporary suspension of whaling while a comprehensive assessment is undertaken for the possible establishment of non-zero catch quotas demonstrates that the attempt was not successful.

It should also be noted that, while Japan's research whaling has been criticized as 'disguised commercial whaling' the research was initiated for the purpose of contributing to the comprehensive scientific assessment demanded by paragraph 10(e).

---

7 Gulland, John. The end of whaling? New Scientist. 29 October 1988, p. 42.

3.2 *Discourse in Presenting and Supporting Moratorium Proposals Leading up to 1982*

In presenting its proposal for a ten year moratorium at the 1974 Annual Meeting, the Commissioner for the United States said 'we are seeking a ten year moratorium based on continuing decrease in catch per unit effort of fin whales and FAO's recognition of the gross uncertainties in the data and questionable assumptions on which present management schemes are based'.[8] Further statements supporting the argument that the moratorium was intended as a conservation and management measure, rather than a permanent prohibition, include statements at the 1979 meeting: by the UK Minister that 'there should be a moratorium on commercial whaling in order to allow a thorough assessment of whale numbers and of their biology' and that 'resumption of whaling should only be considered if evidence of recovery of stocks and improvement in methods of killing justify it'; and by Sweden that 'we strongly favor a moratorium because there are too many gaps in the science but we are prepared to discuss opening of whaling after the moratorium based on scientific results';[9] as well as the statement from France at the 1980 meeting that the 'moratorium is a method of managing whales to be applied at this time to allow their recovery'.[10]

At the 1981 Annual Meeting, Mr. Wiggin, Parliamentary Secretary to the Ministry of Agriculture, Fisheries and Food of the UK said 'we recognize that other countries have a legitimate commercial interest in whaling and if in the future, it could be shown beyond a reasonable doubt that some exploitation of stocks might be safely resumed and that satisfactory methods of killing were available, the lifting of the ban might be considered'. In clarifying their proposal the UK delegation also said 'what we had in mind is a moratorium and not a permanent ban'.[11]

At the 1982 meeting of the IWC's Technical Committee prior to the plenary sessions, the Seychelles proposed a three year phase out of commercial whaling. This was amended by Costa Rica to a two-year phase out which was recommended by the Technical Committee to the plenary. At plenary, the Seychelles

8 International Whaling Commission. 1974. Verbatim record of the 26th IWC Annual Meeting. Available from the IWC Secretariat, Cambridge, UK.

9 International Whaling Commission.1979. Verbatim record of the 31st IWC Annual Meeting. Available from the IWC Secretariat, Cambridge, UK.

10 International Whaling Commission. 1980. Verbatim record of the 32nd IWC Annual Meeting. Available from the IWC Secretariat. Cambridge, UK.

11 International Whaling Commission. 1981. Verbatim record of the 33rd IWC Annual Meeting. Available from the IWC Secretariat, Cambridge, UK.

proposed an amendment to the Costa Rica proposal to return to the three-year phase out of its original proposal and to add the additional clause 'This provision shall be kept under review, based on the best scientific advice, and by 1990 at the latest the Commission shall undertake a comprehensive assessment of the effects of this decision on whale stocks and consider modification of this provision and the establishment of other catch limits'.

In proposing this amendment, the Commissioner for the Seychelles stated: 'We have also taken into account opinions which have been expressed amongst member states here which suggest that it is necessary to provide also for the possibility of review and reassessment and, if the scientific information should so suggest in the interim, perhaps even a resumption of some form of commercial activity'.[12] He also had said in the Technical Committee that: 'I would repeat and remind you that this is a catch limit proposal not a ban or a moratorium'. Spain commented: 'First of all let me advance that I don't consider this as a total ban…but just a temporary interruption of the activity'.[13] The Commissioner from St. Lucia who had supported the 'moratorium' proposal also stated: 'It is with deep regret that I note that there has been a misunderstanding on the proposal of the distinguished delegate from the Seychelles. It is not a proposal for a total ban for commercial whaling, but rather a proposal on catch limits'.[14]

Notwithstanding these statements, some members, including Japan, still opposed the proposal, as they regarded it as 'in substance a total moratorium' noting that 'the Scientific Committee ruled many times in the past that there is no scientific justification or biological need for a total or blanket moratorium'.[15] The vote on the proposal received the required ¾ majority vote (25 votes in favor, with 7 against and 5 abstentions).[16]

4 Perception and the ICJ Judgment

Following the ICJ Judgment on 31 march 2014, the media used headlines such as 'Court demands end to Japanese whaling' (1 April 2014, USA Today); 'U.N.

12 International Whaling Commission. 1982. Verbatim record of the 34th IWC Annual Meeting. Available from the IWC Secretariat, Cambridge, UK.

13 Ibid. at 12.

14 Ibid. at 12.

15 Ibid. at 12.

16 International Whaling Commission. 1983. Chairman's Report of the 34th IWC Annual Meeting. REP. INT. WHAL. COMMN 33, 1983.

court orders Japan to halt Antarctic whaling' (1 April 2014, The Washington Post); 'U.N. court: Antarctic whaling by Japan illegal' (1 April 2014 Chicago Tribune); and 'A Ruling to Protect Whales' (1 April 2014, The New York Times). These and other similar headlines imply that the ICJ ordered a prohibition of all research whaling in the Antarctic, including future research, and that it even concluded that all whaling is illegal, while, in fact, the ICJ judgment is much more limited in scope, as explained below.

These media headlines, enhanced by NGO campaigns, could form the basis for perceptions that the ICJ Judgment DID prohibit present and future research whaling activities in the Antarctic Ocean. As reviewed above, this is analogous to what happened with the adoption of the so-called 'commercial whaling moratorium'.

4.1 *Actual Judgment*

In order to understand the scope of the Judgment accurately, it is necessary to review the pertinent paragraphs of the ICJ Judgment in a holistic manner.

The operative part of the Judgment is paragraph 247 and it states as follows (see Annex 1 for relevant paragraphs of the Schedule to the ICRW).

247. For these reasons,

THE COURT,

(1) Unanimously,
Finds that it has jurisdiction to entertain the Application filed by Australia on 31 May 2010;

(2) By Twelve votes to four,
Finds that the special permits granted by Japan in connection with JARPA II do not fall within the provisions of Article VIII, paragraph 1, of the International Convention for the Regulation of Whaling;

(3) By twelve votes to four,
Finds that Japan, by granting special permits to kill, take and treat fin, humpback and Antarctic minke whales in pursuance of JARPA II, has not acted in conformity with its obligations under paragraph 10 (e) of the Schedule to the International Convention for the Regulation of Whaling;

(4) By twelve votes to four,
Finds that Japan has not acted in conformity with its obligations under paragraph 10 (d) of the Schedule to the International Convention for the Regulation

of Whaling in relation to the killing, taking and treating of fin whales in pursuance of JARPA II;

(5) By twelve votes to four,
Finds that Japan has not acted in conformity with its obligations under paragraph 7 (b) of the Schedule to the International Convention for the Regulation of Whaling in relation to the killing, taking and treating of fin whales in the 'Southern Ocean Sanctuary' in pursuance of JARPA II;

(6) By thirteen votes to three,
Finds that Japan has complied with its obligations under paragraph 30 of the Schedule to the International Convention for the Regulation of Whaling with regard to JARPA II;

(7) By twelve votes to four,
Decides that Japan shall revoke any extant authorization, permit or licence granted in relation to JARPA II, and refrain from granting any further permits in pursuance of that programme.

The decision of the court is that 'Japan shall revoke any extant authorization, permit or licence granted in relation to JARPA II' and 'refrain from granting any further permits in pursuance of that programme' (i.e. JARPA II). The decision does not cover other existing research programs or future research programs, contrary to the views promoted by the media and anti-whaling NGOs. The possible effects of the ICJ Judgment related to future research programs under Article VIII, paragraph 1, will be further elaborated below. Here it should be noted that the Judgment, in its paragraph 246, specifically acknowledges that there may be future whale research programs under Article VIII: 'It is to be expected that Japan will take account of the reasoning and conclusions contained in this Judgment as it evaluates the possibility of granting any future permits under Article VIII, paragraph 1 of the Convention'.

It should be noted that, legally speaking, the ICJ Judgment only binds the Parties to the case and the intervening State; Japan, Australia, and New Zealand. Eighty eight countries currently belong to the IWC and eighty five of them have no legal obligations related to the Judgment. Needless to say, the IWC as an independent international organization, does not have legal obligations related to the ICJ Judgment unless it makes a decision to that effect. On the other hand, it is beyond doubt that the Judgment will have, and already has had, political and diplomatic influence on the discussions at the IWC, as will be discussed below.

The ICJ states its reasoning for the Judgment in the concluding paragraphs from 223 to 227. Especially paragraphs 225, 226, and 227 summarize the reasoning as follows:

> 225. First, the broad objectives of JARPA and JARPA II overlap considerably. To the extent that the objectives are different, the evidence does not reveal how those differences lead to the considerable increase in the scale of lethal sampling in the JARPA II Research Plan. Secondly, the sample sizes for fin and humpback whales are too small to provide the information that is necessary to pursue the JARPA II research objectives based on Japan's own calculations, and the programme's design appears to prevent random sampling of fin whales. Thirdly, the process used to determine the sample size for minke whales lacks transparency, as the experts called by each of the Parties agreed. In particular, the Court notes the absence of complete explanations in the JARPA II Research Plan for the underlying decisions that led to setting the sample size at 850 minke whales (plus or minus 10 per cent) each year. Fourthly, some evidence suggests that the programme could have been adjusted to achieve a far smaller sample size, and Japan does not explain why this was not done. The evidence before the Court further suggests that little attention was given to the possibility of using non-lethal research methods more extensively to achieve the JARPA II objectives and that funding considerations, rather than strictly scientific criteria, played a role in the programme's design.
>
> 226. These problems with the design of JARPA II must also be considered in light of its implementation. First, no humpback whales have been taken, and Japan cites non-scientific reasons for this. Secondly, the take of fin whales is only a small fraction of the number that the JARPA II Research Plan prescribes. Thirdly, the actual take of minke whales has also been far lower than the annual target sample size in all but one season. Despite these gaps between the Research Plan and the programme's implementation, Japan has maintained its reliance on the JARPA II research objectives—most notably, ecosystem research and the goal of constructing a model of multi-species competition—to justify both the use and extent of lethal sampling prescribed by the JARPA II Research Plan for all three species. Neither JARPA II's objectives nor its methods have been revised or adapted to take account of the actual number of whales taken. Nor has Japan explained how those research objectives remain viable given the decision to use six-year and 12-year research periods for different species, coupled with the apparent decision to abandon

> the lethal sampling of humpback whales entirely and to take very few fin whales. Other aspects of JARPA II also cast doubt on its characterization as a programme for purposes of scientific research, such as its open-ended time frame, its limited scientific output to date, and the absence of significant co-operation between JARPA II and other related research projects.
>
> 227. Taken as a whole, the Court considers that JARPA II involves activities that can broadly be characterized as scientific research (see paragraph 127 above), but that the evidence does not establish that the programme's design and implementation are reasonable in relation to achieving its stated objectives. The Court concludes that the special permits granted by Japan for the killing, taking and treating of whales in connection with JARPA II are not "for purposes of scientific research" pursuant to Article VIII, paragraph 1, of the Convention.

The ICJ judgment has also made several important observations that are positive from Japan's perspective. They are also contrary to the views promoted by the media and anti-whaling NGOs.

One of the critical issues surrounding the ICJ Judgment is whether the Judgment changes the basic legal framework and/or the interpretation of the International Convention for the Regulation of Whaling (ICRW).

Australia argued in its Memorial that the ICRW has 'evolved' through its history as a result of adoption of the commercial whaling moratorium, resolutions to promote non-lethal research, and other 'conservation' oriented decisions, and that the objective of the ICRW and the mission of the IWC have changed from management of whaling to protection of whales from such threats as whaling. However, paragraph 56 of the Judgment states as follows:

> 56. ...The objectives of the ICRW are further indicated in the final paragraph of the preamble, which states that the Contracting Parties "decided to conclude a convention to provide for the proper conservation of whale stocks and thus make possible the orderly development of the whaling industry". Amendments to the Schedule and recommendations by the IWC may put an emphasis on one or the other objective pursued by the Convention, but cannot alter its object and purpose.

The Australian argument was explicitly denied by the ICJ. This is significant because one of the main arguments promoted by the anti-whaling side has been that the world has changed, and protection or 'conservation' of whales has become the norm instead of conservation and management (sustainable use) of whales as one of many marine living resources. The ICJ, while recognizing a possible shift of 'emphasis on one or the other objective', declared that 'the orderly development of the whaling industry' – sustainable use in modern terms – remains as one of the objectives of the ICRW. This could have an interesting impact on the future discussions at the IWC meetings.

As to the interpretations of Article VIII of the ICRW, which stipulates the right of Contracting Governments to grant a special permit to kill, take and treat whales for purposes of scientific research, the ICJ Judgment has also made important observations. Australian and New Zealand, based on their anti-whaling position, attempted at the ICJ hearings to restrict (i) the right of Contracting Governments to issue a special permit, (ii) the use of lethal research methods, and (iii) the use and sales of whales taken under a special permit.

On these points, the ICJ has expressed the following views:

> 55. The Court notes that Article VIII is an integral part of the Convention. It therefore has to be interpreted in light of the object and purpose of the Convention and taking into account other provisions of the Convention, including the Schedule. However, since Article VIII, paragraph 1, specifies that "the killing, taking, and treating of whales in accordance with the provisions of this Article shall be exempt from the operation of this Convention", whaling conducted under a special permit which meets the conditions of Article VIII is not subject to the obligations under the Schedule concerning the moratorium on the catching of whales for commercial purposes, the prohibition of commercial whaling in the Southern Ocean Sanctuary and the moratorium relating to factory ships.

This paragraph recognizes the right of Contracting Governments to issue a special permit for '*the killing, taking, and treating of whales in accordance with the provisions of this Article*' and '*notes that Article VIII is an integral part of the Convention*' and '*therefore has to be interpreted in light of the object and purpose of the Convention and taking into account other provisions of the Convention, including the Schedule*'. When read with paragraph 56 above, which reconfirmed the objectives of the ICRW and dismissed the concept of 'evolution' by the adoption of Schedule amendments, paragraph 55 does not add any restrictions to the right of Contracting Governments as long as a special permit is issued for purpose of scientific research.

Furthermore, paragraph 55 states that '*whaling conducted under a special permit which meets the conditions of Article VIII is not subject to the obligations under the Schedule concerning the moratorium on the catching of whales for commercial purposes, the prohibition of commercial whaling in the Southern Ocean Sanctuary and the moratorium relating to factory ships'*. This also reconfirms that special permit research whaling is exempted from other requirements under the Schedule. This is consistent with Japan's arguments at the ICJ.

Regarding the use of lethal research methods, the Court has made some important points from Japan's perspective, for instance:

> 83. Article VIII expressly contemplates the use of lethal methods, and the Court is of the view that Australia and New Zealand overstate the legal significance of the recommendatory resolutions and Guidelines on which they rely. First, many IWC resolutions were adopted without the support of all States parties to the Convention and, in particular, without the concurrence of Japan. Thus, such instruments cannot be regarded as subsequent agreement to an interpretation of Article VIII, nor as subsequent practice establishing an agreement of the parties regarding the interpretation of the treaty within the meaning of subparagraphs (a) and (b), respectively, of paragraph (3) of Article 31 of the Vienna Convention on the Law of Treaties.

Australia and New Zealand maintained that the use of lethal research methods is now strictly restricted to exceptional circumstances, as many IWC resolutions to that effect have been adopted. However, the Court recognizes that '*Article VIII expressly contemplates the use of lethal methods*' and expressed the view that '*Australia and New Zealand overstate the legal significance of the recommendatory resolutions and Guidelines on which they rely*'. The Court is of the view that, because Japan and other members of the IWC did not support those resolutions, '*such instruments cannot be regarded as subsequent agreement to an interpretation of Article VIII, nor as subsequent practice establishing an agreement of the parties regarding the interpretation of the treaty*'.

The second part of this paragraph 83 adds the following view:

> Secondly, as a matter of substance, the relevant resolutions and Guidelines that have been approved by consensus call upon States parties to take into account whether research objectives can practically and scientifically be achieved by using non-lethal research methods, but they do not establish a requirement that lethal methods be used only when other methods are not available.

Even the resolutions and Guidelines adopted by consensus that call upon States to consider '*whether research objectives can practically and scientifically be achieved by using non-lethal research methods*' do not establish a '*requirement that lethal methods be used only when other methods are not available*'. Again, this view is consistent with Japan's positions.

The ICJ even finds that the use of lethal research methods in JARPA II is reasonable:

> 224. The Court finds that the use of lethal sampling per se is not unreasonable in relation to the research objectives of JARPA II".

The anti-whaling side often criticizes Japan's special permit scientific programs as 'commercial whaling in disguise' because, they claim, the whales caught by the programs and their products are sold in the commercial market. The ICJ Judgment states its view on this issue as follows:

> 94. As the Parties and the intervening State accept, Article VIII, paragraph 2, permits the processing and sale of whale meat incidental to the killing of whales pursuant to the grant of a special permit under Article VIII, paragraph 1.
>
> In the Court's view, the fact that a programme involves the sale of whale meat and the use of proceeds to fund research is not sufficient, taken alone, to cause a special permit to fall outside Article VIII.

This paragraph declares that Japan, Australia, and New Zealand accept that the ICRW '*permits the processing and sale of whale meat incidental to the killing of whales pursuant to the grant of a special permit under Article VIII*'. The Court further says that a special permit program involving the sales of whale meat, taken alone, is not outside the scope of Article VIII. While this view would not stop the anti-whaling side from continuing to use the same 'commercial whaling in disguise' criticism, the clear legal position taken by the Court is of help to Japan.

Australia and New Zealand argued that the ICRW, including its objectives, had evolved and interpretation of its provisions should be made on the basis of the evolution. Specifically, they argued that a special permit research program can be granted only in exceptional circumstances, and that Article VIII should be read as restrictively as possible. However, as reflected in the paragraphs above, the ICJ did not accept the evolution argument and accepted the current, or traditional, basic framework and interpretation of Article VIII. In reaching its Judgment, the ICJ did not take an approach involving either changing or restricting the interpretation of Article VIII, observing that: 'neither a restrictive nor an expansive interpretation of Article VIII is justified' (paragraph 58 of the judgment). Rather, they have constructed arguments, especially about sample sizes, to show (in their view) that the plan and implementation of the Second Japanese Whale Research Program in the Antarctic (JARPA II) were not reasonable and, therefore, outside the scope of Article VIII.

5 Effects and Implications of the ICJ Judgment on International Policy Making Regarding Whales and Whaling

5.1 *IWC Situations*

The subject of the ICJ Judgment is in particular the JARPA II. However, it will almost certainly have more general application, affecting the larger picture of the whaling issue and the fundamental conflict in the IWC over whales and whaling—that between those who regard whales as a resource, not different from many other marine living resources that can be utilized in a sustainable manner, and those who promote the protection of whales as a special and charismatic animal. Anti-whaling NGOs see the ICJ Judgment as significant progress toward the total cessation of whaling; and have expressed strong opposition to the statement of Japan expressing their intention to propose a new Antarctic research program that would be designed to meet the requirements and criteria presented by the ICJ Judgment.

Reflecting the fundamental conflicts between the members of the IWC over whales and whaling, the IWC has been almost continually engaged in intense and often emotional exchanges in their discourse on the implementation of research whaling and the management of whaling since the adoption of the commercial whaling moratorium in 1982. Furthermore, a three quarters majority is required in the IWC for the adoption of legally binding decisions. While both sustainable use and anti-whaling sides do not have enough votes to secure a three quarters majority, both sides have more than one quarter of the votes, enough to prevent the adoption of any legally binding decisions. Efforts for building consensus by seeking common interests between the two sides have been rare in the IWC; and as a result the IWC has been dysfunctional as an international organization, repeating unproductive conflicts every year and making no meaningful decisions on conservation and management of whales and whaling except for those related to aboriginal/subsistence whaling. Unfortunately, it is highly unlikely that anything in the ICJ Judgment will change the basic positions of the IWC members. However, the Judgment might promote some dialogue among them, because some legal views are presented by the authoritative ICJ on some aspects of the disputes.

5.2 *Dispute Settlement Approaches*

The IWC has no dispute settlement mechanism. However, in its history the IWC has engaged in a series of quests for a solution acceptable for both sustainable use and anti-whaling sides. For example, there was a compromise proposal from the then Chair of the IWC, Ireland, in 1997. While the Irish proposal failed, IWC's own efforts continued for resolving the difficulties through negotiations for the Revised Management Scheme (RMS) with a view to establishing a monitoring and enforcement scheme for whaling. The RMS consultation was later expanded to the RMS 'package' consultation, which incorporated other issues in dispute (e.g. a sanctuary proposal). The RMS and RMS package consultations also failed. In 2007 the 'Future of the IWC' project, involving renowned diplomatic experts from outside the IWC, was initiated; and a package compromise proposal that would have allowed limited whaling was presented jointly by the Chair and the Vice-Chair in 2010. The Chair and Vice-Chair's proposal was not adopted by the Commission. Despite these extensive efforts, the fundamental conflicts between the sustainable use countries and the anti-whaling countries remain unresolved.

Reflecting this situation in the IWC, and the strong anti-whaling sentiment of its press and its nationals, the whaling issue has become an important domestic policy item in Australia, even becoming one of the important subjects in its Prime Ministerial election campaign. In order to carry out an election campaign pledge, the Australian Government brought Japan to the ICJ in April 2010, claiming that JARPA II was illegal under the ICRW and its Schedule. Politically this means that Australia abandoned the multinational diplomatic negotiations in the IWC for resolving the conflict, and selected a lawsuit as an alternative. It should be noted that Australia brought the case against Japan at the same time as, after the intensive efforts of three years under the Future of the IWC project, a package compromise proposal under the names of the IWC Chair and Vice-Chair was submitted to the IWC. Australia had made a clear political decision; and this approach, replacing international negotiations with a view to legal settlement in an international resource management context, could have wider implications on other situations.

It would be premature to discuss in detail the possible effects of the ICJ Judgment, and of the Australian decision to abandon the IWC efforts for resolving the conflict, on the future discourse at the IWC. The 65th Meeting of the IWC, held in Slovenia in September 2014, has sent out mixed signals for the future direction of the organization. On the one hand, there was more willingness among some IWC members, including Australia and New Zealand, for dialogue, as evidenced by the consensus building discussions on a draft resolution that was proposed for the purpose of incorporating the ICJ Judgment into the IWC system.[17] On the other hand, Latin American states, led by Brazil, took a strong adversarial approach in promoting their South Atlantic Whale Sanctuary proposal. Some of them opposed the draft resolution related to the ICJ Judgment, proposed by New Zealand, as they saw it as a tool legitimizing a special permit research program. Intersessional developments leading to the 66th Meeting of the IWC in 2016 need to be closely observed in order to understand the effects of the ICJ Judgment.

5.3 *Science*

It should be also noted that many of the points in dispute at the ICJ, including the issue of the sample size of JARPA II that forms the basis of the ICJ Judgment, are not legal, but scientific. The ICJ, a legal institution, has produced judgments on the 'reasonableness' of scientific activities, and decided that the research whaling under JARPA II is illegal. This issue also was the basis of arguments in the written opinions of dissenting judges. The meaning of the involvement of the ICJ in science needs to be considered and analyzed in detail. At least it should be noted that the ICJ has made its own scientific judgment that is inconsistent with the views of the IWC Scientific Committee, where around 200 scientists and experts of the world gather annually to discuss whales and whaling. The ICJ Judgment could therefore further jeopardize the *raison d'être* of the IWC, which is already regarded dysfunctional as an organization for conservation and management of resources.

6 Japan's Course of Action

On April 18, 2014, Japan's then Minister of Agriculture, Forestry, and Fisheries, Mr. Yoshimasa HAYASHI issued a policy statement concerning the future whale research programs in light of the ICJ Judgment. The statement provides the following as Japan's basic policy for the whaling issue:

> The ICJ judgment reaffirms that one of the purposes of the International Convention for the Regulation of Whaling (ICRW) is the sustainable exploitation of whale resources. In light of this, Japan has confirmed its basic policy of pursuing the resumption of commercial whaling, by conducting research whaling,...

Based on this basic policy, the statement provides as follows in relation to whale research programs in the fiscal year 2014:

---

17 The consensus building discussions failed and that IWC/65/14 Rev 1 submitted by NZ was adopted as Resolution 2014–5 by vote: 35 in favour, 20 against and 5 abstentions.

(1) In the Antarctic, Japan has decided to cancel the Second Japanese Whale Research Program in the Antarctic (JARPA II) in accordance with the judgment made by the International Court of Justice (ICJ) in the case concerning Whaling in the Antarctic.
(2) In the Northwestern Pacific, Japan will limit its research objectives taking account of the judgment and then reduce the scale of the research activities of the Second Japanese Whale Research Program in the Northwestern Pacific (JARPN II).
(3) In view of the consideration of research programs for FY 2015, Japan will take various necessary measures, including studies of feasibility or practicability of non-lethal methods such as the DNA biopsy samplings in the Northwestern Pacific, taking account of the content of the judgment.

For the fiscal year 2015, the statement says the following:

> Japan will conduct through the cooperation among related ministries an earnest review of the designs of whale research programs in the Antarctic and in the Northwestern Pacific from FY 2015 and with the aim of submitting new research programs to the Scientific Committee of the International Whaling Commission (IWC) by the autumn this year, which reflects the criteria mentioned in the judgment. In doing so, Japan will follow an internationally open and highly transparent process through securing the participation of renowned scientists from Japan and abroad, and through other processes including discussions at the IWC Scientific Committee's workshop and coordination with other institutions conducting relevant studies.

As described above, the ICJ Judgment did not deny a future special permit whale research program *per se* in accordance with Article VIII of the ICRW, and instead sets out a series of criteria for a whale research program that is within the scope of Article VIII. Japan's course of action is therefore consistent with the Judgment. However, anti-whaling countries and NGOs regard the ICJ Judgment as a milestone to total cessation of whaling and a new whale research program in the Antarctic Ocean will receive strong opposition.

7 The 65th Meeting of the IWC

The 65th Meeting of the IWC was held from 15 to 18 September, 2014 in Portoroz, Republic of Slovenia. As this was the first IWC meeting after the ICJ Judgment, it attracted strong attention from the IWC member states and from the media.

Opening statements of Japan, Australia, and New Zealand set out their basic positions toward the IWC65 as follows:

IWC/65/OS Australia

Following the 31 March 2014 judgment delivered by the International Court of Justice, this meeting provides us with an opportunity to embed the principles of the judgment into the work of the Commission.

The full judgment (64 pages), and summary of the judgment (12 pages) can be found on the Court's website. Australia encourages all Commissioners to read the judgment, because the best approach is to let the authoritative judgment speak for itself.

The Court concluded that JARPA II was not a programme for the purposes of scientific research pursuant to Article VIII. Australia welcomes the decision of the Court that Japan 'shall revoke any extant authorization, permit or licence to kill, take or treat whales in relation to JARPA II, and refrain from granting any further permits under Article VIII, paragraph 1, of the Convention, in pursuance of that programme'. The words of the Court, the highest court of international law, speak for themselves. We need just ensure that the IWC, as a responsible international organisation, considers the guidance from the Court and incorporates its learned judgment into the practices of the Commission. Australia therefore commends the NZ Resolution for endorsement by the Commission.

IWC/65/OS New Zealand

The most significant issue since IWC64 has been the decision of the International Court of Justice (ICJ) in the Whaling in the Antarctic case. It is essential to ensure that the reasoning and conclusions contained in the ICJ's judgment are fully taken into account in any future consideration by the IWC of special permit programmes. New Zealand has tabled a draft resolution for the purposes of enabling a discussion within the Commission to that end.

IWC/65/OS Japan

1. The ICJ Judgment and Special Permit Program
On March 31, 2014 the International Court of Justice rendered its Judgment on the whaling case (Whaling in the Antarctic, Australia v. Japan: New

> Zealand intervening), deciding 'that Japan shall revoke any extant authorization, permit or licence granted in relation to JARPA II, and refrain from granting any further permits in pursuance of that programme'. At the same time, the Judgment confirmed that the objective of the ICRW includes 'to provide for the proper conservation of whale stocks and thus make possible the orderly development of the whaling industry' and that '[a]mendments to the Schedule and recommendations by the IWC may put an emphasis on one or the other objective pursued by the Convention, but cannot alter its object and purpose'. The Judgment also states that '[i]t is to be expected that Japan will take account of the reasoning and conclusions contained in this Judgment as it evaluates the possibility of granting any future permits under Article VIII, paragraph 1, of the Convention' and that '[t]he Court finds that the use of lethal sampling per se is not unreasonable in relation to the research objectives of JARPA II' while the ICJ provided several 'reasoning and conclusions' to find 'the special permits granted by Japan in connection with JARPA II do not fall within the provisions of Article VIII, paragraph 1, of the International Convention for the Regulation of Whaling'.
>
> Taking account of the Judgment Japan therefore announced its course of actions on April 18, 2014 through the Statement by Minister for Agriculture, Forestry and Fisheries (http://www.jfa.maff.go.jp/e/pdf/danwa.pdf) that Japan would submit new research plan based upon international law and scientific evidence to the Scientific Committee of the International Whaling Commission (IWC) by the autumn this year, which reflects the criteria mentioned in the Judgment. At this 65th meeting of the IWC Japan is willing to explain its positions on and to share information about the ICJ Judgment and related special permit issues.

There was a common understanding among the Parties to the ICJ case that there was no merit in repeating the ICJ discussions, or in disputing interpretations of the Judgment by other Parties, at the IWC meeting. Accordingly, the interest expressed in the opening statements of Australia and New Zealand was '*to embed the principles of the judgment into the work of the Commission*' by '*ensur[e]ing that the reasoning and conclusions contained in the ICJ's judgment are fully taken into account in any future consideration by the IWC of special permit programmes*' through the draft New Zealand resolution. Japan's interest was to prevent the establishment of wrong perceptions about the ICJ Judgment and to convey the legitimacy of submitting a new research plan in the Antarctic Ocean to the IWC Scientific Committee.

In order to avoid disputes regarding the ICJ judgment at the IWC65, the IWC Bureau meeting held in June 2014 in London decided that a joint presentation would be made at the IWC meeting.

> Australia and Japan proposed to work intersessionally (along with colleagues from New Zealand) to present the Commission with a joint document outlining the implications for IWC.[18]

However, Australia later declined to produce the joint presentation. Australia's position was, as suggested in its opening statement above, that the ICJ judgment should be read as a whole rather than presented through a joint presentation by Australia and Japan '*because the best approach is to let the authoritative judgment speak for itself*'.

Eventually, Japan made a summary presentation[19] with an understanding that Australia and New Zealand could make their own presentations if they found a need to correct, supplement, or dispute Japan's presentation. While both Australia and New Zealand made brief statements after Japan's presentation on the main points of the ICJ Judgment, there was no divisive discussion at the meeting and there were few follow up questions from other members. These somewhat inanimate proceedings at the IWC65 regarding the ICJ Judgment itself could be explained by the contents of the Judgment as described above. If the Judgment had been more one-sided in favor of Australia and New Zealand, the anti-whaling side could have been more aggressive in bringing pressure on Japan to stop whaling.

On the other hand, the IWC65 extensively discussed a draft resolution on Whaling under Special Permit submitted by New Zealand. The proposed resolution was aimed at reflecting the ICJ Judgment in the work and functioning of the IWC. The IWC members strove to achieve consensus on the proposal, and a generally constructive atmosphere prevailed during the discussion. However, the proposal received critical comments from both sides.

Japan argued that a proposed provision making a special permit research program subject to approval by the IWC was contrary to the provisions of Article VIII of the ICRW, and went beyond the ICJ Judgment. On the other hand, some Latin American countries with strong anti-whaling positions criticized the proposal as they saw it as legitimizing a special permit research program. Eventually, New Zealand asked for a vote on the original proposal as no consensus was achieved. The resolution was adopted by 35 votes in support,

18 Chair's Report of the IWC Bureau, Wednesday 4 June 2014, Nobel House, London.

19 Morishita, Joji, ICJ Judgment, IWC/65/22.

20 votes against and 5 abstentions, while receiving no votes from both anti-whaling and pro-whaling countries. As is the case for all resolutions adopted by the IWC, this resolution is legally non-binding.

The final operative paragraph of Resolution 2014–5 on Whaling under Special Permit reads as follows:

> 3. Requests that no further special permits for the take of whales are issued under existing research programmes or any new programme of whale research until:
> (a) the Scientific Committee has reviewed the research programme to enable it to provide advice to the Commission in accordance with the instructions above; and
> (b) the Commission has considered the report of the Scientific Committee and assessed whether the proponent of the special permit programme has acted in accordance with the review process described above; and
> (c) the Commission has, in accordance with Article VI, made such recommendations on the merits or otherwise of the special permit programme as it sees fit.

Upon the adoption of the resolution, Japan announced that it would be submitting a new research plan for the Antarctic Ocean in line with the ICJ Judgment; and that it would implement the research program from the austral summer of 2015/16.

How Resolution 2014–5 will affect future IWC discussions on a special permit whale research program, especially the new plan, is unclear. Because, during the consensus building discussions, there was general common understanding among the IWC members on the proposed resolution, other than the final operative paragraph, the future proceedings at the Scientific Committee could be conducted in a manner consistent with the resolution, but not very different from the past meetings. While no consensus view can be expected from the Scientific Committee, there will be scientific comments both for and against.

As there will be no IWC Commission biannual meeting in 2015, those who supported the resolution may request Japan to allow the Commission to consider the research plan and to make 'recommendations on the merits or otherwise of the special permit programme as it sees fit' in 2016 before initiating the research. This would delay the research for at least a year. Furthermore, because anti-whaling members have the simple majority at the Commission that is required for the adoption of a resolution/recommendation, it would not be realistic to expect the Commission to adopt a positive recommendation to support the research plan, even if Japan were to wait one year to allow the Commission to consider it.

A review of IWC history shows that more than a dozen resolutions against Japan's special permit research programs have been adopted in the past. Those resolutions are legally non-binding just like Resolution 2014–5; and they did not stop Japan's research programs. Japan also announced, at the press conference by the Chief Cabinet Secretary upon the adoption of Resolution 2014–5, that it would implement the new research plan from the austral summer of 2015/16.

However opposition from anti-whaling countries could be stronger than in the past because of the ICJ Judgment, even though the Judgment did not award a new authority to the Commission to effectively deny the right under Article VIII. As Japan remains firmly determined, this could lead to an even stronger confrontation than previously at the IWC. On the other hand, Japan expressed its commitment to follow the ICJ Judgment, and it did demonstrate its good faith through the termination of JARPA II and the preparation of a new research plan to respect the reasoning and conclusions of the ICJ Judgment. If there is sufficient willingness among the IWC members to take advantage of the ICJ Judgment for promoting positive dialogue, rather than confrontation, the IWC65 could mark a turning point in the whaling dispute.

8 New Antarctic Research Plan

On 18 November 2014 the Government of Japan submitted a draft of the New Scientific Research Program in the Antarctic Ocean (NEWREP-A) to the Chairman of the IWC Scientific Committee and the IWC Secretariat. The outline of the proposed research plan is set out in Annex 2, and the full draft document is available at the web site of the Fisheries Agency of Japan (http://www.jfa.maff.go.jp/j/whale/pdf/newrep--a.pdf).

In light of the reasoning and conclusions of the ICJ Judgment, the proposed research plan has incorporated several measures as outlined in Annex 2. For example, the draft plan describes investigations on the feasibility, practicability, and usefulness of several non-lethal research methods.

The submission of the draft research plan has initiated the IWC process to review the plan in accordance with Annex P, Report of the Scientific Committee; Process for the Review of Special Permit Proposals and Research Results from

Existing and Completed Permits (http://iwc.int/permits#guidelines). A small specialist workshop will take place at least 100 days before the annual Scientific Committee meeting in order to review the proposal in light of its stated objectives following the IWC guidelines. The workshop will review and comment on the proposal and produce its report for consideration by the Scientific Committee. The 2015 annual Scientific Committee meeting is to be held from 20 May in San Diego, the United States.

Paragraph 30 of the ICRW Schedule prescribes that the Scientific Committee will review and comment on the proposal. No approval or decision is required by the Scientific Committee or the IWC for the implementation of the plan. However, Japan has expressed its willingness to consider scientific comments to the proposed research plan and to make amendments and improvements if/when they are scientifically appropriate.

9 Conclusions

The whaling dispute is often dictated by perceptions rather than legal views and scientific information. The so-called commercial whaling moratorium adopted in 1982 represents a typical case, as it is understood as providing for the permanent prohibition of commercial whaling; as evidence that all whales are endangered; and as evidence of the world's acceptance of whales as special and charismatic animals. Contrary to this perception, Paragraph 10(e) of the Schedule, that established the moratorium, is a temporary measure to establish zero catch quotas, not a permanent prohibition of commercial whaling; and it itself sets out steps for resuming commercial whaling.

The ICJ Judgment could produce another false perception, namely, that it established a prohibition of all 'scientific whaling', even all whaling, even though the subject of the Judgment is limited to JARPA II and the ICJ clearly assumes possible future special permit programs as discussed above.

The ICJ Judgment is also significant as it could affect the role of science in international policy making. Such questions as 'what should be the role of science when there is a legal dispute' and, conversely, 'what should be the role of law when there is a scientific dispute' need to be further investigated in light of the ICJ Judgment.

The ICJ Judgment will likely affect the future discussions at the IWC. However, at this stage the direction of the change, if any, is unclear. The new Antarctic whale research program proposal is one of the focal points of interest in observing the developments of the whaling issue.

Annex 1: Relevant Paragraphs of the Schedule to the ICRW

7(*b*) In accordance with Article V(1)(*c*) of the Convention, commercial whaling, whether by pelagic operations or from land stations, is prohibited in a region designated as the Southern Ocean Sanctuary. This Sanctuary comprises the waters of the Southern Hemisphere southwards of the following line: starting from 40 degrees S, 50 degrees W; thence due east to 20 degrees E; thence due south to 55 degrees S; thence due east to 130 degrees E; thence due north to 40 degrees S; thence due east to 130 degrees W; thence due south to 60 degrees S; thence due east to 50 degrees W; thence due north to the point of beginning. This prohibition applies irrespective of the conservation status of baleen and toothed whale stocks in this Sanctuary, as may from time to time be determined by the Commission. However, this prohibition shall be reviewed ten years after its initial adoption and at succeeding ten year intervals, and could be revised at such times by the Commission. Nothing in this sub-paragraph is intended to prejudice the special legal and political status of Antarctica.

10(*d*) Notwithstanding the other provisions of paragraph 10 there shall be a moratorium on the taking, killing or treating of whales, except minke whales, by factory ships or whale catchers attached to factory ships. This moratorium applies to sperm whales, killer whales and baleen whales, except minke whales.

30. A Contracting Government shall provide the Secretary to the International Whaling Commission with proposed scientific permits before they are issued and in sufficient time to allow the Scientific Committee to review and comment on them. The proposed permits should specify:

(a) objectives of the research;
(b) number, sex, size and stock of the animals to be taken;
(c) opportunities for participation in the research by scientists of other nations; and
(d) possible effect on conservation of stock.

Proposed permits shall be reviewed and commented on by the Scientific Committee at Annual Meetings when possible. When permits would be granted prior to the next Annual Meeting, the Secretary shall send the proposed permits to members of the Scientific Committee by mail for their comment and review. Preliminary results of any research resulting from the permits should be made available at the next Annual Meeting of the Scientific Committee.

Annex 2

(Provisional Translation)
November, 2014
Fisheries Agency of Japan
Ministry of Foreign Affairs of Japan

Outline of the Proposed Plan for the New Scientific Whale Research Program in the Antarctic Ocean*

1 Research Title

NEWREP-A: *N*ew Scientific *W*hale *R*esearch *P*rogram in the *A*ntarctic Ocean

2 Research Objectives

(1) Improvements in the precision of biological and ecological information for the application of the Revised Management Procedure (RMP) to the Antarctic minke whale.
(2) Investigation of the structure and dynamics of the Antarctic marine ecosystem through building ecosystem models.

3 Research Area

Latitude: South of 60°S, Longitude: 0° to 120°W (the Management Areas III to VI defined by the International Whaling Commission (IWC)) (see Map 8.1).

4 Research Period

12 years (2015/16-2026/27, midterm review after the first six years).

5 Research Methods

(1) Lethal Survey
  a. Whale species: Antarctic minke whales
  b. Sample size: 333 animals
(a) As there is no other means than lethal methods, at this stage, the use of lethal method is indispensable to obtain age data which is necessary for estimating the age-at-sexual maturity (ASM), which makes considerable contribution to achieving the application of the RMP.

* This proposed plan takes account of the reasoning and conclusions contained in the Judgment by the International Court of Justice (ICJ) in the case concerning 'Whaling in the Antarctic' (Australia v. Japan: New Zealand intervening). Japan welcomes outside scientific comments. It will give due regard to such scientific comments and this proposed plan is thus subject to further elaborating and amendment if necessary.

(b) The sample size is limited to the number required for the estimation of the ASM with sufficient accuracy.
(c) Data obtained through lethal sampling will be utilized to the maximum extent to develop improved ecosystem models (Main Objective II).

(2) Non-lethal Surveys

In addition to the non-lethal methods employed by JARPA and JARPA II including sighting surveys for abundance estimation, biopsy sampling of skin tissue and oceanographic observations, the feasibility and practicability of the following non-lethal methods will be examined.
(a) Investigating the feasibility of biopsy sampling from Antarctic minke whales, especially in the offshore area in the Antarctic Ocean.
(b) Investigating the feasibility of age-determination methods other than ear-plug reading by analyzing DNA extracted from biopsy skin samples.
(c) Investigating the feasibility of tracking nutritional status indices by the analysis of retinol and saturated fatty acid extracted from biopsy samples instead of the measurement of body condition such as blubber thickness.
(d) Conducting satellite tagging on Antarctic minke whales to elucidate the location of their breeding grounds and using data-loggers for research on feeding behavior.

(3) Krill abundance survey

Simple surveys for estimating krill abundance using an echosounder will be conducted.

6 Research Vessels to be Used and Personnel to be Involved

Implementing Organization: Institute of Cetacean Research (ICR)
Research vessels: one research base vessel and a few sighting and sampling vessels

7 Backup Plan for Contingency

To minimize any negative influences of disruptions including sabotage activities by an anti-whaling NGO and bad weather conditions and to secure the scientific value of data, this research plan establishes a contingency backup plan including (a) adjustments of research protocols at the scene of disruption, (b) adjustment of the research plan and (c) consideration of alternative analytical methods.

Attachment 1

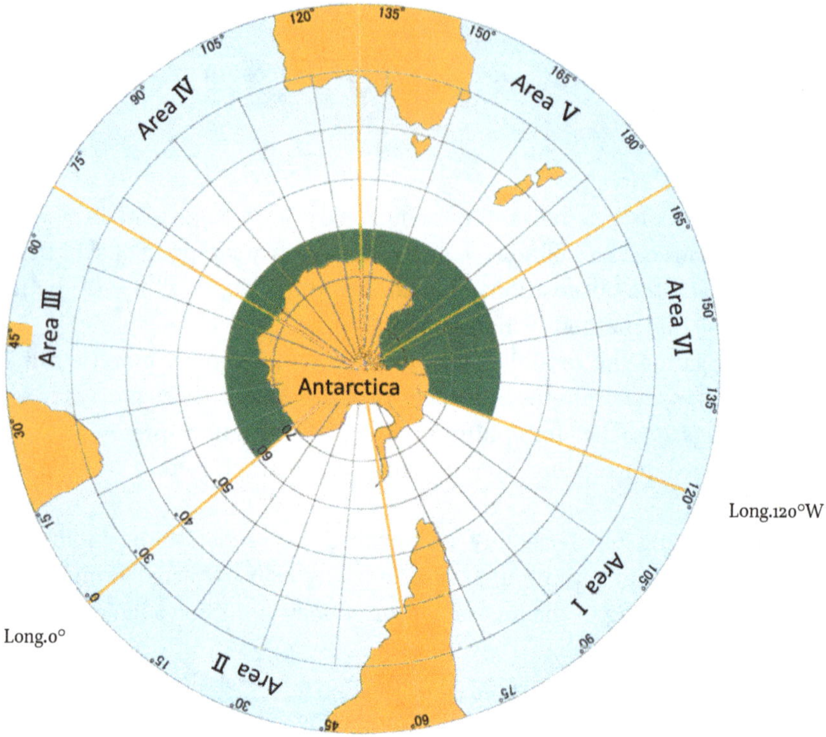

MAP 8.1 *Research Area under NEWREP-A*

8 Participation of Foreign Scientists and Collaboration with Other Researches/Organizations

Participation of foreign scientists will be welcomed and collaboration with other relevant research programs and institutions such as CCAMLR (Commission for the Conservation of Antarctic Marine Living Resources), the National Research Institute of Far Seas Fisheries and the National Institute of Polar Research will be strengthened.

Attachment 2

The New Proposed Plan's – Consideration of the Reasoning and Conclusions of the ICJ Judgment

In light of the reasoning and conclusions of the ICJ Judgment the following measures have been taken through a transparent process including securing the participation of outside experts.

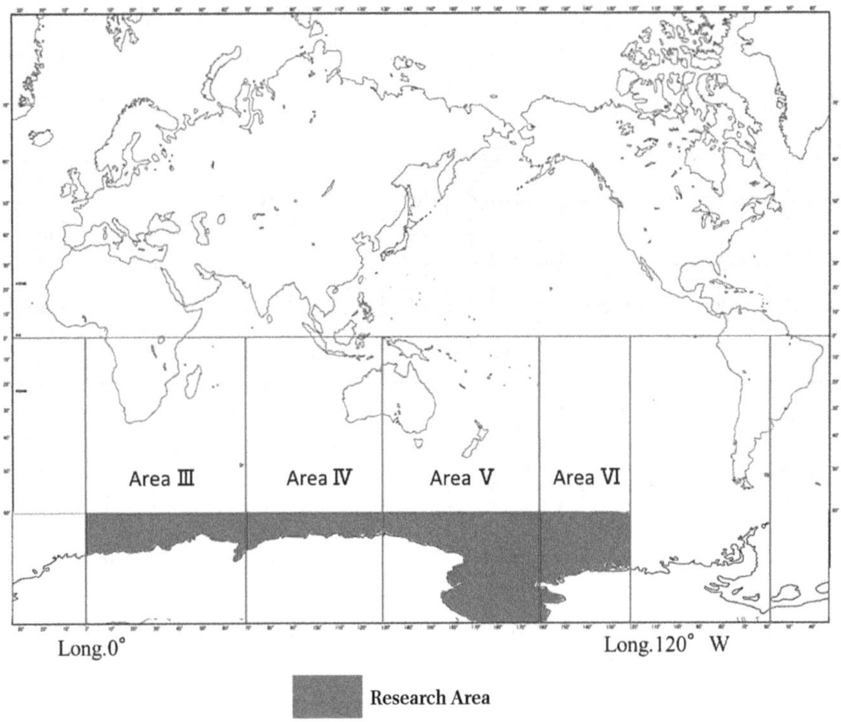

Elements identified in the ICJ Judgment	Measures taken in the new proposed plan
1. *The Decisions regarding the Use of Lethal Methods* The JARPA II Research Plan should have included some analysis of the feasibility of non-lethal methods as a means of reducing the planned scale of lethal sampling.	–As a result of examining necessary data for achieving the two main objectives, it was determined that it is not feasible to obtain age data by the currently available non-lethal research methods. However, the feasibility and practicability of the following non-lethal research methods will be examined. –investigating the feasibility of biopsy sampling to collect genetic samples from Antarctic minke whales. –investigating the feasibility and practicability and usefulness of age-determination and methods for tracking nutritional status using biopsy samples. –conducting trials of satellite tagging and data-logger on Antarctic minke whale.

Elements identified in the ICJ Judgment	**Measures taken in the new proposed plan**
2. *The Scale of Lethal Sampling and the Methodology used to Select Sample Sizes* The evidence relating to whale sample sizes provides scant analysis and justification for the underlying decisions that generate the overall sample size. This raises further concerns about whether the design of JARPA II is reasonable in relation to achieving its stated objectives.	–The sample size of Antarctic minke whales for the lethal research is determined by statistical testing and is limited to the number of samples with which age-at-sexual maturity data, an important element for implementing RMP and not obtainable by the currently available non-lethal research methods, can be estimated with sufficient accuracy.
3. *Discrepancy between the Target Sample Sizes and the Actual Take* Japan's statement that JARPA II can obtain meaningful scientific results based on the far more limited actual take suggests that the target sample sizes are larger than are reasonable in relation to achieving its stated objectives.	–As research activities could be disrupted by both natural and human factors including sabotage activities and bad weather conditions, the proposed plan establishes a contingency backup plan which addresses adjustment at the scene of disruption, adjustment of the research plan and consideration of alternative analytical methods.
4. *Time Frame Associated with a Program* JARPA II's open-ended time frame casts doubt on its characterization as a program for purposes of scientific research.	–This proposed plan has set its research period as 12 years. It has also established 'intermediary targets' with a system of mid-term review by the IWC Scientific Committee after 6 years.
5. *The Program's Scientific Output* In light of the fact JARPA II has been going on since 2005 and has involved the killing of about 3,600 minke whales, the scientific output to date appears limited.	–The scientific output will continue to be presented to the IWC Scientific Committee and there will be increased efforts to publish scientific achievements in peer-reviewed journals. –Scientific data generated from this research will be compiled in a database and be promoted to be widely utilized by outside scientists.

Elements identified in the ICJ Judgment	**Measures taken in the new proposed plan**
6. *Coordination with Related Research Projects* Some further evidence of co-operation between JARPA II and other domestic and international research institutions could have been expected in light of the program's focus on the Antarctic ecosystem and environmental changes in the Antarctic region.	–Deepening collaboration and coordination with relevant research institutions such as the Commission for the Conservation of Antarctic Marine Living Resources (CCAMLR), Japan's National Institute of Polar Research and Japan's National Research Institute of Far Seas Fisheries will be sought and strengthened in the planning, implementation, and data analysis related to surveys for estimating krill abundance. –The participation of foreign scientists in this research will be welcome.

PART 5

Domestic and International Implications of the Judgment

CHAPTER 9

The Whaling Case: An Australian Perspective*

Donald R. Rothwell

Introduction

Australia commenced proceedings in the International Court of Justice (ICJ) in May 2010 after a long running period of diplomatic activity at a bilateral and multilateral level to stop Japan's whaling programs in the Southern Ocean. First with JARPA (1987–2005), and then JARPA II (2005–2013), successive Australian governments sought to raise their concerns over Japan's whaling programs within the International Whaling Commission (IWC), with other conservation-minded States, and directly with Japan. Only after having had these multiple diplomatic efforts rebuffed, did Australia finally commence ICJ proceedings in 2010. Australia took this case to the ICJ against a backdrop of Australia and Japan enjoying strong bilateral relations post World War II, predominantly based upon an economic and trade relationship, but also increased cultural ties. That both countries mutually share the United States as a Pacific security partner also resulted in increased defence and security ties, especially since the 2001 terrorist attacks against the United States. However, notwithstanding the strength of the Australia–Japan bilateral relationship, Australia was prepared to take this dispute all the way to the ICJ in 2010. In that respect, Australia's previous experience before the court as both an applicant and respondent state gave it a degree of confidence in the ICJ as an appropriate forum for international dispute resolution.[1]

The court's 31 March 2014 judgement[2] brought to an end Australia's case against the international legal validity of JARPA II. Japan's subsequent

---

* Based on a paper originally presented at *Whaling in the Antarctic: The ICJ Judgment and Its Implications*, Kobe University Centre for International Law, Kobe University, Kobe, Japan, 31 May–1 June 2014; parts of this chapter are drawn from Donald R. Rothwell "The Antarctic Whaling Case: Litigation in the International Court and the Role Played by NGOs" (2013) 3 *The Polar Journal* 399–414.

1 Prior to the *Whaling Case*, Australia had appeared before the ICJ in three contentious cases once as an applicant in the *Nuclear Test Cases* (Australia v. France) (Interim Measures) [1973] ICJ Reps 99; (Merits) [1974] ICJ Reps 253; and twice as a respondent: *Case Concerning Certain Phosphate Lands in Nauru* (Nauru v. Australia) [1992] ICJ Reps 240; *East Timor* (Portugal v. Australia) (Judgment) [1995] ICJ Reps 90.

2 *Whaling in the Antarctic* (Australia v. Japan; New Zealand Intervening) Judgment of 31 March 2014 [2014] ICJ Reports.

 | DOI 10.1163/9789004313828_011

announcement that it would comply with the ICJ decision, revoke all JARPA II permits, and not undertake any whaling activity based upon special permits issued under Article VIII of the 1946 International Convention for the Regulation of Whaling (ICRW)[3] for the 2014/15 season, was a vindication of Australia's utilisation of the ICJ as a means of dispute settlement. It also represented the conclusion of legal argument that had been in development since 2005. The decision was also a vindication for the non-governmental organisation (NGO) community in Australia and internationally which had been fiercely opposed to both JARPA and JARPA II, and also for the decision of the Labor government of Kevin Rudd to implement its 2007 pre-election policies opposing JARPA II and to explore the commencement of international litigation against Japan if ongoing diplomatic efforts failed. However, Japan's November 2014 announcement of its intention to commence a new whale research program in the Southern Ocean as from 2015/16,[4] has raised new diplomatic and legal issues for Australia that will be assessed in light of the ICJ's decision in the *Whaling Case*.

This chapter will review these issues by first providing background with respect to Australia's historical and contemporary position with respect to whaling, litigation before the Australian courts over the conduct of JARPA II, the development of the Australian case before the ICJ over JARPA II and the role played by NGOs in that process, Australia's objectives and its arguments before the ICJ, and procedural issues before the court. Analysis will be undertaken of the judgment from an Australian perspective, and finally there will be some assessment of Japan's 2014 response to the judgment, and issues and challenges that may raise for Australia.

Australia and Whaling

Australia has a history of having been a whaling nation, with whaling a significant maritime industry during colonial times. Communities along the east, west and southern coasts of Australia reflect and in some cases celebrate the history of whaling. Towns such as Albany (Western Australia), Eden (New South Wales) and Victor Harbour (South Australia) actively promote their whaling history as part of their cultural heritage.[5] When Australia became a

3 International Convention for the Regulation of Whaling 161 UNTS 72.

4 "Proposed Research Plan for New Scientific Whale Research Program in the Antarctic Ocean (NEWREP-A)" (November 2014) available at <www.jfa.maff.go.jp/j/whale/pdf/newsrep--a .pdf> (30 December 2015).

5 See generally Andrew Darby, *Harpoon: Into the Heart of Whaling* (Allen & Unwin, Crows Nest, NSW: 2007).

nation in 1901, whaling remained an important maritime industry. While for a variety of reasons Australia did not became a party to the 1931 International Convention for the Regulation of Whaling, it did become a party to the 1937 International Agreement for the Regulation of Whaling.[6] Following World War II, Australia was a participant in the 1946 Washington conference that concluded the ICRW, signed the Convention on the day it opened for signature on 2 December 1946, ratified the Convention on 1 December 1947, and accordingly was an original party to the Convention when it entered into force on 10 November 1948. The Australian whaling industry continued to operate in the post-war years, albeit within the regulatory constraints imposed by the ICRW. Australia also actively explored developing its own Antarctic whaling industry in the post-war years, and at one stage considered pursuing the transfer of Japan's pre-war whaling fleet to Australia as part of post war reparations.[7] Eventually no Australian whaling interests in the Antarctic were actively pursued in the late 1940s–1950s, and economic factors gradually resulted in the decline of the Australian industry in the 1960s–1970s until such time that there remained only a single whaling station in Albany (Western Australia). In 1978 the Liberal/Country Party coalition government led by Malcom Fraser undertook an enquiry into the whaling industry and the subsequent Frost Inquiry report recommended the closure of the Australian whaling industry, including the Albany Whaling Station.[8] This set in train a process in which Australia joined an international campaign alongside other conservation minded countries to halt commercial whaling. This process focussed on the IWC, and pro-conservation members seeking to adjust the Schedule of the ICRW to reflect conservation objectives.[9] These developments were also reflected in Australian law at the time. The Whaling Act 1960 (Australia) was designed to facilitate and regulate the whaling industry in Australia at that time, framed around the ICRW. However, following the Frost Inquiry report the Fraser government enacted the Whale Protection Act 1980 (Australia) which placed an emphasis upon the protection of whales and did not anticipate the granting of permits for commercial whaling activities.

6 International Agreement for the Regulation of Whaling 190 LNTS 79.

7 David Day, *Antarctica: A Biography* (Knopf, North Sydney, NSW: 2012) 431.

8 Darby, note 5, 114–115; The story associated with the ending of Australia's whaling industry told from an NGO perspective can be found in Chris Pash, *The Last Whale* (Fremantle Press, Fremantle, WA: 2008).

9 The Schedule is attached to the ICRW and contains technical provisions that relate to the manner in which whaling can be conducted, individual provisions regarding certain species, and provisions for closed seasons. Importantly for the present discussion, the ICRW is subject to simple modification via a three-quarter majority of ICRW members states present at an IWC Annual Meeting.

The Development of the Antarctic Whaling Case

Under the ICRW, the state parties to the Convention have traditionally met annually at the IWC which is the principal forum with oversight of the ICRW and which also has capacity to make adjustments and modifications to the Schedule of the Convention. Over time the ICRW Schedule has been subject to considerable adjustment by the IWC and has gradually become more conservation orientated. This is particularly reflected in two provisions found within the ICRW Schedule. The first is paragraph 10(e) which provided for zero catch limits for the take of all whales for commercial purposes as from the 1985/86 pelagic season and the 1986 coastal season. This effectively created a moratorium on all commercial whaling. The second is found in paragraphs 7(a) and (b) which create an Indian Ocean Sanctuary and a Southern Ocean Sanctuary[10] within which all commercial whaling is prohibited. The effect of these two combined provisions is that all commercial whaling activities were prohibited within the Southern Ocean, which extended to those parts of the Southern Ocean where Australia had historic interests as a result of its engagement in Antarctic affairs, its claim to the Australian Antarctic Territory (AAT), and subsequent role in the negotiation of the 1959 Antarctic Treaty and development of the Antarctic Treaty System.[11]

Notwithstanding the moratorium on commercial whaling, reference is made to the lethal take of whales in Article VIII of the ICRW which provides that:

> ...any Contracting Government may grant to any of its nationals a special permit authorizing that national to kill, take and treat whales for purposes of scientific research...and the killing, taking and treating of whales in accordance with the provisions of this Article shall be exempt from the operation of this Convention.

Whaling conducted under Article VIII is commonly referred to as 'special permit' or 'scientific' whaling. Importantly for present purposes, other than through

10 Which became effective from the 1994/5 season. Whaling in the Southern Ocean is conducted during the southern summer and traditionally straddles a period between December–March.

11 For background on Australia's interests in Antarctica see Gillian Triggs, *International Law and Australian Sovereignty in Antarctica* (Legal Books, Sydney: 1986); Donald R. Rothwell and Shirley V. Scott "Flexing Australian Sovereignty in Antarctica: Pushing Antarctic Treaty Limits in the National Interest?" in Lorne K. Kriwoken, Julia Jabour and Alan D. Hemmings (eds) *Looking South: Australia's Antarctic Agenda* (The Federation Press, Annandale: 2007) 7–20.

non-binding Resolutions of the IWC or interpretations of Article VIII by the IWC Scientific Committee, this provision had not been subject to any authoritative interpretation by any court or tribunal.

The introduction of the commercial whaling moratorium proved to be controversial and was subject to objection by Japan, Norway, Peru, and the then Union of Soviet Socialist Republics. Japan, however, withdrew its objection with effect from 1 May 1987 with respect to commercial pelagic whaling and in the same year announced its intention to commence JARPA in the 1987/88 season. The conduct of special permit whaling following the introduction of the moratorium was the subject of a number of IWC Resolutions in 1986,[12] 1987[13] and during the 1990s in which the Commission continually expressed its concern about the issuing by Contracting Governments of special permits. In 2001[14] and 2003[15] the IWC called upon Japan to halt the lethal take of minke whales and revise its research program to focus on non-lethal means of research. As JARPA was conducted in the Southern Ocean, issues were also raised within the IWC as to its consistency with the Southern Ocean Sanctuary.[16]

Japan introduced the Japanese Whale Research Program under Special Permit in the Antarctic (JARPA) in the 1987/88 season and continued this program until the 2004/5 season. JARPA had a principal focus upon minke whales with initially, a sample size of 300 (±10%) being taken each season for research purposes. From the 1995/96 season the sample size was increased to 400 (±10%). At the end of the 2004/2005 Southern Ocean season Japan announced its intention to conduct the Second Phase of the Japanese Whale Research Program under Special Permit in the Antarctic (JARPA II) from the 2005/6 season. Feasibility studies were to be conducted for the first two years with the full-scale program to commence from the 2007/8 season. Projected annual sample sizes for the full-scale research were 850 (±10% allowance) minke whales, 50 humpback whales, and 50 fin whales. The first two years of JARPA II were to be a feasibility study commencing with the sampling each year of a maximum of 850 (±10% allowance) minke whales and 10 fin whales. No humpback whales were scheduled to be taken during the period of the feasibility study. JARPA II was the subject of discussion at the 57th Annual Meeting of the

12 IWC Resolution 1986–1 'Resolution on Special Permits for Scientific Whaling'.
13 IWC Resolution 1987–1 'Resolution on Scientific Research Programs'.
14 IWC Resolution 2001–7 'Resolution on Southern Hemisphere Minke Whales'.
15 IWC Resolution 2003–3 'Resolution on Southern Hemisphere Minke Whales and Special Permit Whaling'.
16 IWC Resolution 1996–7 'Resolution on Special Permit Catches by Japan'; IWC Resolution 1998–4 'Resolution on Whaling under Special Permit'.

IWC held at Ulsan, Republic of Korea, from 20–24 June 2005. The Chair's Summary Report of the meeting notes that 'There was considerable disagreement over the value of this research both within the Scientific Committee and the Commission'.

The announcement by Japan that it intended to substantially expand the take of whales as part of an Article VIII 'special permit' program was the catalyst for more detailed consideration being given to exploring international legal options to challenge the legitimacy of JARPA II.

International Fund for Animal Welfare and International Litigation

The International Fund for Animal Welfare (IFAW) has had a long standing interest in whaling and has been a regular attendee at the IWC. Following Japan's announcement of its intention to commence JARPA II, IFAW began to explore the development of an international legal strategy that pro-conservation governments could pursue through international courts to challenge the international legal validity of JARPA II. It was a strategy that initially had a global reach, but increasingly began to focus on Australia and New Zealand. That Australia decided in 2010 to commence proceedings in the ICJ against Japan, and New Zealand in 2012 sought to intervene in the proceedings in its capacity as an ICRW party, demonstrates how successful IFAW's strategy ultimately was.

Between 2006–2009 IFAW sponsored four international legal panels to consider various aspects of the development of legal argument and legal proceedings against Japan regarding the conduct of JARPA II. They were as follows:

- Paris Panel of May 2006 resulting in the *Report of the International Panel of Independent Legal Experts On: Special Permit ('Scientific') Whaling under International Law*;[17]
- Sydney Panel of December 2006 resulting in a report titled *Japan's Special Permit ('Scientific') Whaling under International Law*;[18]
- London Panel of November 2007 resulting in a report titled *The Taking of Sei and Humpback Whales by Japan: Legal Issues Arising under the Convention of International Trade in Endangered Species of Wild Fauna and Flora* (CITES);[19]

17 International Fund for Animal Welfare, *Report of the International Panel of Independent Legal Experts On: Special Permit ('Scientific') Whaling under International Law* (12 May 2006).

18 International Fund for Animal Welfare, *Japan's Special Permit ('Scientific') Whaling under International Law* (22 December 2006) (on file with author).

19 International Fund for Animal Welfare, *The Taking of Sei and Humpback Whales by Japan: Legal Issues Arising under the Convention of International Trade in Endangered Species of*

- Canberra Panel of January 2009 resulting in a report titled *Japan's 'Scientific' Whaling Program and the Antarctic Treaty System*;[20]

The work of each of the four panels will now be briefly discussed.

The Paris Panel comprised a group of eminent international legal experts[21] who were asked by IFAW to assess what conditions a state party to the ICRW entitled to carry out scientific whaling should have regard to in relation to Article VIII of the Convention and other international instruments, whether scientific whaling being carried out by some members of the IWC was consistent with the ICRW, and the consequences that would follow if the IWC were to adopt a resolution indicating that scientific whaling conducted by a member of the IWC was lawful. The Panel concluded that the whaling conducted by some members of the IWC did not meet the requirements of paragraph 30 of the Schedule and therefore did not meet the exemption provided for in Article VIII of the ICRW and as such was unlawful.[22] It was considered that there was 'strong evidence' that the scientific whaling conducted by some members of the IWC was in violation of the moratorium on commercial whaling.[23] The Panel also assessed the relevance of the 1982 United Nations Convention on the Law of the Sea (LOSC),[24] and identified 11 articles of that instrument where scientific whaling raised 'serious questions of compliance'.[25] Likewise, similar serious questions were raised with respect to compliance with the Convention on Biological Diversity,[26] Convention on International Trade in Endangered Species of Wild Fauna and Flora (CITES),[27] and 1980 Convention on the Conservation of Antarctic Marine Living Resources.[28] In light of those findings it was considered that the conduct of scientific whaling

---

Wild Fauna and Flora (CITES) (1 November 2007) available at <www.ifaw.org/australia/node/1124> (30 January 2015).

20 International Fund for Animal Welfare, *Japan's 'Scientific' Whaling Program and the Antarctic Treaty System* (12 January 2009) (on file with author).

21 The members of the Paris Panel were Professor Laurence Boisson De Chazournes, Professor Pierre-Marie Dupuy, Professor Donald R. Rothwell, Professor Philippe Sands (Coordinator), Ambassador Alberto Székely (Coordinator), William H. Taft IV, and Kate Cook (Rapporteur).

22 International Fund for Animal Welfare, note 17, [2].

23 Ibid.

24 United Nations Convention on the Law of the Sea, 1833 United Nations Treaty Series 397.

25 International Fund for Animal Welfare, note 17, [2].

26 Convention on Biological Diversity, 1760 United Nations Treaty Series 79.

27 Convention on International Trade on Endangered Species of Wild Fauna and Flora, 993 United Nations Treaty Series 243.

28 Convention on the Conservation of Antarctic Marine Living Resources, 1329 United Nations Treaty Series 47; see International Fund for Animal Welfare, note 17, [2].

by some members of the ICRW, including Japan, was not consistent with the ICRW. Particular reference was made to JARPA II.[29] Note was made of the failure by Japan to demonstrate to the IWC Scientific Committee that its special permit whaling had been authorised in exceptional circumstances. Finally, the Paris Panel concluded that a Resolution adopted by the IWC determining that such scientific whaling was lawful would not be capable of altering the legal obligations arising under the ICRW in relation to the prohibition of commercial whaling.[30]

The Paris Panel Report was initially distributed by IFAW to pro-conservation members of the IWC, and advocacy work was undertaken on the report during the 2006 IWC meeting in St. Kitts and Nevis. The Report was subsequently made available to the public, a copy was posted on the IFAW website, and hard copies were printed for distribution. It provided a framework for the legal analysis undertaken by the later Legal Panels and laid the groundwork in identifying the key substantive legal issues that would be relied upon in seeking to contest the legality of Japan's conduct of JARPA II.

The Sydney Panel comprised a group of Australia international law experts,[31] and was asked by IFAW to consider the legality of Japanese whaling in Antarctica and the options available to Australia and New Zealand to pursue international legal action against the Government of Japan in relation to its Antarctic whaling activities under JARPA II. The Panel's report was concluded in December 2006, and delivered to the Australian and New Zealand Governments in January 2007. The Panel noted that the extension of JARPA II to humpback whales was especially significant to Australia given the annual migration of that species along the Australian coastline. The potential economic impact of the killing of humpback whales was also noted given the growth of whale-watching as a tourist and recreational activity.

In its Report, the Panel found that Australia and New Zealand had a number of international legal options open to them to challenge the ongoing conduct by Japan of JARPA II. The legal options canvassed extended from the sponsoring of meetings of scientific, legal and policy experts to review alternative options for the resolution of the dispute within the IWC, to potential international litigation before the ICJ. The Panel also found that it was possible to raise concerns about Japan's scientific whaling program within other international

29 International Fund for Animal Welfare, note 17, [2].

30 Ibid.

31 The members of the Sydney Panel were Dr Natalie Klein, Associate Professor Greg Rose, Professor Donald R. Rothwell (Chair), Professor Ivan Shearer, Dr Tim Stephens (Rapporteur), and Dr Christopher Ward.

forums, including the Antarctic Treaty System. The principal conclusions of the Panel included:

1. That Australia and New Zealand consider sponsoring meetings of legal, scientific and policy experts to discuss options to resolve the dispute with Japan and the future of whale conservation within the International Whaling Commission.
2. That Australia and New Zealand consider requesting Japan to agree to an ad hoc arbitration of the dispute on mutually agreeable terms.
3. That Australia and New Zealand seek to raise their concerns over the conduct of JARPA II before the Antarctic Treaty parties, including at the Commission for the Conservation of Antarctic Marine Living Resources.
4. That Japan's actions under JARPA II may be contrary to the requirement to undertake environmental impact assessment under the provisions of the Convention on Biological Diversity.
5. That Japan's authorisation of JARPA II may lead to breaches of CITES.
6. That Australia and New Zealand consider commencing a legal claim before the International Tribunal for the Law of the Sea (ITLOS) seeking compulsory settlement of a dispute under the LOSC.
7. That Australia and New Zealand also consider commencing a legal claim before the ICJ arguing that Japan's conduct of JARPA II was contrary to the ICRW.

In reaching these conclusions, the Sydney Panel noted that ultimately it was a matter for the Australian and New Zealand governments to weigh up which were the most suitable for pursuing action against Japan in relation to JARPA II.

The London Panel comprised a group of eminent international lawyers[32] who were asked by IFAW to advise on a series of specific questions arising with respect to Japan's take of sei whales from the North Pacific and humpback whales from the Southern Ocean under the provisions of CITES and the remedies available to State Parties to CITES in the event that Japan's take of those whales were found to be illegal. The London Panel, which drew upon the work of the Paris Panel, concluded that the take of either sei or humpback whales would encompass international trade in the form of 'introduction from the sea' under CITES, which would require the prior grant of a certificate.[33] If Japan were to issue such a certificate, given the 'circumstances where the specimens

32 The members of the London Panel were Kate Cook (Rapporteur), Peter H. Sand, Professor Philippe Sands, and Ambassador Alberto Székely (Coordinator).

33 International Fund for Animal Welfare, note 19 [5].

were to be used for primarily commercial purposes', such actions would be invalid and Japan's actions would constitute a breach of CITES.[34] It was also found that there was evidence to suggest that the issuing of any such certificates would be 'detrimental to the survival of the species', which raised separate issues as to Japan's compliance with CITES.[35] The London Panel report considered a range of remedies available to States Parties to CITES, including requesting an investigation by the CITES Standing Committee of an alleged breach, negotiation and arbitration options available under the CITES framework, and potential dispute resolution options available before the ICJ and under the LOSC.[36]

The Canberra Panel comprised a group of Antarctic law and policy experts,[37] convened by IFAW to provide an independent strategic assessment of the options available to Australia, New Zealand and other likeminded states to effect scrutiny of Japan's Antarctic whaling operations through the Antarctic Treaty System (ATS). The Canberra Panel was asked to examine methodologies, timelines and mechanisms for placing the environmental risks of Japan's Antarctic whaling program on the agenda of future Antarctic Treaty Consultative Meetings (ATCM) and other ATS forums. The work of the Canberra Panel directly built upon the Paris, Sydney and London Panel reports. The central recommendation of the Canberra Panel was that the provisions of the Madrid Protocol were relevant to associated environmental impacts arising from the conduct of JARPA II and could be invoked against Japan. Specifically, it was found that the compliance provisions found in Article 13 of the Madrid Protocol oblige Japan to justify to other parties to the ATS that JARPA II did not have adverse impacts on the Antarctic environment, and also allow other parties concerned at such impacts to raise the issue for discussion in the Antarctic Treaty Consultative Meeting (ATCM). Such an approach would bring about increased international attention to the substantial environmental risks connected with JARPA II, hazards that were clearly demonstrated by the explosion and fire on board the whaling factory ship the *Nisshin Maru* in an ice-exposed area of the Ross Sea in 2007. The Canberra Panel took the view that the primary objective in turning to the ATS was to make Japan more accountable for its conduct of JARPA II. This could impose additional safeguards for the Antarctic

34 Ibid.

35 Ibid.

36 Ibid.

37 The members of the Canberra Panel were Dr Alan Hemmings, Professor Stuart Kaye, Joanna Mossop, Professor Donald R. Rothwell (Chair), Dr Tim Stephens (Rapporteur), and Professor Gillian Triggs.

environment in relation to JARPA II in line with improvements achieved in relation to ship-borne tourism operations, which raise many similar environmental concerns.

Significantly for the purposes of this analysis, Australia instituted proceedings against Japan in the ICJ on 31 May 2010 with respect to a dispute concerning Japan's JARPA II program on 'scientific whaling'.[38] The commencement of this claim by Australia can be seen as a result of the role that was played by IFAW in the development of litigation strategies for Australia and other like-minded countries in contesting the legality of JAPRA II before an international court or tribunal. At this point in time the precise impact of IFAW's advocacy on these matters is not possible to identify, and will most likely only be revealed once relevant government Ministers who were involved in the decision-making processes associated with the decision to commence the proceedings against Japan write their biographies, or when Australian government archives make available relevant government documents. Nevertheless, on the basis of the narrative above, the connection between the international legal options considered in the Paris and Sydney Panel reports, and the fact that three members of the Paris Panel were also members of the legal team that represented Australia in the ICJ against Japan,[39] it can be assumed that IFAW's work in developing and promoting the legal case against Japan not only influenced the Australian government to commence proceedings against Japan but also found its way into the actual legal argument that Australia developed and presented to the ICJ during the oral phase of the case in 2013.

Australia's Commencement of Proceedings before the ICJ

Australia's 2010 application commencing proceedings against Japan primarily related to a dispute based upon the interpretation of the ICRW and addressed many of the points that have been noted above. In the Australian Application it was noted that both Australia and Japan had accepted the jurisdiction of the Court by virtue of declarations of acceptance lodged under Article 36(2) of the

38 *Dispute Concerning Japan's JARPA II Program on "Scientific Whaling"* (Australia v. Japan) Application Instituting Proceedings (31 May 2010); details on all the documents filed with the ICJ by Australia and Japan, and the proceedings to date can be found at www.icj-cij.org/docket/ (30 January 2015).

39 Those members of the Paris Panel who were also members of the Australian legal team were Professor Laurence Boisson De Chazournes, Kate Cook, and Professor Philippe Sands.

Statute of the International Court of Justice[40] made respectively by Australia on 22 March 2002 and Japan on 9 July 2007. The Australian Application related the history of the IWC endorsed moratorium on commercial whaling and endorsement of the Southern Ocean Sanctuary. Japan's conduct of JARPA and JARPA II was noted and details given as to the status of minke, fin and humpback whale stocks. Australia also made reference to the refusal of Japan to accept IWC recommendations urging it to halt or vary the JARPA or JARPA II programs. The Australian Application also referred to the ongoing negotiations within the IWC for the so-called 'reform' of the Commission and to the negotiations conducted by the 'Small Working Group' between 2008–2010. As part of that process on 22 April 2010 the IWC Chair and Vice-Chair had produced a 'Consensus Decision to Improve the Conservation of Whales'[41] which was the subject of intense discussions at IWC62 in Agadir, Morocco in June 2010. However, on 23 June 2010 it was announced that further discussions on this IWC agenda item would be suspended. The Australian Application, while lodged before the commencement of IWC62, nevertheless observed that '[i]t has become clear that current and proposed IWC processes cannot resolve the key legal issue that is the subject of the dispute between Australia and Japan, namely the large-scale "special permit" whaling under JARPA II'.[42] The Australian Application concluded by asserting that Japan had breached and was continuing to breach the obligation under paragraph 10(e) of the ICRW Schedule to observe the zero catch limit in relation to the killing of whales for commercial purposes, and its obligation under paragraph 7(b) to act in good faith and refrain from undertaking commercial whaling of humpback and fin whales in the Southern Ocean Sanctuary.[43] The remedies Australia sought included that Japan cease implementation of JARPA II, that it revoke any authorisations, permits or licenses, and provide assurances that it will guarantee that any further action under JARPA II or similar programs be brought into conformity with international law.[44]

An analysis of the decision by Australia to commence legal proceedings against Japan in the ICJ needs to be set against the political backdrop of the relative positions the two major political parties in Australia: the Liberal party and the Australian Labor Party (ALP). John Howard was the Liberal Prime

---

40 Statute of the International Court of Justice 156 United Nations Treaty Series 77.

41 IWC 62/7 Agenda Item 3 (2010).

42 *Dispute Concerning Japan's JARPA II Program on 'Scientific Whaling'* (Australia v. Japan) Application Instituting Proceedings (31 May 2010) [29].

43 *Dispute Concerning Japan's JARPA II Program on 'Scientific Whaling'* (Australia v. Japan) Application Instituting Proceedings (31 May 2010) [36].

44 Ibid. [41].

Minister of Australia from 1996–2007, leading a coalition government comprising the Liberal and National parties. Successive Howard government Environment Ministers had been proactive in national and international debates critical of Japan's conduct of its JARPA program and been vocal at the IWC in opposing the conduct of 'scientific whaling'.[45] However, Japan's 2005 announcement regarding JARPA II significantly changed the discourse with respect to Japan's Southern Ocean whaling programs. This presented an opportunity for NGOs to develop new campaign approaches, and also for political parties to respond to that development. In this regard, the Howard Government did not significantly alter its position with respect to Japan's conduct of JARPA II, with the then Minister for the Environment, Senator Ian Campbell, indicating in 2005 that the government would not pursue international legal options to challenge the conduct of JARPA II.[46] The ALP Opposition, under the leadership of Kevin Rudd, was more open to considering legal options and in November 2005 indicated that 'Australia must take Japan to the International Tribunal for the Law of the Sea to end the barbaric slaughter of whales once and for all'.[47]

With the Howard Government refusing to commit to a JARPA II legal challenge during 2005–2007, further opportunities were presented to the Opposition to directly rely upon the work of the IFAW Legal Panels as the ALP began to settle upon a clear policy position in advance of the 2007 election. Here the work of the Sydney Panel would appear to have been influential. The Sydney Panel concluded its report in December 2006, and IFAW briefings were quickly given to both the Australian and New Zealand governments while the report remained confidential. The Sydney Panel report was then given a wider distribution, including to the then ALP Opposition Environment spokesperson, Peter Garrett, who on 7 May 2007 directly endorsed the Sydney Panel's advice. The ALP indicated that the Sydney Panel's report made 'a forceful case for taking Japan to international courts to stop its illegal whaling'.[48] Further endorsement of the Sydney Panel's work was received from the Australian

45 See Ian Campbell "Whaling-Commercial Whaling" (2007) 26 *Australian Year Book of International Law* 395; Senator Campbell was Minister for the Environment from 2004–2007.

46 AAP "Legal action won't stop whaling, says Campbell" *The Sydney Morning Herald* (November 8, 2005) available at www.smh.com.au/news/world/legal-action-wont-stop-whaling-says-campbell/2005/11/08/1131212030846.html (30 January 2015).

47 Anthony Albanese MP, "Howard Government can stop the whale slaughter, but refuses to act" (8 November 2005) at <www.anthonyalbanese.com.au/file.php?file=/news/870/index.html> (30 January 2015).

48 Peter Garrett MP, "Forceful Case to end Whale Slaughter" Media Statement (7 May 2007) (on file with the author).

Greens who issued a statement calling upon the Howard Government to pursue Japan through the 'international courts'.[49] These developments, especially throughout 2007, show how momentum was building in support of the IFAW Legal Panel reports and especially how the opposition parties in Australia were able to use their endorsement of the work of the Sydney Panel in particular as a point of distinction from the then official Howard government position that international litigation on this matter was not viable.

Throughout the period 2005–2007, IFAW was therefore able to build a sustained campaign in response to Japan's announcement that JARPA II would commence as a full scale scientific research program as from the 2007/2008 season. IFAW was able to highlight the distinctions between JARPA and JARPA II not only with respect to the anticipated take of whales, but also Japan's announced intention to expand the take from minke whales to include fin and humpback whales. Given the annual migration of humpback whales along the eastern Australian coastline in particular, and the growth of the whale watching industry, Japan's intention to take humpback whales under JARPA II provided more of a direct link for the Australian public between Japan's whaling program in the Southern Ocean and whales that are observed during the annual migration along the Australian coast.[50] IFAW was able to skilfully use the media during this period to promote the work of the Legal Panels, especially the Sydney Panel,[51] to highlight the potential for an alternate approach towards the whaling issue other than the continuation of debates within the IWC. That IFAW's campaign was taking place in a volatile pre-election environment gave added oxygen to the campaign in the media, which was further enhanced by the position taken by the Howard government of opposing international legal action against Japan in favour of a continuation of diplomacy.[52] The endorsement by the ALP of the IFAW Legal Panel's work meant that following the November 2007 Federal election there was a strong expectation that the new Rudd ALP government would honour its pre-election commitments and immediately commence legal action against Japan. Ultimately, a more conservative position was taken as reflected in the joint statement by the

49 Australian Green Party "International experts: Japanese whaling unlawful" (1 June 2006) (on file with the author).

50 See Darby, note 5, 194–209 discussing humpback whales and their migration along the Australian coastline.

51 International Fund for Animal Welfare, "Australian Govt can stop Japan Whaling" News Release (7 May 2007) (on file with author); ABC Online "Govt urged to take Japan to court over whale hunt" (6 May 2007) available at www.abc.net.au/news/newsitems/200705/s1915705.htm (30 January 2015).

52 See Andrew Darby, "Court fight urged to save whales" *The (Melbourne) Age*, 8 May 2007, p. 7.

Minister for the Foreign Affairs, Stephen Smith, and Peter Garrett as Minister for the Environment on 19 December 2007,[53] though it was made clear the government was actively considering its legal options.

Between December 2007 and March 2010 Australia continued to pursue a number of strategies with respect to its campaign against JARPA II. Diplomatically, Australia appointed a 'Whales Envoy' whose role was separate from the position of IWC Commissioner and who sought to explore options for the diplomatic settlement of what was now clearly emerging as a disagreement between Australia and Japan over JARPA II. Australia also advanced various positions within the IWC in which it sought to promote non-lethal research into whales, while also contributing to ongoing IWC debates about the reform of the Commission and whether alternate positions could be endorsed with respect to the status of the moratorium and agreement being reached upon a 'Revised Management Plan' for whaling.[54] Australia separately also sought additional legal advice regarding its options to challenge JARPA II before an international court or tribunal. To that end, Australia dispatched an Australian government customs vessel to the Southern Ocean in January–February 2008 to monitor JARPA II activities. Australia did not interfere with Japan's conduct of JARPA II but did collect photographic evidence of Japan's Article VIII program which it indicated could be used in the event of international litigation being commenced.[55] By 2010 Australia's position on whaling gradually became clearer with Prime Minister Rudd announcing that Australia was looking at commencing proceedings in the ICJ. The Australian position was made clear on 28 May 2010 at a joint press conference of the Minister for Environment Protection, Heritage and the Arts, the Attorney-General, and Minister for Foreign Affairs in which it was indicated that Australia would in

53 See statement by Minister for Foreign Affairs, Mr Stephen Smith, and Minister for the Environment, Heritage and the Arts, Mr Peter Garrett reproduced in "Australian Practice in International Law 2007: Whaling" (2009) 28 *Australian Year Book of International Law* 377–378.

54 See statements made by Minister for the Environment, Heritage and the Arts, Mr Peter Garrett reproduced in "Australian Practice in International Law 2008: Whaling – International Whaling Commission (IWC)" (2010) 29 *Australian Year Book of International Law* 359–360; "Australian Practice in International Law 2009: Whaling – International Whale Conservation" (2012) 30 *Australian Year Book of International Law* 334–335.

55 See statement made by Minister for Foreign Affairs, Mr Stephen Smith, the Minister for Home Affairs, Mr Bob Debus, and the Minister for the Environment, Heritage and the Arts, Mr Peter Garrett reproduced in "Australian Practice in International Law 2008: Whaling – International Whaling Commission (IWC)" (2010) 29 *Australian Year Book of International Law* 361.

fact commence proceedings against Japan in the ICJ over JARPA II.[56] The Australian application was filed before the ICJ on 31 May 2010.

While the above analysis has focussed upon Australia, it should not be forgotten that the IFAW campaign also included New Zealand and the work of the Sydney Panel in particular identified individual and joint legal options for both Australia and New Zealand.[57] While IFAW's campaign on this issue in New Zealand did not match the scale of the Australian campaign, New Zealand was clearly supportive of Australia's approach in its exploration of international legal strategies and ultimately elected to intervene in the ICJ proceedings in 2012 in support of the Australian interpretation of the ICRW that Japan's conduct of JARPA II was not consistent with the convention.

Australia's Application and Objectives before the ICJ

Australia announced on 28 May 2010 that it would commence legal proceedings against Japan in the ICJ over JARPA II,[58] and the Australian Application instituting proceedings was filed on 31 May.[59] At the time of Australia's 28 May announcement, the Minister for Foreign Affairs, Attorney-General, and the Minister for Environment Protection jointly indicated that:

> The decision underlies the Government's commitment to bring to an end Japan's program of so-called 'scientific' whaling in the Southern Ocean. It also demonstrates our commitment to do what it takes to end whaling globally. ...the Australian Government will keep working tirelessly to

56 See statement by Minister for Foreign Affairs, Mr Stephen Smith, and Minister for the Environment Protection, Heritage and the Arts, Mr Peter Garrett, and the Attorney-General, Mr Robert McClelland reproduced in "Australian Practice in International Law 2010: Whaling – International Court of Justice" (2013) 31 *Australian Year Book of International Law* 246–247.

57 Andrew Darby "Tensions ease in anti-whaling fight" *The* (Melbourne) *Age*, February 14, 2007, available at <www.theage.com.au/news/national/tensions-ease-in-antiwhaling-fight/2007/02/13/1171128973983.html> (30 January 2015).

58 See statement by Minister for Foreign Affairs, Mr Stephen Smith, and Minister for the Environment Protection, Heritage and the Arts, Mr Peter Garrett, and the Attorney-General, Mr Robert McClelland reproduced in "Australian Practice in International Law 2010: Whaling – International Court of Justice" (2013) 31 *Australian Year Book of International Law* 246–247.

59 *Dispute Concerning Japan's JARPA II Program on "Scientific Whaling"* (Australia v. Japan) Application Instituting Proceedings (31 May 2010).

> achieve an end to whaling in the Southern Ocean, and we will use all legal and diplomatic avenues to achieve our goal.[60]

The Australia Application comprised 20 pages, with an accompanying Annex.[61] In addition to certain pro forma procedural matters such as notification of the appointment of Agent and Co-Agent, the Australian Application addressed matters relating to the Court's Jurisdiction, the Content of the Dispute, Obligations Breached by Japan, and Remedies sought by Australia. The matter of jurisdiction is only briefly addressed, indicating that the Court has jurisdiction under Article 36 (2) of the Statute as a result of the declarations made by Australia and Japan on 22 March 2002 and 9 July 2007 respectively.[62] The Content of the Dispute relates to matters that have already been noted above with respect to the moratorium on commercial whaling, the Southern Ocean Sanctuary, JARPA and JARPA II, IWC negotiations, and Japan's refusal to comply with requests to halt JARPA II. Preliminary material is also presented as to the status of minke, fin and humpback whale stocks.[63] Australia asserted that Japan had breached and was continuing to breach its obligations under the ICRW with respect to the zero catch limit under paragraph 10(e) of the Schedule, and paragraph 7(b) of the Schedule with respect to the Southern Ocean Sanctuary. Australia also asserted that:

> ...having regard to the scale of the JARPA II program, to the lack of any demonstrated relevance for the conservation and management of whale stocks, and to the risks presented to targeted species and stocks, the JARPA II program cannot be justified under Article VIII of the ICRW.[64]

Australia further asserted that Japan's conduct was in breach of obligations under the 1973 Convention on International Trade in Endangered Species of Wild Fauna and Flora,[65] and the 1992 Convention on Biological Diversity.[66]

60 Smith, Garrett, and McClelland, note 58.

61 *Dispute Concerning Japan's JARPA II Program on "Scientific Whaling"* (Australia v. Japan) Application Instituting Proceedings (31 May 2010). The Annex is a 21 December 2007 Aide Memoire from 30 IWC members, including Australia, to Japan raising their concerns over the conduct of JARPA II.

62 Ibid. [4].

63 Ibid. [13]–[16].

64 Ibid [37].

65 Ibid [38].

66 Ibid.

With respect to remedies sought, Australia asked the ICJ to 'adjudge and declare that Japan is in breach of its international obligations in implementing the JARPA II program in the Southern Ocean' and additionally that Japan be ordered to:

(a) cease implementation of JARPA II;
(b) revoke any authorisations, permits or licences allowing the activities which are the subject of this application to be undertaken; and
(c) provide assurances and guarantees that it will not take any further action under the JARPA II or any similar program until such program has been brought into conformity with its obligations under international law.[67]

Parallel to the formal Australian Application, it is also relevant to reflect upon the 28 May 2010 joint Ministerial Media Release in which Ministers Smith, Garrett and McClelland make reference to the Australia/Japan bilateral relationship:

> Australia and Japan share a comprehensive strategic, security and economic partnership. We share a substantial commercial relationship built over many decades, growing strategic and security linkages, and work together closely in key international forums such as the G20, United Nations, the World Trade Organisation and APEC. The Government's action today reflects a disagreement in one element of a relationship that is deep, broad and multi-dimensional. Both Australia and Japan have agreed that, whatever our differences on whaling, this issue should not be allowed to jeopardise the strength and the growth of our bilateral relationship.[68]

Australia's objectives were therefore relatively limited and focussed upon legal argument and relief framed around an interpretation of JARPA II, but were aimed at ensuring that if the ICJ accepted its arguments JARPA II would not be permitted to continue in its existing form or in any related form until such time as Japan's conduct was brought into conformity with international law. This was substantially then an application by Australia in which it sought to ensure that JARPA II was brought to an end, thereby removing the immediate controversy that had developed between the two countries over the conduct of Japan's Southern Ocean whaling program. But the Australian application did

67 Ibid. [40].
68 Smith, Garrett, and McClelland, note 58.

implicitly acknowledge that Japan may seek to undertake a future Southern Ocean whaling program, and to that end orders that Australia sought to ensure that 'JARPA II or any similar program' be 'brought into conformity with its obligations under international law' is significant. As such, Australia understood when making its application to the ICJ that it had a limited ability to bring about a complete cessation of Japanese whaling in the Southern Ocean, but that it could seek to ensure that any future conduct by Japan was in conformity with international law.

In this respect it needs to be recalled that Australia's 2010 application not only referred to Japan's obligations under the International Convention for the Regulation of Whaling (ICRW), but also to the potential obligations of Japan under the Convention on Biological Diversity (CBD), and the Convention on International Trade in Endangered Species of Wild Fauna and Flora (CITES). Ultimately, Australia did not pursue substantive argument before the ICJ based on the CBD and CITES, and chose to substantively frame its argument around Article VIII of the ICRW. However, even with the decision to focus the legal argument upon Article VIII, Australia was well aware that the ICJ was not a constitutional court capable of striking Article VIII out of the ICRW. Article VIII would always remain in the ICRW following the ICJ's decision and Australia was aware that it would have to deal with that reality and that there was the potential that if the Australian application succeeded the court would have provided sufficient guidance for Japan for it to reassess the conduct of JARPA II and seek to develop an alternate whaling program framed around the court's judgment and interpretation of the ICRW and Article VIII in particular.

Finally, it can be observed that Australia did not seek to raise before the court Japan's conduct of JARPN II, Japan's whale research program in the North Pacific which had also commenced in 2005. In principle there was nothing to stop Australia seeking to have made such an application as the foundation of the Australian application was not that Japan was conducting whaling in waters within Australian jurisdiction, or adjacent to Australian territory, but that Australia sought to contest Japan's interpretation of the ICRW via its conduct issuing Article VIII special permits and that Australia's standing to do so before the ICJ was based upon Australia's position as a party to the ICRW. Australia could, if it had wished to do so, have expanded its application to challenge both JARPA II and JARPN II. This would have significantly expanded the scope of the Australian case, most likely have resulted in delays in the progress of the case, added additional complexities with respect to not only the legal argument but also the scientific evidence, and created challenges for the ICJ in having to deal with two related but different Article VIII special permit programs. An expansion of the case to also include JARPN II would also have raised issues within Australia as to the national interest in challenging Japan's North Pacific whaling program where Australia does not have a direct interest. This is not the case with respect to JARPA II which has been conducted within the 'Australian Whale Sanctuary' offshore the Australian Antarctic Territory. While Japan does not recognize either Australia's territorial claim to Antarctica, or the Australian Whale Sanctuary underpinned by the Environment Protection and Biodiversity Conservation Act 1999, both provide a legal and national interest for Australia with respect to JARPA II which is absent in the case of JARPN II.

Australian Argument before the ICJ

In assessing the principal Australian arguments before the ICJ, attention will be given to those arguments that eventually proved to be decisive in the court's judgment. Accordingly, the discussion that follows will consider Australia's arguments with respect to jurisdiction, Article VIII of the ICRW, the test for interpreting Article VIII, and matters relating to the evidence presented to the court regarding science. With respect to jurisdiction, Australia had invoked the Court's jurisdiction based on declarations made by the parties under Article 36(2) of the Court's Statute. The Japanese Declaration was made on 9 July 2007 and was without reservation, other than the need for reciprocity. The Australian Declaration, made on 22 March 2002, accepted the jurisdiction of the court based on reciprocity with the exception of:

> (b) any dispute concerning or relating to the delimitation of maritime zones, including the territorial sea, the exclusive economic zone and the continental shelf, or arising out of, concerning, or relating to the exploitation of any disputed area of or adjacent to any such maritime zone pending its delimitation.[69]

While Japan raised the issue of jurisdiction during the written phase, Australia's response to Japan's argument on that point were principally made during the oral phase of the case. The Australian argument essentially addressed three points: (1) that the reservation only addressed disputes between Australia and another country with respect to a maritime claim that overlaps that of Australia; (2) the Australian reservation does not cover a dispute covering the

69 Declaration under the Statute of the International Court of Justice concerning Australia's acceptance of the jurisdiction of the International Court of Justice [2002] Australian Treaty Series 5.

validity or otherwise of JARPA II; and (3) the dispute concerning compliance of JARPA II with the ICRW exists whether or not Australia asserts a maritime claim adjacent to Antarctica.[70] Australia sought to characterise the reservation within its Declaration as to be understood against the backdrop of maritime boundary delimitations with New Zealand and Timor Leste, and any exploitation of resources that may arise as a result of that delimitation.[71]

With respect to the provisions of the ICRW and Article VIII, Australia adopted an approach before the court which argued that Article VIII needed to be read alongside the context of the other provisions of the convention, in which the moratorium on whaling and the Southern Ocean Sanctuary were of significance.[72] Given that context, Australia argued that Article VIII cannot be relied upon so as to 'have the effect of undermining the effectiveness of the regulatory regime as a whole',[73] and accordingly urged a restrictive interpretation.[74] A critical issue for Australia, upon which significant aspects of the case turned, was the amount of discretion that rested with a state issuing Article VIII special permits. Australia argued that Article VIII set out 'a standard of an objective nature to which the State of nationality had to conform'.[75] In this regard, Australia argued for a standard of review with respect to Article VIII that went beyond good faith and looked at objective elements in determining whether a special permit had been issued for the purposes of scientific research. Australia particularly referred to the 'design and implementation of the whaling programme, as well as any results obtained'.[76] Australia sought to raise two particular issues that were relevant in distinguishing between a legitimate Article VIII programme of scientific research, and one that was in violation of the Schedule moratorium on commercial whaling. The first went to the quantity of whale meat that was generated as a result of an Article VIII programme, which could 'cast doubt on whether the killing, taking and treating of whales is for purposes of scientific research'.[77] The second was the pursuit of policy goals that went beyond scientific objectives, to which Australia pointed to providing employment and maintaining a whaling infrastructure as factors relevant to the conduct of JARPA II.[78]

---

70 *Whaling in the Antarctic*, note 2 [34–35].
71 Ibid. [35].
72 Ibid. [53].
73 Ibid.
74 Ibid. [57].
75 Ibid. [61].
76 Ibid. [63].
77 Ibid [91].
78 Ibid [95].

A critical issue for the Australian case was how 'scientific research' was to be interpreted by the court. To that end, Australia placed emphasis on the evidence presented before the court by two scientific experts called by Australia, Mr Mangel and Mr Gales.[79] Placing particular reliance upon the views of Mr Mangel, Australia maintained that 'scientific research' had four characteristics:

1. defined and achievable objectives that aim to contribute to knowledge important to the conservation and management of stocks;
2. 'appropriate methods', including the use of lethal methods only where the objectives of the research cannot be achieved by any other means;
3. peer review; and
4. the avoidance of adverse effects on stock.[80]

The evidence presented to the court by these experts called on behalf of Australia, and the manner in which Australia subsequently developed their views into legal argument as to what constituted 'scientific research' for the purposes of the ICRW was an important dimension of the Australian case.

The Judgment

The analysis of the court's judgment will focus upon those matters that Australia principally relied upon in argument. The first was jurisdiction, which also was the first substantive matter the court addressed, eventually finding unanimously in favour of jurisdiction.[81] Here the court referred to its observations in *Fisheries Jurisdiction* (Spain v. Canada)[82] that the intention of the reserving state was not only to be deduced from the text of the reservation but

79 See evidence-in-chief and cross examination of these two witnesses, including questions from the bench at *Public sitting held on 27 June 2013, at 10a.m., at the Peace Palace, President Tomka presiding, in the case concerning* Whaling in the Antarctic (Australia v. Japan: New Zealand intervening) VERBATIM RECORD CR2013/9, pp. 39–71 (Mr Mangel); *Public sitting held on 27 June 2013, at 3 p.m., at the Peace Palace, President Tomka presiding, in the case concerning* Whaling in the Antarctic (Australia v. Japan: New Zealand intervening) VERBATIM RECORD CR2013/10, pp. 14–33 (Mr Gales).

80 *Whaling in the Antarctic,* note 2 [74].

81 Ibid [247].

82 *Fisheries Jurisdiction* (Spain v. Canada), Jurisdiction of the Court, Judgment [1998] ICJ Rep 454 [48].

also the context and from an examination of the evidence.[83] The court determined that the reservation in paragraph b was to read in unity and that the disputes it referred to were those that concern 'maritime delimitation in an area where there are overlapping claims or the exploitation of such an area or of an area adjacent thereto'.[84] As the parties were of the view that the case was not about maritime delimitation, the question the court was faced with was whether JARPA II involved the exploitation of an area that was the subject of a dispute relating to delimitation. However, while Japan had contested Australia's maritime claims in the Antarctic, Japan did not claim sovereign rights over the area with the result that the delimitation of the maritime areas constituted a dispute between the parties. As there were no overlapping claims between the parties, the reservation in paragraph (b) was not applicable.[85] The court finally observed on this point that Australia did not seek to contend that JARPA II was unlawful because it took place within maritime zones over which Australia asserts sovereign rights.[86] In this respect the clear distinction that Australia had sought to draw throughout the duration of the dispute between the application of international law and the provisions of the ICRW with respect to JARPA II, and the application of Australian law being a matter for Australian courts was significant.[87]

As to its interpretation of the ICRW, recalling that this was the first occasion the ICJ had been asked to undertake this task, the court made brief reference to the principles of treaty interpretation found in the Vienna Convention on the Law of Treaties,[88] and then turning to Article VIII observed that as it was an integral part of the convention it needed to be interpreted in light of the object and purpose of the convention.[89] However the court did not accept Australia's submission that Article VIII was to be interpreted restrictively, observing that neither 'a restrictive nor an expansive interpretation' was justified.[90] As to the review of special permits issued under Article VIII, the court determined that its approach would be to: (1) determine whether the programme under which the activities occurs involves scientific research, and (2) if the programme which involves the use of lethal methods is 'reasonable in relation to achieving

83 *Whaling in the Antarctic*, note 2 [36].

84 Ibid [37].

85 Ibid [39].

86 Ibid [40].

87 See also comments on jurisdiction in Ibid, Separate Opinion of Judge Owada [3]; Dissenting Opinion of Judge Abraham [3–14] (though in the majority on this point).

88 Convention on the Law of Treaties, 1155 United Nations Treaty Series 331.

89 *Whaling in the Antarctic*, note 2 [55].

90 Ibid [58].

its stated objectives'.[91] In this respect the court's most significant statement as to its interpretative approach towards Article VIII related to its observations at paragraph 88, where it blended aspects of the arguments presented by the parties to indicate the elements it would assess in determining whether JARPA II's design and implementation was reasonable in relation to its stated scientific objectives. The elements to be taken into account were:

- decisions regarding the use of lethal methods;
- the scale of the programme's use of lethal sampling;
- the methodology used to select sample sizes;
- a comparison of the target sample sizes and the actual take;
- the time frame associated with a programme;
- the programme's scientific output; and
- the degree to which a programme coordinates its activities with related research projects.

On the question of what constitutes 'scientific research', the court was not persuaded to adopt the four criteria presented by Australia, which was ultimately characterised as 'well-conceived scientific research, rather than serving as an interpretation of the term as used in the Convention'.[92] Only the fourth criterion that Australia put before the court, that scientific research must not have an adverse effect on whales stocks, was seen by the court to be persuasive, partly because both Japan and New Zealand made similar arguments. Nevertheless, the court observed that Australia had nonetheless conceded that even if Japan had been able to meet the sample sizes set under JARPA II there would have been no adverse effect on the relevant stocks.[93]

Finally, it can be observed that while Australia put forward two scientific experts to a single expert from Japan, the judgment suggests that while the court identified certain differences between the experts, there was a considerable level of broad agreement between them. For example, the court noted the degree of agreement between the experts on the question of hypotheses in scientific research generally,[94] and that lethal methods can have a place in scientific research.[95] Notwithstanding the weight given by Australia (and also Japan) to the opinions of the scientific experts, and the interest demonstrated

91 Ibid [67].
92 Ibid [86].
93 Ibid [85].
94 Ibid [77].
95 Ibid [82].

by the judges in their evidence as reflected in the questions posed by the judges of the experts, the court nevertheless observed that '[t]heir conclusions as scientists, however, must be distinguished from the interpretation of the Convention, which is the task of the Court'.[96]

The Judgment and the Australian Arguments

In summary the critical elements of the court's Orders from an Australian perspective were:

1. That JARPA II Special Permits do not fall within the provisions of Article VIII;
2. That JARPA II does not fall within Japan's obligations under paragraphs 10(e) and (f) of the ICRW Schedule to not engage in commercial whaling, and use factory ships;
3. That JARPA II does not fall within Japan's obligations to refrain from conducting commercial whaling within the IWC's declared Southern Ocean Sanctuary.[97]

In addition, the court also decided that:

> Japan shall revoke any extant authorization, permit or licence granted in relation to JARPA II, and refrain from granting any further permits in pursuance of that programme.[98]

The most immediate consequence of these orders was that JARPA II was to cease immediately. However, as the judgment was delivered a few weeks after the conclusion of the 2013–2014 JARPA II season it had no immediate implications for any Southern Ocean whaling activities that may have been ongoing at the time. In that respect the timing of the judgment is notable in that the Court was able to avoid injecting itself into any controversies by not having the decision handed down during the actual conduct of JARPA II activities which may have forced Japan to immediately react to the judgment, and may have encouraged the Sea Shepherd Conservation Society to engage in more robust protest activities with the Japanese whaling fleet.

96 Ibid.
97 Ibid [247].
98 Ibid.

Impact of the Judgment for Australia

Within Australia, there was a mixed response to the judgment. Officially, the government of Australia's response was rather muted. The Australian Attorney-General, Senator George Brandis, gave a Press Conference soon after the judgment was handed down in which it was observed that:

> The Australian government welcomes the decision. We note that both Australia and Japan have stated on a number of occasions that both countries would accept and respect the decision of the Court. ...the decision seems to dispose of the issue of whaling. ... The fact that Australia and Japan were able to be in dispute on this narrow issue at the International Court of Justice, and agree to differ, but to nevertheless maintain an excellent relationship not withstanding that difference, I think demonstrates very clearly that what a strong relationship and what an enduring relationship this is.[99]

There was no further official response by Australia to the decision, and while the Attorney-General's Department's website contains details on the case and its outcome, there is no accompanying commentary and the statements made are very neutral with respect to the political dimensions of the case.[100]

The official Opposition, however, took a more celebratory tone in their comments following the decision. The former Attorney-General, Mark Dreyfus, who was a member of the Australian legal team who argued the case before the ICJ, observed that:

> Australian's en masse believe that whaling should stop in the Southern Ocean. That's the point we took to the world court, and we've been vindicated. We've shown what Australia can achieve as a nation if we are willing to stand up and flight for what we believe in.[101]

99 George Brandis, Attorney-General for Australia, Minister for the Arts "Transcript – Press Conference, Perth" (31 March 2014) available at <www.attorneygeneral.gov.au/transcripts/Page/31March2014-TranscriptPressConferencePerth.aspx> (30 January 2015).

100 See Attorney-General's Department (Australian Government) "Whaling in the Antarctic litigation" at <www.ag.gov.au/Internationalrelations/InternationalLaw/Pages/WhalingintheAntarcticLitigation.aspx> (30 January 2015).

101 RN Breakfast "Mark Dreyfus hails ICJ whaling ruling" (1 April 2014) available at <www.abc.net.au/radionational/programs/breakfast/japanese-whaling-ruled-unlawful/5358114> (30 January 2015).

In considering the consequences of the judgment, Dreyfus also observed that: 'What they [the court] haven't done is to completely order cessation for all time, and that's something that our new government is going to have to take up with Japan'.[102]

These different political responses within Australia can be explained by the fact that the Abbott government did not initiate the case before the ICJ, and to a degree inherited the proceedings which by the time it took office in September 2013 had reached the stage where the court was deliberating and there was nothing the new government could have done to influence the outcome. The Abbott government was also actively seeking to conclude a Free Trade Agreement with Japan in 2014, and its response no doubt reflected a desire to be respectful of Japan and to do nothing that could in any way detract from reaching a successful conclusion to those negotiations or unsettling the bilateral relationship between the two countries. This view of the Abbott government is confirmed when comments made by the Prime Minister immediately following the judgment are considered. Prime Minister Abbott observed that Japan should 'appropriately reflect' on the judgment, but that trade remained Australia's priority with Japan and would remain the focus of a planned April 2014 visit to Tokyo.[103] On the other hand, the Labor Opposition would have been keen to ensure that they were acknowledged for having commenced the case in 2010 and being the political party in government that ultimately successfully argued the case in The Hague.

The medium term impact of the court's orders related to the conduct of JARPA II in the 2014–2015 season. Here Australia was also reassured by Japan's immediate response to the ICJ judgement when it indicated that all JARPA II related activities would cease immediately and that Japan would not undertake any Southern Ocean whaling in 2014/2015. In that respect, Australia's immediate objective of seeking to bring about an end to JARPA II had been a success. Nevertheless, there are some dimensions of the court's judgement and Japan's subsequent conduct that would have concerned Australia. The first was that the ICJ in paragraph 246 made clear that Article VIII of the ICRW remains intact and that it remains open to Japan to issue further special permits consistently with Article VIII. The court observed that:

> It is to be expected that Japan will take account of the reasoning and conclusion contained in this Judgment as it evaluates the possibility of

102 Ibid.

103 "Japan should 'reflect' on whaling ruling, says Tony Abbott" *the guardian* (Australia) (2 April 2014).

granting any future permits under Article VIII, paragraph 1 of the Convention.

The second is that soon after the judgment the Japanese whaling fleet commenced JARPN II for the 2014 season.[104] While the ICJ's judgement does not apply to JARPN II, that Japan continued with an activity based upon Article VIII Special Permits so soon after its processes relating to the issuing of such permits in the case of JARPA II were found deficient was a cause for concern.

Procedural Issues

The *Whaling* case raised a number of interesting procedural issues, some of which were commented on in the court's Declaration regarding the New Zealand intervention, and also in the court's judgment. The first related to the appointment of the Australian judge ad hoc. In its application instituting the proceedings Australia had indicated its intention to designate an ad hoc judge consistent with Article 31 of the Statute of the court and Article 35(1) of the Rules.[105] In indicating its intention to designate an ad hoc judge Australia was mindful that Judge Owada was a national of Japan and that his term of appointment to the court coincided with the anticipated duration of the proceedings. Australia subsequently elected to nominate Professor Hilary Charlesworth as its judge ad hoc to sit in the case.[106] In this respect it can be observed that Australia had some familiarity with the appointment of and the role of a judge ad hoc, having previously made similar appointments in the *French Nuclear Tests* case[107] and the *East Timor* case.[108] It can also be observed that up until the 2014 election to the court of Professor James Crawford, the only other

104 Andrew Darby "Japan strikes back on whaling" *The Sydney Morning Herald*, April 19, 2014 available at <www.smh.com.au/environment/whale-watch/japan-strikes-back-on-whaling-20140419-36x70.html> (30 January 2015).

105 *Dispute Concerning Japan's JARPA II Program on "Scientific Whaling"* (Australia v. Japan) Application Instituting Proceedings (31 May 2010) [42].

106 *Whaling in the Antarctic*, note 2 [4].

107 *Nuclear Test Cases* (Australia v. France) (Interim Measures) [1973] ICJ Reps 99; (Merits) [1974] ICJ Reps 253 where the Australian nominated judge ad hoc was Sir Garfield Barwick.

108 *East Timor* (Portugal v. Australia) (Judgment) [1995] ICJ Reps 90 where the Australian nominated judge ad hoc was Sir Ninian Stephen. Note that Australia did not nominate a judge ad hoc in the *Case Concerning Certain Phosphate Lands in Nauru* (Nauru v. Australia) [1992] ICJ Reps 240, and neither did Nauru.

Australian national who had served on the court was Sir Percy Spender who was judge from 1958–1967.[109]

The appointment of Professor Charlesworth eventually proved to be contentious between the parties following the New Zealand intervention of 20 November 2012. It became contentious because there was a judge of New Zealand nationality on the court, Sir Ken Keith, whose term as judge was also conterminous with the proceedings in the *Whaling* case. The issue that was then raised as a result of the New Zealand intervention was whether Australia was gaining a 'litigation advantage' by having a judge ad hoc of its own nationality appointed to the court, and also having a judge of New Zealand nationality sitting on the court when it was known that New Zealand's position in the proceedings was sympathetic to the Australian position regarding the interpretation of Article VIII and also that New Zealand was opposed to the conduct of JARPA II.

These issues were directly addressed by the court in its 6 February 2013 Order regarding the New Zealand intervention. The court noted that while Japan had not objected to the New Zealand intervention, Japan had sought to draw to the attention of the court 'certain serious anomalies that would arise from the admission of New Zealand as an intervenor' with respect to the equality of the parties.[110] The Japanese submissions on this point went to whether the New Zealand intervention would result in 'procedural inequality under the Statute and Rules of the Court'[111] as a result of both Australia and New Zealand having the same interest in the case. The court's response was to observe that the New Zealand intervention under Article 63 of the Statute was limited to submitting observations 'on the construction of the convention in question and does not allow the intervenor, which does not become a party to the proceedings, to deal with any other aspect of the case before the Court' and that accordingly the equality of the parties to the dispute was not affected.[112] In this respect the observations of Judge Owada, who voted in favour of the Order, are of interest:

> It is regrettable that a State Party to a case before the Court and a State seeking to intervene in that case pursuant to Article 63 of the Statute

109 This it to be contrasted with Japanese nationals who have served on the court: Judge Tanaka (1961–1970), Judge Oda (1976–2003), and Judge Owada (2003–).

110 *Whaling in the Antarctic* (Australia v. Japan) Declaration of Intervention of New Zealand (Order) 6 February 2013 [17].

111 Ibid [17].

112 Ibid [18].

> should engage in what could be perceived as active collaboration in litigation strategy to use the Court's Statute and the Rules of Court for the purpose of promoting their common interest, as is candidly admitted in their Joint Media Release of 15 December 2010.[113]

The other procedural matter that arose was the Australian decision to not proceed to a second round of written pleadings. In this respect, the court had fixed 9 May 2011 and 9 March 2012 respectively as the dates by which the Australian Memorial and Japanese Counter-Memorial were to be submitted.[114] Following submission of the Japanese Counter-Memorial the President of the Court met with the Agents of the parties on 23 April 2012 in which Australia indicated that it did not consider it necessary to organize a second round of pleadings, while Japan did make such a request. The court decided against a second round of written pleadings and the parties were so advised on 2 May 2012.[115]

For Australia this decision on the second round of pleadings immediately resulted in a certain acceleration of the proceedings and meant that it was possible to schedule the oral proceedings at an earlier stage, which in turn resulted in the judgement being delivered in a shorter time frame than if a second round of written pleadings had been ordered. In this respect it needs to be recalled that the ALP had taken pursuit of Japan over its conduct of JARPA II to the 2007 Australian election, and likewise its policies on this issue and track record commencing litigation before the ICJ in the *Whaling* case was a part of its 2010 re-election strategy. For the ALP government of the then Prime Minister, Julia Gillard, continuing to pursue this matter before the ICJ in a timely fashion and seeking an early resolution of the case would have been seen as advantageous for domestic political purposes and also for the bi-lateral relationship as it ensured that the uncertainty arising from the outcome of the case would not impact upon the short term relationship between Australia and Japan.

In this respect, the comments on this issue by some judges are of interest. Judge Owada made brief allusion to the absence of a second round of pleadings in the context of the parties not being able to develop their arguments fully with respect to jurisdiction, but effectively conceded that matter was not

---

113 Ibid, Declaration of Judge Owada [5]; see Kevin Rudd and Murray McCully, "Australia and New Zealand agree on strategy for whaling legal case" (Joint Media Release, 15 December 2010), available at <foreignminister.gov.au/releases/Pages/2010/kr_mr_101215.aspx> (30 January 2015).

114 *Whaling in the Antarctic*, note 2 [4].

115 Ibid [4].

determinative in terms of his final view on the issue of jurisdiction.[116] Judge Greenwood in his Separate Opinion noted that the Rules of Court make it clear that a second round of written pleadings is 'by no means automatic'.[117] Observing that unless the parties are agreed the court has a discretion as to whether to order a second round of further pleadings, Judge Greenwood identified that factors the court could consider in such orders were: (1) if this is necessary, because the court does not have sufficient information on a particular matter, (2) that the court has an obligation to ensure proceedings do not become unnecessarily protracted, and (3) that a distinction exists between an Applicant not seeking a second round of pleadings and a Respondent.[118] On the last point it was observed that:

> If, having seen the Counter-Memorial, the Applicant considers that it does not need a second round of written pleading, it is difficult to see on what grounds the Respondent can claim to need such a second round. It has already had the last word and no injustice is done by denying it the opportunity to rehearse or add to its case.[119]

Judge Greenwood did concede that one clear point of contention between the parties that had been highlighted in the Japanese Counter-Memorial was that of jurisdiction, however, in this case '[c]onsideration of the objection required no documents or other evidence beyond what had been submitted by Japan and what was already freely and publicly available and Japan, represented by very experienced counsel, can have been in little doubt what form Australia's response would take'.[120]

Japan's Response from an Australian Perspective

Prior to the ICJ's judgement there was considerable speculation in Australia as to what Japan's response would be if Australia was successful and also the impact upon the bilateral relationship. To that end, Australia was reassured by Japan's immediate response to the judgement through statements made in The

116 Ibid, Separate Opinion of Judge Owada [3].
117 Ibid, Separate Opinion of Judge Greenwood [33].
118 Ibid [34–36].
119 Ibid [36].
120 Ibid [37].

Hague on 31 March 2014,[121] and also by the Japanese government.[122] Confirmation from Japan that it would respect the judgment of the court, and that it would cease any future conduct of JARPA II activity for the 2014/2015 season ensured that Australia was confident that its initial goal of bringing JARPA II to a halt had succeeded. Concern was raised, however, when in April and May 2014 members of the Abe government suggested that Japan was closely considering the Court's judgment and assessing options by which it could seek to resume Article VIII whaling at some point in the future. A policy statement issued on 18 April 2014 by Yoshimasa Hayashi, the Japanese Minister of Agriculture, Forestry and Fisheries, noted that the ICJ had reaffirmed that one of the purposes of the ICRW was the 'sustainable exploitation of whale resources' and that as a result of this 'Japan has confirmed its basic policy of pursuing the resumption of commercial whaling, by conducting research whaling…'.[123] Notwithstanding this development, Prime Minister Abe was enthusiastically received in Australia in July 2014 during which time he was honoured with an invitation to make a joint address to the Australian Parliament.[124] Australia and Japan also significantly advanced their economic relationship in 2014 by concluding a Free Trade Agreement.[125] These factors certainly indicated that the Liberal government of Prime Minister Abbott was keen to ensure that the Australia/Japan bilateral relationship remained strong and that the issue of whaling would not be allowed to divert the two countries from enhancing not only their economic but also their security relationship.

---

121 See "Remarks by the Agent of Japan, Koji Tsuruoka: International Court of Justice – Whaling in the Antarctic – (Australia v. Japan: New Zealand intervening)" Ministry of Foreign Affairs of Japan (March 31, 2014) available at <www.mofa.go.jp/ecm/fsh/page2e_000012.html> (30 January 2015) where it is observed 'Japan is disappointed and regrets that the Court ruled that JARPA II by Japan did not fall within Article VIII, paragraph 1, of the ICRW. However, Japan will abide by the Judgment of the Court as a State that places great importance on the international legal order and the rule of law as a basis of the international community'.

122 See statement by the Chief Cabinet Secretary: Prime Minister of Japan and His Cabinet "International Court of Justice (Australia v. Japan: New Zealand intervening) – Statement by Chief Cabinet Secretary, the Government of Japan" (Monday, March 31, 2014) available at <japan.kantei.go.jp/96_abe/decisions/2014/ICJ.html> (30 January 2015).

123 Policy Statement by Mr. Yoshimasa Hayashi, Minister of Agriculture, Forestry and Fisheries, April 18, 2014.

124 Parliament of Australia "Visit to Australia by His Excellency Mr Shinzo Abe Prime Minister of Japan – Address to Members & Senators in HOR Chamber" (July 8, 2014) available at <parlview.aph.gov.au> (30 January 2015).

125 Agreement between Australia and Japan for an Economic Partnership (Canberra, 8 July 2014) [2014] Australian Treaties Not in Force 14.

Based on the comments made by Japan in the months following the ICJ Judgment, Japan's 18 November 2014 announcement that it was proposing to commence a new Article VIII special permit research program in the Southern Ocean and Antarctica would not have come as a surprise to Australia. The 'Proposed Research Plan for New Scientific Whale Research Program in the Antarctic Ocean (NEWREP-A)' sought to directly refer back to the ICJ judgment in the *Whaling* case, and the parameters the court placed around how Article VIII should be interpreted, including the critical phrase 'for the purposes of scientific research'. In that respect NEWREP-A directly seeks to build upon the court's observations in paragraph 246 that 'It is to be expected that Japan will take account of the reasoning and conclusions contained in this Judgment as it evaluates the possibility of granting any future permits under Article VIII, paragraph 1, of this Convention'.

A number of preliminary observations can be made with respect to NEWREP-A. The first is that the proposal is more extensive than the documentation that was publicly released with respect to JARPA II. In this regard, Japan appears to have taken into account a number of the observations made by the ICJ in the judgement with respect to how a legitimate Article VIII program of scientific research needs to be justified. Direct reference is made throughout the proposal to Japan's consideration of the reasoning of the ICJ and how it incorporated those factors into the development of NEWREP-A. In that respect, Japan specifically identified seven elements drawn from the judgement. Those were:

1. 'decisions regarding the use of lethal methods';
2. 'the scale of the programme's use of lethal sampling';
3. 'the methodology used to select sample sizes';
4. 'a comparison of the target sample sizes and the actual take';
5. 'the time frame associated with a programme';
6. 'the programme's scientific output'; and
7. 'the degree to which a programme co-ordinates its activities with related research projects'.[126]

The second is that there is a strong emphasis throughout the proposal with the principle of cooperation between and amongst parties to the ICRW, how that cooperation would assist in the improvement of the Revised Management Program, and also how the proposal would assist in cooperation between the

126 Government of Japan, *Proposed Research Plan for New Scientific Whale Research Program in the Antarctic Ocean* (NEWREP-A) (Tokyo: 2014) 8–11.

ICRW regime and that found in CCAMLR. The third observation is that there is considerable initial justification given for how the proposed take of 333 minke whales per season was arrived at. In this respect, the Appendices attached to the Proposal provide additional material and justification for how that quota was calculated. The final observation is that the proposed plan seeks to incorporate comments from the IWC Scientific Committee, and other 'comments on this proposed plan that are based on scientific consideration'.[127] This will raise issues for how Japan seeks to incorporate those comments from the Scientific Committee within its proposed time frame for the commencement of NEWREP-A in the 2015/2016 season, and also whether it intends to allow for comment by the IWC in plenary. In this respect, IWC Resolution 2014–5 adopted at the 2014 IWC plenary may take on additional significance through its call for proponents of Article VIII 'special permit' programs to not issue further special permits until 'the Commission has, in accordance with Article VI, made such recommendations on the merits or otherwise of the special permit programme as it sees fit'.[128] That the IWC is not scheduled to meet again until 2016, therefore creates a challenge for Japan as to how it will manage those processes, mindful of the ICJ's references to states seeking to cooperate with the IWC as part of the process associated with issuing special permits.

There was no formal response from Australia to Japan's announcement of NEWREP-A, other than a statement issued on behalf of the Minister for the Environment, Greg Hunt. In that statement it was observed that, Australia:

> ...remains opposed to all forms of commercial whaling and believes that all information necessary for the contemporary conservation and management of whales can be obtained using non-lethal methods. Australia remains of the view that lethal research on whales is unnecessary and that non-lethal techniques are the most effective and efficient method of studying all cetaceans.[129]

Members of the Australian Greens party urged the Australian government to take a more forceful approach, with Senator Peter Whish-Wilson stating that

127 Ibid 3.

128 IWC Resolution 2014–5 "Resolution on Special Permit Whaling" [3c].

129 Oliver Milman, "Australian government to scrutinise new Japanese whaling plan" *the guardian* (Australian edition) 19 November 2014 www.theguardian.com/environment/2014/nov/19/australian-government-to-scrutinise-new-japanese-whaling-plan (30 January 2015).

Australia should 'pile on the diplomatic pressure' to back up the judgment of the ICJ and prevent Japan undertaking further whaling.[130]

Beyond these comments, there has been no further formal response by the Government of Australia to Japan's NEWREP-A proposal, though an opportunity will arise to do so in the context of the 2015 IWC Scientific Committee meeting, and eventually at the 2016 plenary of the IWC. In the meantime, it could be anticipated that the Abbott government will come under some political pressure to take a harder diplomatic and public stance on efforts by Japan to resume any form of whaling program in the Southern Ocean and its response will be closely monitored by NGOs and others in Australia and also by Tokyo.

Concluding Remarks

Australia's commencement of proceedings against Japan in the ICJ over JARPAII represented the culmination of Australia's evolution from a whaling nation to a pro-conservation nation, which was prepared to go to great lengths to seek to uphold its position. The Australian legal position on JARPA and eventually JARPAII took some time to develop. Australia's initial response to JARPA in the 1980s was to work within the IWC to seek to uphold the moratorium on commercial whaling, and while that campaign had success in recruiting other like-minded States to the conservation cause, it had little apparent impact upon Japan. That campaign increased in 2005 when Japan announced its intention to commence JARPAII, but again Australia's response was purely diplomatic during which it preferred to argue its position before the IWC. The only concession made by Japan during this period was its decision in 2007, following pressure from the US, to suspend the take of humpback whales and limit its conduct of JARPAII to the take of minke and fin whales.

While at this point in time it is not possible to comment with any certainty as to the legal advice the Australian Government possessed prior to 2007, there can be no denying that the advocacy and legal advice provided by IFAW was influential in shaping the policy position of the ALP during its time in Opposition in 2005–2007. These views were taken into government in December 2007 and the Rudd Government over the course of the next three years took an increasingly assertive stance with respect to Japan's conduct of JARPAII, including the sending of an Australian government Customs vessel to the Southern Ocean in 2008 to monitor the activities of the Japanese whaling

130 Ibid.

fleet. Australia adopted a 'multi-track' approach during this time in which it sought to advance its pro-conservation position in the IWC, promote non-lethal means of whale research, and considered its legal options. In early 2010, Australia began to increase its diplomatic pressure and rhetoric, effectively giving Japan an ultimatum that unless JARPAII was halted Australia would commence international litigation. Eventually by the end of May 2010, following the collapse of IWC reform talks in which Australia and Japan had been critical participants, Australia elected to commence proceedings in the ICJ. In doing so, Australia understood the risks associated with taking a major trading partner and close Asian ally before an international court, that the Australian case was without precedent given that Article VIII of the ICRW had never before been the subject of determination by an international court or tribunal, and that a loss could have significantly undermined the pro-conservation position within the IWC which in turn would have had wider diplomatic consequences for Australia. When all of these factors are considered together, Australia's positon on whaling and particularly Japan's conduct of JARPAII would have been considered by some to have been 'brave'. However this assessment needs to be balanced against the history of Australia pursuing environmental causes before international courts as occurred in *Nuclear Tests* and *Southern Bluefin Tuna*, and Australia's probable knowledge that it had the backing of at least New Zealand in commencing litigation and that there was every probably that New Zealand would either join Australia or seek to intervene in the case.

Ultimately once proceedings had commenced before the ICJ, Australia took a relatively conservative position. It abandoned arguments that it had initially sought to advance under CITES and the CBD, electing instead to focus upon legal argument framed around the ICRW. Australia did rely heavily on experts, as was inevitably going to be the case where the phrase 'for the purposes of scientific research' was so critical for both sides, and where there was a range of scientific opinion as to how JARPAII could be characterised in the context of Article VIII. In that respect, the procedural issues that arose before the ICJ were of some significance in this instance. They related not only to how the evidence of experts was to be presented before the court, but also how expert testimony was to be tested by way of cross-examination. This in turn provided an opportunity for the Judges to ask questions of the witnesses, which assisted to elucidate some aspects of the Australian position and also clarify thinking as to how JARPAII was properly characterised. Other procedural issues, while they appeared small in the context of the overall case, proved to be significant. These included the court's 2013 ruling in favour of the New Zealand intervention, which effectively saw a challenge to Australia's capacity to appoint an ad hoc judge in the case dismissed. Australia was also able to succeed in bringing the written phase of the case to an end after one round of written pleadings, which resulted in the oral proceedings taking place only three years after the case had been commenced which generally speeded up the proceedings and resulted in the March 2014 judgement being handed down less than four years after the case had commenced.

Given Australia's success in the ICJ finding that Japan's conduct of JARPAII was contrary to Article VIII of the ICRW, Australia would be well satisfied with the final result of the case. Article VIII remains embedded in the ICRW however, and Australia would have commenced these proceedings fully aware that it was not within the remit of the ICJ to strike down Article VIII in the way that a municipal court may strike down an unconstitutional provision in a statute. All the ICJ could do was determine the parameters of Article VIII, and then assess whether Japan's conduct of JARPAII was consistent with that interpretation. While Australia succeeded in striking down JARPAII, it did not succeed in striking down the prospect of Japan or any other party to the ICRW relying upon the ICJ's judgement to propose and ultimately conduct another Article VIII 'special permit' whaling program either in the Southern Ocean or elsewhere. If Japan continues with its proposal to undertake NEWREP-A, or a variation thereof, at some time in the future Australia will be faced with a range of options. Will it respond through the use of diplomatic means, or will it again consider its legal options to challenge a new Japanese whaling program in the Southern Ocean? Will Australia's position depend on the government of the day, or is Australia now so implacably opposed to all forms of whaling other than that permitted via an IWC quota that it will instinctively seek to oppose and challenge any resumption of a Japanese whaling program? Ultimately, these are matters for the future. What can be observed is that Australia's actions in challenging Japan's conduct of JARPAII in the ICJ created a benchmark for how states may seek to legally challenge the conduct of whaling in international law, and how Article VIII of the ICRW is to be interpreted. It also reflects a peaceful resolution of an international dispute between two States which had only a generation ago been at war, which in itself should be greatly reassuring for the capacity of international law to bring about reconciliation and harmonious and peaceful relations between States.

CHAPTER 10

After the *Whaling in the Antarctic Judgment*: Its Lessons and Prospects from a Japanese Perspective

Hironobu Sakai

Introduction

There is no doubt that the *Whaling in the Antarctic Judgment*, which was delivered on 31 March 2014,[1] is deeply disappointing to the Japanese Government as well as to those involved in whaling activities. In the *Whaling in the Antarctic Case*, the International Court of Justice ('the Court' or 'the ICJ') held Japan to be in violation of three provisions of the operative Schedule to the International Convention for the Regulation of Whaling ('the ICRW') relating to: the taking of fin whales; the use of factory ships for the research hunt; and the global moratorium on commercial whaling. As a result, the Court ordered Japan to revoke any permit granted under, and to refrain from granting any further permits in relation to, the *Japanese Whale Research Program under Special Permit in the Antarctic Phase II* ('JARPA II').[2]

Conflicts around Japan's whaling activities under the ICRW flared up between the whaling countries and anti-whaling countries within the

1 *Whaling in the Antarctic (Australia v. Japan: New Zealand intervening), Judgment, I.C.J. Reports 2014*, p.226.

2 JARPA II started in the seasons 2005/06 following the end of the *Japanese Whale Research Program under Special Permit in the Antarctic* ('JARPA'), which was conducted between the austral summer seasons of 1987/88 and 2004/05 with the four main objectives: (a) estimation of biological parameters to improve the stock management of the Southern Hemisphere minke whale; (b) elucidate the role of whales in the Antarctic marine ecosystem; (c) elucidation of the effect of environmental change on cetaceans; and (d) elucidation of the stock structure of Southern Hemisphere minke whales to improve stock management. For JARPA II, its objectives are the following: (a) monitoring the Antarctic ecosystem (whale abundance trends and biological parameters; krill abundance and the feeding ecology of whales; effects of contaminants on cetaceans; cetacean habitat); (b) modelling competition among whale species and future management objectives (constructing a model of competition among whale species; new management objectives including the restoration of the cetacean ecosystem); (c) elucidation of temporal and spatial changes in stock structure; and (d) improving the management procedure for Antarctic minke whale stocks. Institute of Cetacean Research, *Scientific Contribution from JAPRA/JAPRA II*. http://www.icrwhale.org/pdf/ScientificContributionJARPA.pdf.

 | DOI 10.1163/9789004313828_012

International Whaling Commission ('the IWC') in particular since 1982, when it adopted a moratorium, or zero catch-limit, on commercial whaling, following which, nevertheless, Japan continued to engage in scientific research whaling.[3] This whaling discord may be seen to have had multi-dimensional aspects: legal, political, ethical, cultural and so on.[4] It is to be noted that the case brought before the Court by Australia, a typical anti-whaling country,[5] is also a part of those wider conflicts.[6] Therefore, it stands to reason that the Court, as a judicial organ, exercised restraint, and, as will be considered later, concentrated mainly on the questions relating to the interpretation of Article VIII of the ICRW for the resolution of the case. However, the problem is whether the Court rightly employed this method at an appropriate stage of the proceedings.

3 Maria Clara Maffei, 'The International Convention for the Regulation of Whaling', *International Journal of Maritime and Coastal Law*, Vol. 12 (1997), pp. 293–294. Despite the moratorium, scientific research whaling continued to be permitted. Howard S. Schiffman, 'Scientific Research Whaling in International Law: Objectives and Objections', *ILSA Journal of International & Comparative Law*, Vol. 8 (2002), pp. 474–476. As an example of critical views in relation to Japan's scientific research whaling activities, Reuben B. Ackerman, 'Japanese Whaling in the Pacific Ocean: Defiance of International Whaling Norms in the Name of "Scientific Research", Culture, and Tradition', *Boston College Law Review*, Vol. 25 (2002), pp. 323–341.

4 Joji Morishita, 'Multiple Analysis of the Whaling Issue: Understanding the Dispute by a Matrix', *Marin Policy*, Vol. 30 (2006), pp. 802–808, who properly analyses the structure of the whaling dispute within the IWC. For the concise history of the conflicts in particular from the view of international environmental law, Gregory Rose & Saundra Crane, 'The Evolution of International Whaling Law', in Philippe Sands (ed.), *Greening International Law* (Earthscan Publications Ltd, 1993), pp. 159–181. With reference to the Japanese activities on whaling, Atsuko Kanehara, 'Japanese Practices concerning the International Regulation of Whaling', *Japanese Annual of International Law*, No. 46 (2003), pp. 127–148.

5 Australia filed an Application instituting proceedings against Japan in the ICJ on 31 May 2010. http://www.icj-cij.org/docket/files/148/15951.pdf. For the background of Australia's filing, Donald R. Rothwell, 'Australia v. Japan: JARPA II Whaling Case before the International Court of Justice', *The Hague Justice Portal, 2 July 2010*. http://www.haguejusticeportal.net/index.php?id=11840.

6 Some Judges seemed to be confused in dealing with this case and with the Parties' arguments before the Court. Dissenting Opinion of Judge Yusuf, *I.C.J. Reports 2014*, p.383, para. 2, who warned that '[m]any of us are distributed by the killing of these iconic and intelligent animals and by the manner in which they are killed. However, these perfectly justified emotional reactions should not make us overlook that it is only by reference to the law that the issues before this Court can be resolved. The judicial settlement of disputes between States cannot made on emotional or purely ethical grounds'.

There are many issues, both procedural and substantive, in the sphere of international law, which the *Whaling in the Antarctic Case* has brought up. This chapter picks out, and considers, only a few of them, in particular from a Japanese perspective.

1 The Questions of Jurisdiction of the Court and Admissibility of the Application

1.1 *Why Didn't Japan Submit Any Preliminary Objections before the Proceedings on the Merits?*

Australia requested the Court to find that JARPA II was incompatible with the ICRW, relying upon the compulsory jurisdiction according to Article 36, paragraph 2, of the ICJ Statute. Japan maintained, against Australia's legal arguments, pursuant to the principle of reciprocity, that the jurisdiction of the Court had not been established—invoking, in this respect, Australia's reservation (b) to its declaration under the 'optional clause'.[7] Australia's reservations to this clause are as follows:

> The Government of Australia declares that it recognises as compulsory *ipso facto* and without special agreement, in relation to any other State accepting the same obligation, the jurisdiction of the International Court of Justice in conformity with paragraph 2 of Article 36 of the Statute of the Court, until such time as notice may be given to the Secretary-General of the United Nations withdrawing this declaration. This declaration is effective immediately.
>
> This declaration does not apply to:[...]
>
> (b) any dispute concerning or relating to the delimitation of maritime zones, including the territorial sea, the exclusive economic zone and the continental shelf, or arising out of, concerning, or relating to the exploitation of any disputed area of or adjacent to any such maritime zone pending its delimitation;[...][8]

7 Counter-Memorials of Japan, 9 March 2012, pp. 23–53. http://www.icj-cij.org/docket/files/148/17384.pdf.

8 Australia's Declaration was deposited with the United Nations Secretary-General on 22 March 2002. http://www.icj-cij.org/jurisdiction/index.php?p1=5&p2=1&p3=3&code=AU. For Japan, it was on 9 July 2007 that it deposited with the United Nations Secretary-General its Declaration, which 'does not apply to any dispute in respect of which any other party to the dispute has accepted the compulsory jurisdiction of the International Court of Justice only in relation to or for the purpose of the dispute; or where the acceptance of the Court's compulsory

Japan's jurisdictional challenge aimed at characterizing the taking of whales under JARPA II as constituting 'the exploitation of any disputed area of or adjacent to any such maritime zone pending its delimitation', because the area in which the JARPA II programme was implemented, lay in Australia's claimed EEZ seaward of Antarctica, and Japan had constantly opposed Australia's Antarctic maritime claims. Thus, Japan asserted that this case falls within the meaning of a dispute 'arising out of, concerning, or relating to the exploitation of any disputed area of or adjacent to any such maritime zone pending its delimitation', and that Australia's reservation should operate so as to exclude the jurisdiction of the Court.[9]

The Court rejected Japan's assertion by interpreting a declaration accepting the compulsory jurisdiction 'in harmony with a natural and reasonable way of reading the text, having due regard to the intention of the declaring State'.[10] And as in its past cases, the Court did not hesitate 'to place a certain emphasis on the intention of the depositing State'.[11] The Court found the wording of the second part of Australia's reservation (b) ('arising out of, concerning, or relating to the exploitation of any disputed area of or adjacent to any such maritime zone pending its delimitation') to be closely linked to that of the first part ('concerning or relating to the delimitation of maritime zones, including the territorial sea, the exclusive economic zone and the continental shelf'), and noted that 'the reservation thus has to be read as a unity'. Thus the Court denied Japan's view that the second part, introduced by the second conjunction 'or', is separate from the first part, with the result that the reservation would apply both to disputes on delimitation and to other kinds of dispute involving the exploitation of maritime zones or adjacent areas pending delimitation. According to the *Judgment*, the reservation requires a maritime delimitation dispute, which simply did not exist in this situation.[12] Thus, the Court concluded

---

jurisdiction on behalf of any other party to the dispute was deposited or notified less than twelve months prior to the filing of the application bringing the dispute before the Court'. http://www.icj-cij.org/jurisdiction/index.php?p1=5&p2=1&p3=3&code=JP.

9 On a detailed argument for the lack of jurisdiction, CR 2013/12, pp. 22–39, pars. 1–34 (Pellet).

10 *Whaling in the Antarctic*, *I.C.J. Reports 2014*, p.244, para. 36. The Court quotes some terms in this part from its *Judgment* on the preliminary objections in the *Anglo-Iranian Oil Co. Case*. *Anglo-Iranian Oil Co. (United Kingdom v. Iran), Preliminary Objection, Judgment, I.C.J. Reports 1952*, p. 104.

11 *Fisheries Jurisdiction (Spain v. Canada), Jurisdiction of the Court, Judgment, I.C.J. Reports 1998*, p. 454, para. 48.

12 Cameron Jefferies, 'Whaling in the Antarctic: Part III – Jurisdiction', University of Alberta Faculty of Law Blog. http://ualbertalaw.typepad.com/faculty/2014/04/whaling-in-the-antarctic-part-iii-jurisdiction.html.

that '[t]he fact that Japan questions those maritime entitlements does not render the delimitation of these maritime areas under dispute as between the Parties', and that '[t]here are no overlapping claims of the Parties to the present proceedings which may render reservation (b) applicable'.[13]

It is well known that Australia established the Australian Antarctic Territory in 1933, and then, since the latter part of the 1970's, shifted its policy to favour the protection of whales by enacting certain legal regulations on whaling. Finally, Australia extended the reach of its related acts even to its Exclusive Economic Zone (EEZ) offshore the Australian Antarctic Territory,[14] notwithstanding that Australia is a Contracting Party to the Antarctic Treaty, Article 4, paragraph 2, of which provides that '[N]o acts or activities taking place while the present treaty is in force shall constitute a basis for asserting, supporting or denying a claim to territorial sovereignty in Antarctica or create any rights of sovereignty in Antarctica.[...]' As a matter of general practice, the Australian government has deliberately refrained from exercising its enforcement of EEZ jurisdiction off Antarctica against non-nationals,[15] and this was also true, even when the Federal Court declared that Japanese whaling in Australia's Antarctic water is unlawful under the *Environment Protection and Biodiversity Conservation Act 1999* and granted an injunction restraining it in the *Human Society International Inc. v. Kyodo Senpaku Kaisha Ltd.* case.[16]

13 *Whaling in the Antarctic, I.C.J. Reports 2014*, pp.245–246, para. 39.

14 For Australia's setting the maritime zones in Antarctica and exercising its jurisdiction there, Ruth Davis, 'Enforcing Australian Law in Antarctica: The HIS Litigation', *Melbourne Journal of International Law*, Vol. 8 (2007), pp. 150–154.

15 James Crawford, 'The Antarctic Treaty after 50 Years', in Duncan French, Matthew Saul & Nigel D. White (eds.), *International Law and Dispute Settlement. New Problems and Techniques* (Hart Publishing, 2010), pp. 294–295.

16 *Human Society International Inc. v. Kyodo Senpaku Kaisha Ltd.* [2008] FCA 3. Stuart Kaye, Michael Johnson & Rachel Baird, 'Law', in Marcus Haward & Tom Griffiths (eds.), *Australia and the Antarctic Treaty System: 50 Years of Influence* (University of New South Wales Press, 2011), pp. 113–114; Natalie Klein & Nikolas Hughes, 'National Litigation and International Law: Repercussions for Australia's Promotion of Marine Resources', *Melbourne University Law Review*, Vol. 33 (2009), pp. 201–203. On this case, see also, Tim Stephens, 'The Japanese Whaling Case: Implications for Australia and Beyond', *Asia Pacific Journal of Environmental Law*, Vol. 11 (2008), pp. 137–146; Rachel Baird & Chantal Le Feuvre, 'They Said They'd Never Win: Humane Society International Inc v Kyodo Senpaku Kaisha Ltd', *Asia Pacific Journal of Environmental Law*, Vol. 11 (2008), pp. 147–158; Donald K. Anton, 'Australian Jurisdiction and Whales in Antarctica: Why the Australian Whale Sanctuary in Antarctic Waters Does Not Pass International Legal Muster and is also a Bad Idea as Applied to Non-Nationals', *Asia Pacific Journal of Environmental Law*, Vol. 11 (2008), pp. 159–192.

From a procedural viewpoint, it is interesting, and even incomprehensible, that Japan did not challenge the jurisdiction of the Court as a preliminary objection in order to ask the Court to consider the jurisdiction question apart from other issues in the proceedings on the merits. Japan would have had much more time to focus on, and to scrutinize, certain questions other than the jurisdiction of the Court and the admissibility of Australia's Application, if it had submitted preliminary objections so that the proceedings on the merits could have been suspended in accordance with Article 79, paragraph 1 of the Rules of the Court.[17]

Given that the Court has a recent tendency to consider, with the consent of the parties, issues relating to the jurisdiction of the Court and the admissibility of the Application during the proceedings on the merits, Japan's plans to settle the case without submitting any preliminary objections to the Court probably represented its aim to resolve the matter as definitely and quickly as possible;[18] and they may also have corresponded with the interests of the Court itself, in resolving smoothly as many as possible of the cases on its docket. Nevertheless, a respondent is entitled to make full use of any procedural means to ask the Court to reject the claims by the applicant, and so Japan could have chosen a policy which would have enabled it to take time to carefully consider the jurisdictional questions at the stage of preliminary objections. More profound reflection on the jurisdictional issues, including the matter of Australia's reservation (b), and even (a),[19] might have enabled better presentation of the reasonable grounds for Japan's arguments, so that the Court might have declined to consider the claims by Australia,[20] although it might have been difficult to

---

17 Article 79, paragraph 5 of the current Rules of the Court provides that '[u]pon receipt by the Registry of a preliminary objection, the proceedings on the merits shall be suspended [....]'. This suspension occurs automatically whenever a preliminary objection should be raised. Shabtai Rosenne, *The Law and Practice of the International Court 1920–2005, Vol. II, Fourth Edition* (Martinus Nijhoff Publishers, 2006), pp. 862–863.

18 Australia also, in the *East Timor* case, has declined to raise any separate preliminary objection so as to resolve the matter as soon as possible. Henry Burmester, 'Australia's Experience in International Litigation', in Natalie Klein (ed.), *Litigating International Law Disputes. Weighing the Balance* (Cambridge University Press, 2014), p. 75.

19 Australia's reservation (a) states that 'any dispute in regard to which the parties thereto have agreed or shall agree to have recourse to some other method of peaceful settlement'.

20 Dissenting Opinion of Judge Owada, *I.C.J. Reports 2014*, p.301, para. 3, who states that 'under the somewhat unfortunate procedural circumstances, the Parties were not provided in the proceedings with ample opportunities to develop their respective arguments on the issue of jurisdiction, [...]'.

convince the Court to accept the objections by Japan, even at the stage of preliminary objections, bearing in mind that the Court unanimously rejected Japan's arguments on the jurisdictional question in the *Judgment* on the merits.[21]

1.2 *Did Australia Establish Its* Locus Standi *in the* Whaling in the Antarctic Case?

Australia did not appear to claim that it was the direct victim of an injurious act committed by Japan, but instead maintained that Japan had violated its international obligations under the ICRW, to which Australia is also a Contracting party and hence entitled to claim cessation of those violations.[22] There seems to remain a question as to a party's quality or interest in relation to the issue of its personal admissibility. In fact, the issue of this right of appearance before the Court, *locus standi*, was not covered in detail by the Court in this case. However, surely it must be an important issue in relation to characterizing the Court's treatment of this case as a kind of judicial control by the Court over violations of international law.

In practice, an applicant normally needs to justify its *locus standi* by pointing to some violation of its own rights.[23] In the *South West Africa Case* (the Second Phase), the Court dealt with the question whether the applicants had any legal rights or interest regarding the subject-matter of their claim—the Applicants' *locus standi ratione materiae*[24]—, regarding it as a matter of the merits of the case; and the Court denied that the Applicants possessed any 'separate self-contained right which they could assert, independently of, or additionally to, the rights of the League.'[25]

In the *Whaling in the Antarctic Case*, what right of Australia was violated by the acts of Japan? It can be argued that no subjective rights of Australia were infringed, nor any damage to that State caused, because no whales which were allegedly hurt by Japanese whaling activities belonged to Australia, nor did any

---

21 *Whaling in the Antarctic*, *ibid.*, pp.298–299, para. 247(1).

22 CR 2013/18, p.18, para.19 (Burmester). 'Australia does not claim to be an injured State because of the fact that some of the JARPA II take is from waters over which Australia claims sovereign rights and jurisdiction'.

23 Robert Kolb, *The International Court of Justice* (Hart Publishing, 2013), p. 270.

24 Shabtai Rosenne, *The Law and Practice of the International Court 1920–2005, Vol. III, Fourth Edition* (Martinus Nijhoff Publishers, 2006), p. 1171, which points out that 'the Court made a distinction between the applicants' *locus standi in judicio ratione personae* – the subject of the 1962 judgment on the Preliminary Objections, and their *locus standi ratione materiae* – the subject of the Second Phase judgment of 1966'.

25 *South West Africa, Second Phase, Judgment, I.C.J. Reports 1966*, p. 29, para. 33.

Japanese activities ever violate any Australian territory. This concern seems to have been shared by one of the Judges on the Court, who posed a question to the Applicant during the oral proceedings: 'What injury, if any, has Australia suffered as a result of Japan's alleged breaches of the ICRW through JARPA II?'[26]

If any right was violated in this case, it must have been a right of Australia as a Contracting party to the ICRW that was allegedly infringed by Japan's violation of its obligations under the ICRW. In such a case, when an applicant who is not injured by a violation nevertheless seeks relief for that violation from the Court, the applicant must testify that the obligation breached is owed to a group of States including the applicant State, and that it was established for the protection of a collective interest of the group. As a matter of fact, Australia responded to the above-mentioned question as follows:

> L'Australie, comme tous les autres Etats parties à la convention de 1946, a un intérêt commun à ce que l'intégrité du regime découlant de la convention soit maintenue.[27]

This statement brings to mind Article 48 of the Articles on State Responsibility for Internationally Wrongful Acts ('the ARSIWA'),[28] which have considerably broadened the scope of *jus standi* as a provision serving the collective interests of a group or the interest of the international community as a whole.[29] In particular 'a collective interest' referred by Australia appears in Article 48, paragraph 1(a), which concerns the case where the obligation breached is owed 'to a group of States' and 'the obligation is established for the protection of a collective interest of the group'. In other words, the collective interest is relevant to the category of multilateral treaties establishing obligations which in any specific circumstance are owed to all the other States party to the treaty.[30] In the present case, the obligation at issue is one owed to all Contracting parties to the ICRW, an obligation *erga omnes partes*, on the premise that there may be a collective interest of the Contracting parties secured by the ICRW.[31]

26 CR 2013/13, p. 73 (Judge Bandhari).

27 CR 2013/18, p. 33, para. 18 (Boisson de Chazournes).

28 UN Doc.A/RES/56/83, Annex.

29 Ch.Tomuschat, 'Article 36', in A. Zimmermann et al. (eds.), *The Statute of the International Court of Justice. A Commentary. Second Edition* (Oxford University Press, 2012), p. 705.

30 Giorgio Gaja, 'States Having an Interest in Compliance with the Obligation Breached', in James Crawford, Alain Pellet & Simon Olleson (eds.), *The Law of International Responsibility* (Oxford University Press, 2010), p. 959.

31 James Crawford, 'Responsibility for Breaches of Communitarian Norms: An Appraisal of Article 48 of the ILC Articles on Responsibility of States for Internationally Wrongful

In fact, both Australia and New Zealand explored the 'collective interest' arguments in the oral proceedings.[32]

The Court referred to obligations *erga omnes* in the *Barcelona Traction Case* in 1970,[33] and has since, in its recent jurisprudence, heard several cases in which the obligations breached were of an *erga omnes* character,[34] which paved the way for the development of Article 48 of the ARSIWA.[35] Moreover, this provision was expressly mentioned in the *Responsibilities and Obligations Case* by the Seabed Disputes Chamber of the International Tribunal for the Law of the Sea ('the ITLOS'), which affirmed the *erga omnes* character of the obligations respecting the preservation of the environment of the high seas and the Area in its 2011 Advisory Opinion.[36] Finally, in its 2012 *Judgment* in the *Obligation to Prosecute or Extradite Case*, the Court observed that the common interest in compliance with the relevant obligations under the Convention against Torture and Other Cruel, Inhuman or Degrading Treatment or Punishment ('the CAT') implied the entitlement of each State party to that Convention to make a claim concerning the cessation of an alleged breach by

---

Acts', in Ulrich Fastenrath, Rudolf Geiger, Daniel-Erasmus Khan, Andreas Paulus, Sabine von Schorlemer & Christoph Vedder (eds.), *From Bilateralism to Community Interest. Essays in Honour of Judge Bruno Simma* (Oxford University Press, 2011), p. 236, who argues that '[t]he remedies sought by Australia, which invokes Japan's obligations *erga omnes partes* under the Whaling Convention [...], coincide with the remedies available under Article 48(2) of the ILC Articles'. See also, Ruth Davis, 'The Whaling Dispute in the South: An Australian Perspective', *Journal of East Asia and International Law*, Vol. IV (2011), p. 440.

32 Henry Burmester, Counsel for Australia, also stated that '[e]very party has the same interest in ensuring compliance by every other party with its obligations under the 1946 Convention. Australia is seeking to uphold its collective interest, an interest it shares with all other parties'. CR 2013/18, p. 28, para. 19 (Burmester); *ibid.*, pp. 33–34, paras. 19–20 (Boisson de Chazournes); CR 2013/17, p. 33, para. 63 (Finlayson).

33 *Barcelona Traction, Light and Power Company, Limited, Judgment, I.C.J. Reports 1970*, p. 32, para. 33.

34 *East Timor (Portugal v. Australia), Judgment, I.C.J. Reports 1995*, p. 102, para. 29; *Application of the Convention on the Prevention and Punishment of the Crime of Genocide, Preliminary Objections, Judgment, I.C.J. Reports 1996*, p. 616, para. 31; *Legal Consequences of the Construction of a Wall in the Occupied Palestinian Territory, Advisory Opinion, I.C.J. Reports 2004*, p. 199, paras. 155–157; *Armed Activities on the Territory of the Congo (New Application: 2002) (Democratic Republic of the Congo v. Rwanda), Jurisdiction and Admissibility, Judgment, I.C.J. Reports 2006*, pp. 31–32, para. 64.

35 James Crawford, *State Responsibility. The General Part* (Cambridge University Press, 2013), p. 552.

36 *Responsibilities and Obligations of States Sponsoring Persons and Entities with respect to Activities in the Area, Advisory Opinion, 1 February 2011, ITLOS Reports 2011*, p. 59, para. 180.

another State party, and held that any State party to the CAT 'may invoke the responsibility of another State party with a view to ascertaining the alleged failure to comply with its obligations *erga omnes partes* [...]'.[37] It is true that this illustrates that there is a contemporary tendency in cases for international courts and tribunals to recognize and to apply the concept of obligations *erga omnes* (*partes*).[38]

At the heart of the *Whaling in the Antarctic Case*, in this respect, is what may be alleged as common interests of the Contracting parties to the ICRW, as in the Genocide Convention or the CAT. It is, however, easier to consider the prevention of genocide or of torture as a common interest of the parties to the Genocide Convention and the CAT, respectively. The Contracting parties to those Conventions hold shared values and a common interest in assuring the prevention of genocide or torture.[39] But is this true of the ICRW in the *Whaling in the Antarctic Case*? The International Law Commission ('the ILC'), when elaborating the draft ARSIWA, makes it clear that 'collective obligations' might concern, 'for example, the environment or security of a region (e.g. a regional nuclear free zone treaty), or a regional system for the protection of human rights'.[40] The ICRW, however, cannot fall within this category of treaty, in particular the category of environmental protection treaty, even if some claim that the ICRW, or whales themselves, constitute a kind of symbol for environmental protection.[41] Seemingly, the Court hesitated to characterize the ICRW solely as an environmental protection treaty, refraining from referring to the

37 *Questions relating to the Obligations to Prosecute or Extradite* (*Belgium v. Senegal*), *Judgment, I.C.J. Reports 2012*, p. 450, para. 69.

38 Gleider I Hernández, *The International Court of Justice and the Judicial Function* (Oxford University Press, 2014), pp. 226–228.

39 The Court, in the *Genocide Convention Reservation Case*, stated that 'the contracting States do not have any interests of their own; they merely have, one and all, a common interest, namely, the accomplishment of those high purposes which are the *raison d'être* of the Convention'. *Reservations to the Convention on the Prevention and Punishment of the Crime of Genocide, Advisory Opinion, I.C.J. Reports 1951*, p. 23. With respect to the CAT, the Court observed that '[t]he States parties to the Convention have a common interest to ensure, in view of their shared values, that acts of torture are prevented and that, if they occur, their authors do not enjoy impunity'. *Questions relating to the Obligations to Prosecute or Extradite (Belgium v. Senegal), Judgment, I.C.J. Reports 2012*, p. 450, para. 68.

40 James Crawford, *The International Law Commission's Articles on State Responsibility. Introduction, Text and Commentaries* (Cambridge University Press, 2002), p. 277, para. (7).

41 Malgosia Fitzmaurice, 'The International Court of Justice and International Environmental Law', in Christian J. Tams & James Sloan (eds.), *The Development of International Law by the International Court of Justice* (Oxford University Press, 2013), p. 373, who writes that '[b]ecause of the general and longstanding public interest in the protection of marine

precautionary principle or approach in the interpretation of the ICRW,[42] and irrespective of strong support for such characterization by some Judges, who took the position that the ICRW is concerned with environmental protection.[43]

Otherwise, might the Court not have to find that protection of the Antarctic ecosystem (the research of which is one of the aims pursued by the JARPA II activities) amounts to a common interest of the Contracting parties to the ICRW, so that the ICRW might give rise to an obligation of a Contracting party, owed to all the other Contracting parties, to protect the ecosystem in the Antarctic? It would mean that the ICRW would no longer be for the management of the whaling industry but rather for environmental protection. On this issue, the Court's interpretation of the object and purpose of the ICRW may have a vital influence.

However, it is precisely on what is the common interest of the Contracting parties to the ICRW that, whenever the issue of implementation of the ICRW is raised, the views between the pro-whaling States and anti-whaling States have most sharply conflicted; and it is this which has produced the paralysis of the functions of the IWC. It is in this situation, in which, since the conclusion of the ICRW in 1946, there has been no agreement among the Contracting Governments to the ICRW on what is or should be their common interest, that Australia, an anti-whaling State, makes use of the Court so as to persuade it to regard as the common interest under the ICRW the protection of whales, irrespective of the quantity of whale resources.[44] Is it appropriate that the Court gave its judgment in favour of the almost unconditional protection of whales as being the common interest of the Contracting parties to the ICRW in this respect? Moreover, it follows that the Court, having in the present case established the *locus standi* of the Applicant as a Contracting Government to the ICRW solely on account of the

---

mammals, this case is likely to define the public perception of the Court as a protector of environmental concerns'.

42 This means that the Court did not concur with an argument that the IWC's policy of maintaining the moratorium on commercial whaling is an application of the precautionary principle, which is explored in A.W. Harris, 'The Best Scientific Evidence Available: The Whaling Moratorium and Divergent Interpretation of Science', *William & Mary Environmental Law & Policy Review*, Vol. 29 (2005), pp. 375–450.

43 Separate Opinion of Judge Cançado Trindade, *I.C.J. Reports 2014*, pp.374–375, paras. 70–71 Separate Opinion of Judge *ad hoc* Charlesworth, *ibid.*, pp.455–456, paras. 6–10. See also, Jeffrey J. Smith, 'Evolving to Conservation?: The International Court's Decision in the Australia/Japan Whaling Case', *Ocean Development & International Law*, Vol. 45 (2014), p. 318.

44 As for the decision-making of the unilateral resort to the Court of the dispute by Australia, see, Shirley V. Scott, 'Australia's Decision to Initiate *Whaling in the Antarctic*: Winning the Case versus Resolving the Dispute', *Australian Journal of International Affairs*, Vol. 68, No. 1 (2014), pp. 1–16.

alleged violations of obligations *erga omnes partes* by another Contracting Government, will be obliged to take on the role of monitoring the implementation of other multilateral treaties similar to the ICRW, some of the parties to which may claim the *raison d'être* of those treaties to be the realization of the common interest of the international community as a whole. Is it, indeed, possible, or appropriate, for the Court, of which the main role is to resolve disputes between States relating to alleged violations of their own rights, as a matter of the judicial policy of the Court, to deal with cases where the common interest of the parties to the treaties, even of the international community as whole, is in dispute[45]?

As a matter of fact, the Court proceeded to the merits on the premise of the existence of *locus standi* of the applicant. It is worth noting, in this respect, that the Court avoided referring to the term 'obligation *erga omnes partes*' or even to 'Article 48, paragraph 1(a)', though it substantively recognized the functions of that provision to establish the standing of Australia, which had suffered no damage, only in relation to the alleged violations of the ICRW by another Contracting Government. Interestingly, or perhaps, rather curiously, Japan refrained from raising any objections on the *locus standi* of the Applicant in this case.[46] However, one may question whether it might not have been better for the Respondent to have done so, at least 'in the alternative', even if other arguments might have been contradictory to its objections on the jurisdiction. In any case, it may be fortunate for the Court not to scrutinize this matter because Japan did not argue against the *locus standi* of the Applicant on this stage.[47] It is, however, not good—it could even be dangerous—that, on the

45 For consideration concerning the ICJ's judicial policy on public interest in the context of its provisional measures, Hironobu Sakai, 'New Development of the Orders on Provisional Measures by the International Court of Justice', *Japanese Yearbook of International Law*, Vol. 52 (2009), pp. 231–280.

46 Frans A. Nelissen & Steffen van der Velde, 'Australia Attempts to Harpoon Japanese Whaling Program. Proceedings Instituted by Australia against Japan: Application Instituting Proceedings, 31 May 2010', *The Hague Justice Portal*, 5 July 2010, p. 4, who had assumed that 'Japan might argue that Australia has no legal interest in Japanese whaling practices and that, therefore, the case should be declared inadmissible'. http://www.haguejusticeportal.net/Docs/Commentaries%20PDF/HJP_Nelissen_Aust_Japan_EN.pdf See also, Natalie Klein & Tim Stephens, 'Whaling in the Antarctic: Protecting Rights in Areas beyond National Jurisdiction through International Litigation', in Clive Schofield, Seokwoo Lee & Moon-Sang Kwon (eds.), *The Limits of Maritime Jurisdiction* (Brill, 2014), p. 539.

47 Shotaro Hamamoto, Council for Japan in the *Whaling in the Antarctic Case*, states that 'highlighting the fact that JARPA II was taking place *out of* Australia's EEZ might have diminished the impact of the jurisdictional argument that JARPA II was taking place in the relevant maritime zone or in area *adjacent to* it'. Shotatro Hamamoto, 'Procedural Questions in the Whaling Judgment: Admissibility, Intervention and Use of Experts',

basis alone that the Court seemed to recognize the existence of obligations *erga omnes partes* and their legal effects in the proceedings in the *Whaling* case, it should become prevalent in subsequent similar cases before the Court, and that it should exercise some influence on the attitude of other judicial organs.

2 The Issues Concerning the Court's Reasoning for Its Findings of Violations of the ICRW by Japan

2.1 *Did the Court Adopt an Appropriate Method for the Interpretation of the ICRW?*

Australia claimed that Japan was in breach of its international obligations in authorizing and implementing JARPA II in the Southern Ocean in violation of paragraphs 10(e),[48] 7(b),[49] 10(d),[50] and 30[51] of the Schedule to the ICRW, and

pp. 2–3. *ICJ Judgment on Whaling in the Antarctic: Its Significance and Implications. The Honorable Shigeru Oda Commemorative Lectures in the 2014 Annual Conference of the Japanese Society of International Law.* [filed with the author].

48 Paragraph 10(e) provides: Notwithstanding the other provisions of paragraph 10, catch limits for the killing for commercial purposes of whales from all stocks for the 1986 coastal and the 1985/86 pelagic seasons and thereafter shall be zero. This provision will be kept under review, based upon the best scientific advice, and by 1990 at the latest the Commission will undertake a comprehensive assessment of the effects of this decision on whale stocks and consider modification of this provision and the establishment of other catch limits.

49 Paragraph 7(b) provides: In accordance with Article V(1)(c) of the Convention, commercial whaling, whether by pelagic operations or from land stations, is prohibited in a region designated as the Southern Ocean Sanctuary. This Sanctuary comprises the waters of the Southern Hemisphere southwards of the following line: starting from 40 degrees S, 50 degrees W; thence due east to 20 degrees E; thence due south to 55 degrees S; thence due east to 130 degrees E; thence due north to 40 degrees S; thence due east to 130 degrees W; thence due south to 60 degrees S; thence due east to 50 degrees W; thence due north to the point of beginning. This prohibition applies irrespective of the conservation status of baleen and toothed whale stocks in this Sanctuary, as may from time to time be determined by the Commission. However, this prohibition shall be reviewed ten years after its initial adoption and at succeeding ten year intervals, and could be revised at such times by the Commission. Nothing in this sub-paragraph is intended to prejudice the special legal and political status of Antarctica.

50 Paragraph 10(d) provides: Notwithstanding the other provisions of paragraph 10 there shall be a moratorium on the taking, killing or treating of whales, except minke whales, by factory ships or whale catchers attached to factory ships. This moratorium applies to sperm whales, killer whales and baleen whales, except minke whales.

51 Paragraph 30 provides: A Contracting Government shall provide the Secretary to the International Whaling Commission with proposed scientific permits before they are

that JARPA II was not a program for purposes of scientific research within the meaning of Article VIII of the ICRW.[52] Against Australia's claims, Japan argued that it 'has acted and is continuing to act in full conformity with the ICRW and for the furtherance of its object and purpose',[53] and that the implementation of special permit whaling should be consistent with the object and purpose of the ICRW, as is the case with the JARPA II, which has been conducted 'for purposes of scientific research'. According to the Respondent, 'once it is determined that a whaling expedition is a special permit whaling expedition authorized by a Contracting Government, the ICRW is no longer relevant to questions of the legality of the expedition', and '[n]o other State or body is given the power to overturn decisions taken by a Contracting Government in exercise of its right to authorize special permit whaling.'[54] Thus, one of the main subjects in this case is whether the whaling activities pursuant to JARPA II constitute scientific research within the meaning of Article VIII, paragraph 1, of the ICRW. Article VIII, paragraph 1, provides that

> Notwithstanding anything contained in this Convention any Contracting Government may grant to any of its nationals a special permit authorizing that national to kill, take and treat whales for purposes of scientific research subject to such restrictions as to number and subject to such other conditions as the Contracting Government think fit, and the killing, taking, and treating of whales in accordance with the provisions of this article shall be exempt from the operation of this Convention.

Just after the resort to the Court by Australia, it was assumed that the most crucial issue in the case might be the interpretation of this Article and its

---

issued and in sufficient time to allow the Scientific Committee to review and comment on them. The proposed permits should specify:

(a) objectives of the research;

(b) number, sex, size and stock of the animals to be taken;

(c) opportunities for participation in the research by scientists of other nations and;

(d) possible effect on conservation of stock.

Proposed permits shall be reviewed and commented on by the Scientific Committee at Annual Meetings when possible. When permits would be grated prior to the next Annual Meeting, the Secretary shall send the proposed permits to members of the Scientific Committee by mail for their comment and review. Preliminary results of any research resulting from the permits should be made available at the next Annual Meeting of the Scientific Committee.

52 *Whaling in the Antarctic*, *I.C.J. Reports 2014*, pp.239–240, para. 25.

53 Counter-Memorial of Japan, pp. 4–5, para. 13.

54 *Ibid.*, pp. 326–327, para. 7.16.

application to JARPA II;[55] and in fact, the Court rightly assessed the Application as involving the question of interpretation and application of Article VIII of the ICRW, which is central to the case.

The Court, in its *Judgment*, points out that 'in contrast to the 1931 and 1937 treaties, the text of the ICRW does not contain substantive provisions regulating the conservation of whale stocks or the management of the whaling industry',[56] which are to be found in the Schedule, to emphasize a distinction between the ICRW and the two preceding treaties. The Court also makes it clear that an amendment by the International Whaling Commission ('the IWC') of the Schedule, which forms an integral part of the ICRW, becomes binding on a State party unless it presents an objection. Thus, the Court draws attention to the way in which the functions conferred on the IWC, which has amended the Schedule many times, have made the ICRW an 'evolving' instrument.[57]

Japan certainly does not deny that law may evolve.[58] However, the Court's idea that the recommendations of the IWC on any matters which relate to whales or whaling and to the objectives and purposes of the ICRW, which are not binding on any, or all, Contracting Governments, may be relevant for the interpretation of the ICRW or its Schedule, 'when these are adopted by consensus or by a unanimous vote',[59] is hardly convincing. Here, while the Court rejects the idea that recommendations adopted even by a majority vote may have the relevance which Australia has contends, it does seem to support such an effect in relation to recommendation adopted in some other way: by consensus or by a unanimous vote. Nevertheless, an agreement is different from consensus. The interpretation of a treaty should be in light of the rules on treaty interpretation, expressed in Article 31 of the Vienna Convention on the

55 C. Brighton, 'Speculation and Significance: Japan's Scientific Whaling before the International Court of Justice', *New Zealand Yearbook of International Law*, Vol. 9 (2011), pp. 16–30.

56 *Whaling in the Antarctic, I.C.J. Reports 2014*, pp.247–248, para. 45.

57 *Ibid.* This wording recalls the Court's statement in the *Namibia Advisory Opinion*. In its view, 'the Court must take into consideration the changes which have occurred in the supervening half-century, and its interpretation cannot remain unaffected by the subsequent development of law, through the Charter of the United Nations and by way of Customary law. Moreover, an international instrument has to be interpreted and applied within the framework of the entire legal system prevailing at the time of the interpretation'. *Legal Consequences for States of the Continued Presence of South Africa in Namibia (South West Africa) notwithstanding Security Council Resolution 276 (1970), Advisory Opinion, I.C.J. Reports 1971*, p. 31, para. 53.

58 CR 2013/12, p. 20, para. 27 (Tsuruoka).

59 *Whaling in the Antarctic, I.C.J. Reports 2014*, p.248, para. 46.

Law of Treaties ('the VCLT'), and should be based upon an agreement of the State parties to the treaty, not upon consensus.

In relation to the object and purpose of the ICRW, the Court confirmed that the ICRW embodies the purposes of ensuring the conservation of all species of whales while allowing for their sustainable exploitation, as set out in its preamble.[60] The Court also states that amendments to the Schedule and recommendations by the IWC may put an emphasis on one or the other objective pursued by the ICRW, but that they cannot alter its object and purpose. Thus, the Court rejects both Japan's and Australia's arguments and adopts a middle way between them, observing that 'neither a restrictive nor an expansive interpretation of Article VIII is justified'.[61]

The position adopted by the Court seems to be correct when it affirms that 'Article VIII gives discretion to a State party to the ICRW to reject the request for a special permit' while adding that 'whether the killing, taking and treating of whales pursuant to a requested special permit is for purposes of scientific research cannot depend simply on that State's perception'.[62] Nevertheless, as a matter of fact, the Court appears to support Australia's argument that a special permit is to be treated as exceptional, not only by setting a standard of an objective nature to which the issuing State has to conform, but also by introducing a relatively high standard of review in relation to the term 'for purposes of scientific research', as will be further mentioned below.

2.2 *Where Do the Two Elements Come from and Why Should They be Examined Separately in Article VIII, Paragraph 1, of the ICRW?*

The Court limited its own role in relation to the interpretation of the ICRW in this case to focus, specifically, on the words in Article VIII, 'for purposes of scientific research'. According to the Court, this term consists of two elements, 'for purposes of' and 'scientific research', which are cumulative.[63] This view was originally suggested by the Australian analysis in its Memorial of the meaning of this term,[64] but the appropriateness of such dichotomy was itself never discussed

60 For these objectives of the ICRW, 'reconciliation of the two forms of objective has proved highly problematic'. Malgosia Fitzmaurice, 'The International Convention for the Regulation of Whaling and International Whaling Commission – Conservation or Preservation – Can the Gordian Knot be Cut (or Untangled)?' *The Yearbook of Polar Law*, Vol. V (2013), p. 461.

61 *Whaling in the Antarctic, I.C.J. Reports 2014*, p.252, para. 58.

62 *Ibid.*, p.253, para. 61.

63 *Ibid.*, pp.254–255, para. 70.

64 Memorial of Australia, 9 May 2011, pp. 155–156, para. 4.38. http://www.icj-cij.org/docket/files/148/17382.pdf.

before the Court. In any event, this distinction between the two elements is rather artificial, but is still decisive for the *Judgment* of the Court.

Though adopting this distinction to interpret the term 'for purposes of scientific research' in Article VIII, the Court, in reality, looks mainly into 'for purposes of', but not into 'scientific research', so that it may avoid entering into scientific questions, which should be distinguished from the interpretation of the ICRW. As for 'scientific research', the Court rejected the idea of its four essential characteristics, endorsed by Australia; and declined to devise alternative criteria or to offer a general definition of it. Moreover, respecting the experts' opinion that lethal methods, notwithstanding the difference of views on the conditions for their use, can have a place in scientific research, and based on the information before it, the Court found that 'the JARPA II activities involving the lethal sampling of whales can broadly be characterized as "scientific research"'.[65]

On the other hand, the Court, more closely examining the meaning of the term 'for purposes of', considered whether seven elements of a programme's design and implementation were reasonable in relation to its scientific objectives. In order to decide reasonableness,[66] the Court looked at 'decisions regarding the use of lethal method; the scale of the programme's use of lethal sampling; the methodology used to select sample size; a comparison of the target sample sizes and the actual take; the time frame associated with a programme; the programme's scientific output; and the degree to which a programme co-ordinates its activities with related research projects',[67] and the Court finally found that the design and implementation of JARPA II, in particular the target sample size, were not reasonable in relation to achieving its objectives.[68] In other words, the Court concluded that the special permits

65 *Whaling in the Antarctic*, *I.C.J. Reports 2014*, p.267, para. 127.

66 The *Whaling in the Antarctic Case* is not the first one where the Court approaches its analysis on the basis of applying a court-articulated 'reasonableness' standard to specific facts. Examples of this include *Nicaragua Case*, *Sovereignty over Pedra Branca/Pulau Batu Puteh Case*, and *Navigational and Related Rights Case*. See, Anastasia Telesetsky, Donald K. Anton & Timo Koivurova, 'ICJ's Decision in Australia v. Japan: Giving up the Spear or Refining the Scientific Design?' *Ocean Development & International Law*, Vol. 45 (2014), p. 333. Claire Brighton, 'Unravelling Reasonableness: A Question of Treaty Interpretation', *Australian Year Book of International Law*, Vol. 32 (2014), p. 133, who points out that 'without the foundation provided by a textual interpretation, the majority's reliance on 'unreasonableness' appears entirely unconnected to the text and the object and purpose of the ICRW'.

67 *Whaling in the Antarctic*, *I.C.J. Reports 2014*, p.258, para. 88.

68 *Ibid.*, p.258, para. 88, p.292, para. 224.

granted by Japan in connection with JARPA II were not 'for purposes of scientific research', by considering only the meaning of the words 'for purposes of'.

This methodology adopted by the Court also appears to be strange. Firstly, the term 'for purposes of' may express the intention or the objectives of a programme, which must be of a subjective nature, while the Court maintained that this standard of review was an objective one. If the objective criteria were only to be related to the term 'scientific research', and the term 'for purposes of' were to be subjective, the Court would, at the stage of considering the term 'for purposes of', have, as Judge Abraham pointed out, respected the intention of the Japanese Government. Thus, he said:

> 《En vue de》, selon moi, renvoie à l'intention, aux fins recherchées, aux buts réellement poursuivis (qui peuvent être différents de ceux qui ont été proclamés). Pas selon l'arrêt. La Cour insiste au contraire sur le fait que son critère est 《objectif》 (paragraphe 67), autrement dit qu'il ne s'agit pas pour elle de déceler les intentions réelles du Japon, de chercher la réalité des fins poursuivies derrière les apparances.[69]

Secondly, it has to be seriously doubted whether the distinction between the two elements, itself, is reasonable in relation to the Court's application of the objective standard of review in this case. This dichotomy seems to be much too artificial. The issue in this case was whether JARPA II activities involved scientific research under Article VIII of the ICRW. Japan argues that the test of whether a State's decision is objectively reasonable should have been considered as a standard of review of the expression 'for purposes of scientific research' as a whole, not of the expression 'for purposes of' separately. Thirdly, the *Judgment* is not clear on the point of the relationship between 'scientific research' in a broader sense and 'scientific research' pursuant to Article VIII. Bearing in mind that the Court found that Japan had complied with its obligations under paragraph 30 of the Schedule to the ICRW, can it not then be said that JARPA II activities have been recognized as 'scientific research' by the Court? If 'scientific whaling, even if with flaws, remains scientific in nature',[70] can JARPA II activities still be illegal 'scientific research'?[71] In addition, to the

69 Opinion dissidente de M.le juge Abraham, *C.I.J. Recueil 2014*, p.329, par. 32.

70 Separate Opinion of Judge Xue, *ibid.*, p.430, para. 35.

71 In 2014 IWC Assembly, Japan pointed out that the second sentence of paragraph 227 of the *Judgment*, which states that the special permits granted by Japan for JARPA II are not for scientific research as defined by Article VIII, was contradictory to the Court's statement in its paragraph 127, which reads that 'the JARPA II activities involving lethal sampling of

extent that the Court maintained another dichotomy, namely, between scientific research whaling and commercial whaling, the view that illegal scientific research whaling must be regarded as commercial whaling is unavoidable. However, is it correct that scientific research whaling with some defects under Article VIII of the ICRW will inevitably change into commercial whaling, which is contrary to paragraph 10(e) of the Schedule? As Judge Owada rightly pointed out, '[e]ven if JARPA II contained some defects as a programme for purposes of scientific research, that fact in itself would not turn these activities into activities for commercial whaling.'[72]

2.3 *Doesn't the Court Apply 'Standard of Review' Much Too Strictly in Relation to the Conduct of the Japanese Authorities?*

While the Court considers that 'Article VIII gives discretion to a State party to the ICRW to reject the request for a special permit or to specify the conditions under which a permit will be granted', it also notes that 'whether the killing, taking and treating of whales pursuant to a requested special permit is for purposes of scientific research cannot depend simply on that State's perception'.[73] For these activities to fall within Article VIII, according to the Court, the whaling programme must involve scientific research, and at the same time they are required to be 'for purposes of' scientific research, which is evidenced by 'examining whether, in the use of lethal methods, the programme's design and implementation are reasonable in relation to achieving its stated objectives'. Thus, the Court observes that '[t]his standard of review is an objective one.'[74]

With regard to this method adopted by the Court, two things may be noted. Firstly, there seems something of a lack of legal support for the Court's decision to evaluate the facts related to Japan's whaling activities, as well as to the design and implementation of JARPA II, in accordance with a standard of review based on a legally defined standard of reasonableness. In other words,

---

whales can broadly be characterized as "scientific research"'. The IWC Chair's Report of the 65th Meeting, p. 26, para. 239.

72 Dissenting Opinion of Judge Owada, *I.C.J. Reports 2014*, p.320, para. 49.

73 *Whaling in the Antarctic*, *ibid.*, p.253, para. 61.

74 *Ibid.*, p.254, para. 67. In this paragraph, the Court adopts the distinct standards for the two steps along which the review unfolded. Enzo Cannizzaro, 'Margin of Appreciation and Reasonableness in the ICJ's Decision in the Whaling Case', in Pierre d'Argent, Béatrice Bonafé & Jean Combacau (eds.), *Les limites du droit international. Essais en l'honneur de Joe Verhoeven* (Bruylant, 2015), p. 452, who states that '[t]he first relates to whether "the program under which these activities occur involves scientific research"', and that '[t]he second relates to whether "the program's design and implementation are reasonable in achieving its stated objectives"'.

the *Judgment* endorsed a reasonable standard without ever explaining why it was appropriate.[75] Judge Owada asserted in this respect that 'it was not possible nor useful to try to apply this concept of "reasonableness" in a general way as the standard of substantive assessment'.[76]

Secondly, even if this reasoning in the *Judgment* were to be followed, the Court should have reviewed the way in which Japan exercised its discretion under Article VIII, and the results of that exercise, so as to ask whether the issue of a special permit by Japan under JARPA II amounted to an abuse of rights, was *ultra vires*, or involved bad faith by a Contracting Government to the ICRW.[77] Put differently, the Court should have focused on the question whether Japan's action in issuing a special permit under JARPA II, not JARPA II itself, was legal and reasonable objectively.[78] This is because it is not the task of the Court to review and evaluate the design and implementation of a research plan for scientific whaling. Judge Yusuf maintained in his dissenting opinion that:

> the Court should have focused its analysis on the lawful use by Japan of its discretionary power under Article VIII, in light of the object and purpose of the Convention, in issuing a special permit for JARPA II and whether or not Japan has violated or is violating its obligations under the ICRW in authorizing and implementing JARPA II. ......[79]

75 Sonia E. Rolland, 'Whaling in the Antarctic (Australia v. Japan: New Zealand Intervening)', *American Journal of International Law*, Vol. 108 (2014), p. 500, who argues the possibility that the Court might have considered alternative standards, indicating that 'for example, a "necessary" test or a "but for" test (entailing a sine qua non relationship), or requiring that the activity consist primarily (or even exclusively) of scientific research, or determining the least harmful way to achieve the result, or balancing the harm to the whale population and the scientific value of the research'.

76 Dissenting Opinion of Judge Owada, *I.C.J. Reports 2014*, pp.316–317, para. 39.

77 Before the referral of this case to the Court by Australia, some scholars had already put forward the idea that the Japanese scientific permit whaling programme is an abuse of rights. Eldon V. Greenberg, Paul S. Hoff & Michael I. Goulding, 'Japan's Whale Research Program and International Law', *California Western International Law Journal*, Vol. 32 (2002), pp. 177–185; Joanna Mossop, 'Australia v Japan: Whaling in the International Court of Justice', *New Zealand Yearbook of International Law*, Vol. 7 (2009), pp. 173–174.

78 Separate Opinion of Judge Xue, *I.C.J. Reports 2014*, p.425, para. 16, who stated that '[t]he standard of review as agreed by the Parties that tests whether a State's decision to grant special permits is objectively reasonable, supported by coherent reasoning and respectable scientific evidence, should therefore primarily relate to special permits rather than the programme in general'.

79 Dissenting Opinion of Judge Yusuf, *ibid.*, p.388, para. 17.

However, the Court made no reference to the question of abuse of rights, even though many Judges expressly or implicitly addressed the good faith issue in their respective individual opinions, and submitted their different understandings of whether good faith was at issue.[80] The Court instead considered the reasonableness of JARPA II on the basis of the so called 'the standard of review',[81] and confirmed the it did 'not need to address other arguments invoked by Australia in support of its claims',[82] though it rejected the idea that Article VIII gives a State total discretion to grant permits and to direct the taking of whales for scientific research.[83]

It is supposed that the Court's introduction of the concept 'standard of review' may have been suggested by the arguments of both Parties, as well as by the current practice of the World Trade Organization ('the WTO') and its Dispute Settlement Body ('the DSB') in the field of multilateral trade, in which the concept of a deferential standard of review has been established due to the sensitivity of national sovereignty in the policy-making of the Member States.[84] Compared with such a standard of review in the WTO, however, it should be asked whether it is appropriate or useful in this case for the Court to adopt a seemingly higher level of standard of review. If there is a strong presumption that a Contracting party has made its determination to grant special permits 'in light of careful consideration that the activities are carried

80 Dissenting Opinion of Judge Owada, *ibid.*, p.388, para. 21; Opinion dissidente de M.le juge Abraham, *ibid.*, pp.327–328, paras. 28–29; Dissenting Opinion of Judge Yusuf, *ibid.*, pp.401–402, para. 54; Separate Opinion of Judge Greenwood, *ibid.*, pp.416–417, para. 29; Separate Opinion of Judge Sebutinde, *ibid.*, p.436, para. 18; Separate Opinion of Judge Cançado Trindade, *ibid.*, p.354, para. 17; Separate Opinion of Judge *ad hoc* Charlesworth, *ibid.*, pp.457–458, para. 13.

81 See also, Gillian Triggs, 'Japanese Scientific Whaling: An Abuse of Right or Optimum Utilisation?' *Asia Pacific Journal of Environmental Law*, Vol. 5 (2000), p. 52; Natalie Klein, 'Whales and Tuna: The Past and Future of Litigation between Australia and Japan', *Georgetown International Environmental Law Review*, Vol. 21 (2009), pp. 201–202. Nevertheless, Phillip J. Clapham, 'Japan's Whaling Following the International Court of Justice Ruling: Brave New World – Or Business as Usual?' *Marine Policy*, Vol. 51 (2015), p. 239, which states that the Court's 'ruling lent independent support to what many on the IWC Scientific Committee had been saying for years: that Japan's whaling was an abuse of Article VIII'.

82 *Whaling in the Antarctic, I.C.J. Reports 2014*, p.298, para. 243.

83 As for this idea, Nobuyuki Yagi, 'The Status of Scientific Research Whaling in International Law', *ILSA Journal of International & Comparative Law*, Vol. 8 (2002), p. 493.

84 Caroline E. Foster, 'International Adjudication – Standard of Review and Burden of Proof: *Australia-Apples* and *Whaling in the Antarctic*', *Review of European Community & International Environmental Law*, Vol. 21 (2012), pp. 80–81.

out for purposes of scientific research', such determination should be accorded maximum deference in the ICRW regime. That is not to say that the Court cannot question the determination by the Contracting party. Japan also did not argue that there should be complete deference to the determinations made by a State.[85] It is just that a *de novo* review, or substantially its equivalent, should be avoided. As Judge Owada rightly pointed out in his dissenting opinion:

> The function of the Court as a court of law gives it the power to interpret and apply the provisions of the Convention from a legal point of view. Given the nature and the specific characteristics of the regulatory framework created by the Convention, however, this power of the Court has to be exercised with a certain degree of restraint, the extent that what is involved is (a) related to the application of the regulatory framework of the Convention, and (b) concerned with the techno-scientific task of assessing the merits of scientific research assigned by the Convention to the Scientific Committee.[86]

2.4 *Is the* Actori Incumbit Probatio *No Longer Valid in a Future 'Standard of Review' Case in the Court?*

The 'standard of review' method adopted and the avoidance of introducing good faith, or *bona fide,* arguments by the Court in this case seem to have had a significant influence on the procedural aspects of the proceedings in favour of the Applicant.

The Court observed that 'the dispute before it arises from a decision by a State party to the ICRW to grant special permits under Article VIII of that treaty', and then that '[i]nherent in such a decision is the determination by the State party that the programme's use of lethal methods is for purposes of scientific research'. Thus, the Court asked 'the authorizing State, which has granted

---

85 CR 2013/15, p. 19, para. 28 (Lowe). Japan also agrees with Australia and New Zealand 'in regarding the test as being whether a State's decision is objectively reasonable, or 'supported by coherent reasoning and respectable scientific evidence and [...] in this sense, objectively justifiable'. CR 2013/22, p. 60, para. 21 (Lowe). *Whaling in the Antarctic, I.C.J. Reports 2014,* p.254, para. 66, which evaluates this statement as refining Japan's position regarding the standard of review to be applied in this case.

86 Dissenting Opinion of Judge Owada, *ibid.,* p.308, para. 20. But see, Robert Kolb, 'Short Reflections on the ICJ's Whaling Case and the Review by the International Courts and Tribunals of "Discretionary Powers"', *Australian Year Book of International Law,* Vol. 32 (2014), pp. 139–140.

special permits, to explain the objective basis for its determination'.[87] By recalling this observation, the Court seems to have affirmed that any scientific whaling pursuant to a special permit under Article VIII is an exception to the moratorium on commercial whaling, so that the Court may, in fact, place the burden of proof not so much on the State claiming violations of Article VIII of the ICRW by a State granting a special permit under that provision as on the granting State itself.

However, it is by no means free from doubt that the Court seems to construe the legal effect of the ICRW so that the principal obligation is the moratorium on commercial whaling, with Article VIII, paragraph 1, as an exception to that obligation. Such construction ignores the fact that Article allows a Contracting party to the ICRW a wide discretion in implementing scientific research whaling. A Contracting party enjoys the power to take whales for purposes of scientific research under the ICRW, though it is required to use that power in accordance with the conditions based on Article VIII. In such an instance, a State is presumed to exercise its right to grant any of its nationals a special permit to engage in what is legal conduct *bona fide*—an issue to which the Court made no reference in the *Judgment*. Generally speaking, when one party to a dispute claims that another has acted in bad faith, it is accepted that the latter's good faith is presumed. Thus, a respondent, whose conduct is alleged to be illegal by an applicant, must be the beneficiary of a very strong presumption in its favour;[88] and it is the applicant arguing the illegality of the respondent's acts who has to override that presumption. In the present case, however, the Court did not require Australia to demonstrate that Japan had engaged in whaling activities based on Article VIII in bad faith.

This has the effect, in fact, of shifting the burden of proof, from the viewpoint of procedural law, and of introducing a concept of accountability, from the viewpoint of substantive law, in the sense that the Court calls upon 'the authorizing State, which has granted special permits, to explain the objective basis for its determination'. This follows from the Court's adoption of the reasonableness test in evaluating the conformity of JARPA II with Article VIII,

87 *Whaling in the Antarctic, I.C.J. Reports 2014*, p.254, para. 68.

88 Opinion dissidente de M.le Juge Abraham, *ibid.*, pp.327–328, par. 28. But Julian Wyatt, 'Should We Presume that Japan Acted in Good Faith? Reflections on Judge Abraham's Burden of Proof Based Analysis', *Australian Year Book of International Law*, Vol. 32 (2014), p. 163, who, criticizing the Dissenting Opinion by Judge Abraham, points out that the juxtaposition of his admission that Japan did not have the discretionary right to unilaterally determine whether its whaling came within the scope of Article VIII of the ICRW and his imposition of a heavy burden of proof on Australia in light of a presumption of Japanese good faith is difficult to reconcile.

paragraph 1, of the ICRW.[89] However, the burden of proof lies traditionally with the applicant;[90] and thus, according to this rule, it should have been Australia who was called upon to establish, by credible evidence, that the special permits granted by Japan in relation to JARPA II were not based on Article VIII, and that the determination to grant the special permits to its nationals by Japan was not objectively reasonable—since the maxim *actori incumbit probatio* means that the burden of proof lies on the applicant.[91]

Furthermore, the fact that the Court did not refer to the precautionary principle or approach in its reasoning is also critical. While, arguably, the reversal of the burden of proof should occur to give effect to the precautionary principle if necessary for the due administration of justice,[92] the Court did not

89 Anastasia Telesetsky, Donald K. Anton & Timo Koivurova, *supra* note 66, p. 336, who point out that '[i]n creating the reasonableness test, the International Court ultimately shifts the burden of proof from the applicant having to prove that the respondent acted in bad faith to the respondent having to demonstrate that its approach to the design and implementation of its research plan was reasonable'. Voir aussi, Pascale Martine-Bidou, 'L'Affaire de la chasse à la baleine dans l'Antarctique: Droit des traités ou protection des espèces?' *Revue general de droit international public*, Tome 119 (2015), p. 412.

90 Caroline E. Foster, 'Burden of Proof in International Courts and Tribunals', *Australian Year Book of International Law*, Vol. 29 (2008), p. 27. The Court has also constantly confirmed that the applicant must establish its case and that a party asserting a fact must establish it. *Pulp Mills on the River Uruguay (Argentina v. Uruguay), Judgment, I.C.J. Reports 2010*, p. 71, para. 162; *Maritime Delimitation in the Black Sea (Romania v. Ukraine), Judgment, I.C.J. Reports 2009*, p. 86, para. 68; *Sovereignty over Pedra Branca/Pulau Batu Puteh, Middle Rocks and South Ledge (Malaysia/Singapore), Judgment, I.C.J. Reports 2008*, p. 31, para. 45; *Application of the Convention on the Prevention and Punishment of the Crime of Genocide (Bosnia and Herzegovina v. Serbia and Montenegro), Judgment, I.C.J. Reports 2007*, p. 128, para. 204; *Military and Paramilitary Activities in and against Nicaragua (Nicaragua v. United States of America), Jurisdiction and Admissibility, Judgment, I.C.J. Reports 1984*, p. 437, para. 101.

91 Ad hoc Editorial Chamber, 'The Whaling Dispute in the South Pacific: A Japanese Perspective', *Journal of East Asia and International Law*, Vol. 4 (2011), p. 451.

92 Separate Opinion of Judge ad hoc Charlesworth, *I.C.J. Reports 2014*, p.456, para. 10, who asserts the practical reverse of the burden of proof in favour of Australia based on the precautionary principle. See also, Caroline E. Foster, *Science and the Precautionary Principle in International Courts and Tribunals: Expert Evidence, Burden of Proof and Finality* (Cambridge University Press, 2011), pp. 272–280. In contrast to the argument on the relationship between precautionary principle and the shifting of the burden of proof, John M. Macdonald, 'Appreciating the Precautionary Principle as an Ethical Evolution in Ocean Management', *Ocean Development & International Law*, Vol. 26 (1995), p. 263, who states that the precautionary principle would require States to take measures to prevent harm and to reduce and control pollution, and that '[t]his interpretation would not so

address this argument, making no reference to that principle in this case. Indeed, as in the *Pulp Mills Case,*[93] the Court declined to reverse the burden of proof in favour of the Applicant.

There is no doubt that Japan became accountable for the reasonableness of its own conduct when the Court introduced the method of 'standard of review' to evaluate their actions in the present case.[94] Nevertheless, it is to be stressed that Japan's decisions and policies on the whaling should be legally assessed on the strong presumption that they would have been reasonable under Article VIII of the ICRW—a presumption which Australia must overturn to persuade the Court to find Japan in violation of the ICRW. In this context, it is not clear whether or not Australia would have been successful in establishing the unreasonableness of the conduct of Japan based on JARPA II. In fact, Judge Owada asserted that 'the Applicant has failed to establish that the activities carried out pursuant to JARPA II are not "reasonable" scientific activities'.[95]

2.5 *Why Is the Duty to Co-operate with the IWC and the Scientific Committee Imposed upon Japan as Involving a Legal Obligation to Give Due Regard to IWC Resolutions and Guidelines?*

As for Article VIII of the ICRW, there seems to be something strange in the reasoning of the *Judgment*: how the Court can derive certain legal effects from the terms of Article VIII. The Contracting parties to the ICRW, according to the Court, 'have a duty to co-operate with the IWC and the Scientific Committee and thus should give due regard to recommendations calling for an assessment of the feasibility of non-lethal alternatives'.[96] Moreover, the Court, in another paragraph of the *Judgment*, refers to 'Japan's obligation to give due regard to

much require a shifting of the burden of proof, but rather a commitment by nations to pursue stronger and more environmentally friendly technologies or management regimes when engaged in activities that have the potential to harm the marine environment'.

93 *Pulp Mills on the River Uruguay (Argentina v. Uruguay), Judgment, I.C.J. Reports 2010*, p. 71, para. 164, which states that 'while a precautionary approach may be relevant in the interpretation and application of the provisions of the [1975] Statute, it does not follow that it operates as a reversal of the burden of proof'.

94 In the *Navigational and Related Rights Case*, the Court has observed that 'a court examining the reasonableness of a regulation must recognize that the regulator [...] has the primary responsibility for assessing the need for regulation and for choosing, on the basis of its knowledge of the situation, the measure that it deems most appropriate to meet that need'. *Dispute regarding Navigational and Related Rights (Costa Rica v. Nicaragua), Judgment, I.C.J. Reports 2009*, p. 253, para. 101.

95 Dissenting Opinion of Judge Owada, *I.C.J. Reports 2014*, p.318, para. 45.

96 *Whaling in the Antarctic, ibid.*, p.257, para. 83.

IWC resolutions and Guidelines'.[97] Thus, the Court first seems not to have considered the duty to co-operation as binding, because of the use of the terms, 'duty' and 'should'; but later, seems to have found a legally binding effect in it in the form of the obligation to give due regard. Consequently, the Court found Japan in violation of this obligation, since Japan had not fully analysed the feasibility of using non-lethal methods in order to reduce or eliminate the need for lethal sampling.

Truly, Japan accepts that 'there is a duty on the part of the Contracting Governments to consider a recommendation in good faith and, if requested, to explain their action or inaction';[98] but, at the same time, Japan emphasizes that the IWC resolutions and Guidelines have no binding effect.[99] As for the Court, it also clearly confirms that the resolutions and the Guidelines without the concurrence of Japan:

> cannot be regarded as subsequent agreement to an interpretation of Article VIII, nor as subsequent practice establishing an agreement of the parties regarding the interpretation of the treaty within the meaning of subparagraphs (a) and (b), respectively, of paragraph (3) of Article 31 of the Vienna Convention on the Law of Treaties.[100]

The duty of co-operation by a State party with the IWC was also referred to by the Court, not only in relation to the Recommendations of the IWC and its Scientific Committee, but also in the context of a State party's duty of communication with the IWC,[101] as provided, in particular, in paragraph 30 of the Schedule, which provides that a Contracting Government issuing a special permit is under the obligation to provide the IWC Secretary with proposed scientific permits before they are issued. The Court observed that:

---

97 *Ibid.*, p.271, para. 144.

98 Counter-Memorial of Japan, p. 373, para. 8.63.

99 *Ibid.*, pp. 371–372, paras. 8.58–8.59.

100 *Whaling in the Antarctic, I.C.J. Reports 2014,* p.257, para. 83. For example, 'subsequent practice' in Article 31(3)(b) of the VCLT, 'even if only some parties participated in it, must be accepted by all the parties, *ie* the parties as a whole'. Oliver Dörr & Kirsten Schmalenbach (eds.), *Vienna Convention on the Law of Treaties. A Commentary* (Springer, 2012), p. 559. And if some events do not fall within the meaning of such 'subsequent practice', '*A fortiori*, they cannot have given rise to an "agreement between the parties regarding the interpretation of the treaty or the application of its provisions"' in Article 31(3)(a). *Kasikili/ Sedudu Island (Botswana/Namibia), Judgment, I.C.J. Reports 1999*, p. 1087, para. 63.

101 Philippe Weckel, 'Chasse à la baleine dans l'Antarctique (Australie c. Japon; Nouvelle-Zélande (intervenant))', *Revue général de droit international public*, tome 118 (2014), p. 402.

> paragraph 30 and the related Guidelines regarding the submission of proposed permits and the review by the Scientific Committee [...] must be appreciated in light of the duty of co-operation with the IWC and its Scientific Committee [...]. [...] the implementation of JARPA II differs in significant respects from the original design of the programme that was reflected in the JARPA II Research Plan. Under such circumstances, consideration by a State party of revising the original design of the programme for review would demonstrate co-operation by a State party with the Scientific Committee.[102]

The Scientific Committee, in reviewing the new proposal of JARPA II in 2005, had already admitted that '[t]he proposal provides the information required under Paragraph 30 of the Schedule.'[103] In this respect, the Court also, finally, recognized that 'Japan has met the requirements of paragraph 30 as far as JARPA II is concerned',[104] and observed that Japan has complied with its obligations under paragraph 30 of the Schedule. Thus, while the Court found that Japan did not comply with its obligations under paragraphs 7(b), 10(d), and 10(e) of the Schedule, it held that Japan had complied with its obligations under paragraph 30 of the Schedule. Although Judge Sebutinde argued that paragraph 30 imposed 'more than a formal or procedural obligation to notify the Scientific Committee of certain information',[105] the Court properly regarded the duty of co-operation under this paragraph as an obligation characteristic solely of procedure, and confirmed that Japan lawfully submitted the JARPA II to the Scientific Committee in accordance with the procedure under paragraph 30 of the Schedule.

102 *Whaling in the Antarctic, I.C.J. Reports 2014,* p.297, para. 240.

103 Report of the Scientific Committee 2005, p. 50.

104 *Whaling in the Antarctic, I.C.J. Reports 2014,* p.297, para. 242.

105 Separate Opinion of Judge Sebutinde, *ibid.,* pp.434–435, para. 15. Some Judges did not agree with the finding in operative paragraph 6 of the *Judgment*, and argued that Japan has not complied with the duty of co-operation with the Scientific Committee and thus that it has breached paragraph 30 of the Schedule. Separate Opinion of Judge Bhandari, *ibid.,* pp.438–444, paras. 3–19; Separate Opinion of Judge *ad hoc* Charlesworth, *ibid.,* pp.456–459, paras. 11–17. Judge Greenwood, though voted in favour of the Court' finding on this respect, saying that the duty of co-operation 'means that a State is not free to adopt a formalistic approach to paragraph 30, and confirmed that Japan has not provided any information regarding whether, or how, JARPA II has been adapted to take account of the changed circumstances, which leads him to argue that '[i]t must, therefore, be open to question whether there has been a full compliance with the duty of co-operation'. Separate Opinion of Judge Greenwood, *ibid.,* pp.417–418, paras. 30–31.

3 Legal Effects and Ramifications of the *Whaling in the Antarctic Judgment*

3.1 *Can Japan Keep Engaging the Whaling Activities in the Antarctic as Well as in the North Pacific Ocean after this* Judgment?

The Court, while holding that special permits in the Antarctic in connection with JARPA II did not fall within Article VIII, paragraph 1, of the ICRW, and that Japan should revoke any extant permit to kill, take or treat whales in relation to JARPA II, and should refrain from granting any further permits under Article VIII, paragraph 1, in pursuance of that programme, did not deny Japan the right to submit a new whaling programme in accordance with Article VIII. In confirming that, the Court said that it:

> sees no need to order the additional remedy requested by Australia, which would require Japan to refrain from authorizing or implementing any special permit whaling which is not for purposes of scientific research within the meaning of Article VIII. That obligation already applies to all States parties. It is to be expected that Japan will take account of the reasoning and conclusions contained in this Judgement as it evaluates the possibility of granting any future permits under Article VIII, paragraph 1, of the Convention.[106]

Immediately after the *Judgment* was delivered, Mr.Fumio Kishida, Minister for Foreign Affairs of Japan, stated definitely, 'Japan is a country which places great importance on the international legal order and the rule of law. Therefore, the Government of Japan will abide by the Judgment of the Court', although he added that the decision was very regrettable and that he was deeply disappointed by it.[107] Then, as the Minister for Foreign Affairs promised, Japan suspended whaling practices in the Antarctic for the 2014–2015 season, and was ready to submit a new research programme, which must be consistent with the *Judgment*, by this autumn to the IWC.[108] Finally, on 18 November 2014, the

---

106 *Whaling in the Antarctic, ibid.,* p.298, para. 246.

107 Press Conference by Minister for Foreign Affairs Fumio Kishida, 1 April 2014, http://www.mofa.go.jp/press/kaiken/kaiken4e_000059.html (last visit on 24 May 2014).

108 Minister for Agriculture, Forestry and Fisheries also announced in its statement that Japan would submit a new research plan based upon international law and scientific evidence to the Scientific Committee of the IWC by the autumn this year, which reflects the criteria mentioned in the *Judgment*. Policy towards the Future Whale Research Program. Statement by Minister for Agriculture, Forestry and Fisheries, the Government of Japan, 18 April 2014. http://www.jfa.maff.go.jp/e/pdf/danwa.pdf.

Japanese Government proposed a new research whaling plan in the Antarctic and submitted it to the IWC Scientific Committee. This is called New Scientific Whale Research Program in the Antarctic Ocean ('NEWREP-A'),[109] which was formulated taking into account the standards indicated in the *Judgment* of the Court in this case, and which could be amended when judged necessary, based on discussions at the Scientific Committee.[110]

The process of getting a new approval for the programme may not be so easy, because Japan must not only convince the IWC and the Contracting Governments to the ICRW that its intent to undertake a new whaling programme complies with the *Judgment* of the Court, but must also present the programme for review and analysis by the Scientific Committee, which would require a higher level of scrutiny. Moreover, the IWC adopted the Resolution on Whaling under Special Permit ('Resolution 2014–5') in September 2014, which espouses the standard of review and the reasonableness test confirmed by the *Judgment* of the Court.[111] Thus, the circumstances around Japan in the IWC are, and will remain, so difficult that whether NEWREP-A will be accepted in the Scientific Committee as well as in the IWC remains unpredictable.

---

109 The whole of this new plan can be seen on the website of the Fishery Agency of the Japanese Government. http://www.jfa.maff.go.jp/j/whale/pdf/newrep--a.pdf.

110 Press Conference by Foreign Minister Fumio Kishida, 18 November 2014. http://www.mofa.go.jp/press/kaiken/kaiken4e_000126.html.

111 Michael Johnson, '*Whaling in the Antarctic* – the ICJ Decision and its Consequences for Future Special Permit Whaling', *Australia Year Book of International Law*, Vol. 32 (2014), p. 96. Resolution 2014–5, para.1 provides that the IWC instructs the Scientific Committee, in its review of new and existing special permit research programmes, to provide advice to the Commission on:
(a) whether the design and implementation of the programme, including sample sizes, are reasonable in relation to achieving the programme's stated research objectives;
(b) whether the elements of the research that rely on lethally obtained data are likely to lead to improvements in the conservation and management of whales;
(c) whether the objectives of the research could be achieved by non-lethal means or whether there are reasonably equivalent objectives that could be achieved non-lethally;
(d) whether the scale of lethal sampling is reasonable in relation to the programme's stated research objectives, and non-lethal alternatives are not feasible to either replace or reduce the scale of lethal sampling proposed; and
(e) such other matters as the Scientific Committee considers relevant to the programme, having regard to the decision of the International Court of Justice, including the methodology used to select sample sizes, a comparison of the target sample sizes and actual take, the timeframe associated with a programme, the programme's scientific output; and the degree to which a programme coordinates its activities with related research projects. https://archive.iwc.int/pages/view.php?ref=3723&search=%21collection72&order_by=relevance&sort=DESC&offset=0&archive=0&k=&curpos=0.

In any case, it is to be stressed that as the ICRW allows the Contracting parties to issue special permits to kill, take and treat whales for purposes of scientific research, Japan is entitled to keep engaging in whaling in the Antarctic so long as it conforms to the decision of the Court.

The *Judgment*, of course, has no binding effect except between Australia and Japan in respect of the whaling pursuant to JARPA II in the Antarctic, because of the effect of *res judicata*.[112] However, for Japan, it is to be noted that this *Judgment* may have a certain legal influence on its whaling activities in other sea areas, in particular the *Japanese Whale Research Program under Special Permit in the North Pacific Phase II* ('JARPN II').[113] The statement by the Court that Japan's obligation to refrain from authorizing any special permit whaling which is not for purposes of scientific research within the meaning of Article VIII 'already applies for all States parties' means that the Court recognizes an obligation *erga omnes partes* of the Contracting parties to the ICRW. Therefore, any Contracting party to the ICRW may institute proceedings before the Court against Japan for any alleged violations of that obligation if the jurisdiction of the Court can be established through a declaration under Article 36, paragraph 2, of the Statute of the Court etc. In this sense, Japan must comply with the *Judgment* in relation to any future programme in the Antarctic as well as to JARPN II.

Furthermore, the *Whaling in the Antarctic Judgment* has also had a considerable influence upon the IWC discussions and has had after effects with respect to the regulation of whaling pursuant to Article VIII of the ICRW even outside the Antarctic. In the above-mentioned Resolution 2014–5, the criteria to be considered by the Scientific Committee for the review of the special permits

112 The principle of *res judicata*, according to the Court, signifies that 'the decisions of the Court are not only binding on the parties, but are final, in the sense that they cannot be reopened by the parties as regards the issues that have been determined, save by procedures, of an exceptional nature, specially laid down for that purpose'. *Application of the Convention on the Prevention and Punishment of the Crime of Genocide (Bosnia and Herzegovina v. Serbia and Montenegro), Judgment, I.C.J. Reports 2007*, p. 90, para. 115.

113 JARPN II started with two feasibility surveys in 2000 and 2001. The first full survey started in 2002. The objectives of JARPN II are the following: (a) feeding ecology and ecosystem studies (prey consumption by cetaceans, prey preference of cetaceans, ecosystem modeling); (b) monitoring environmental pollutant in cetaceans and the marine ecosystem (pattern of accumulation of pollutants in cetaceans, bioaccumulation process of pollutants through the food chain, relationships between chemical pollutants and cetacean health), and (c) stock structure of large whales (common minke whale, Bryde's whale, sei whale and sperm whale). Institute of Cetacean Research, *Scientific Contribution from JARPN/JARPN II*. http://www.icrwhale.org/pdf/ScientificContributionJARPN.pdf.

can be applied to all whaling activities under special permit programmes without geographical limits. In the discussions on the draft resolution, New Zealand, a proposing country, stated that while the *Judgment* was only binding on the parties to the dispute, 'all members of the IWC must pay close attention to the ruling and consider its wider implications.'[114] Against this assertion, Japan argued that the *Judgment* 'relates to the particular case of JARPA II' and that 'it does not change the legal framework of the IWC'.[115] Although some other Member states expressed concerns in relation to the proposal,[116] the Resolution was finally adopted, not by consensus but by vote (35 in favour, 20 against and 5 abstentions).[117] Accordingly, the Scientific Committee, as a subsidiary organ of the IWC, must comply with the Resolution, which incorporates the main elements of the *Judgment*, and proponents of future whaling programmes, in Antarctic or in any other ocean, would be expected to follow it. Thus, the practical impact of the *Whaling in the Antarctic Judgment* may go beyond the geographical and personal limits of its purely legal effect, and would prompt the development of criteria for the review of scientific research whaling of a restrictive nature.

3.2 *Will the Court Actively Apply 'Standard of Review' to the Future Cases Solely Concerning Violations of International Legal Rules?*

As mentioned above, the Court has recently confirmed the existence of this kind of obligation, and has provided redress to the Applicant for violations of such obligations in its 2012 *Judgment* in the *Obligation to Prosecute or*

114 The IWC Chair's Report of the 65th Meeting, p. 17, para. 141, although it is superfluous to say that New Zealand was rather inaccurate in saying that 'the parties to the dispute, i.e., Australia, Japan and New Zealand', because New Zealand 'avails itself of the right conferred upon it by Article 63, paragraph 2, of the Statute to intervene as a non-party in the proceedings' in its Declaration, which the Court ordered to be admissible. *Declaration of Intervention of the Government of New Zealand*, p. 20, para. 35. However, New Zealand also confirms that 'by availing itself of its right to intervene, it accepts that the construction given by the judgment in this case will be equally binding upon it'. *Ibid.*, p. 6, para. 9.

115 Japan also stressed that 'changes suggested would require amendments to Article VIII of the Convention or paragraph 30 of the Schedule and that a Resolution would therefore not be appropriate'. The IWC Chair's Report of the 65th Meeting, p. 17, para. 144.

116 For example, Norway noted that the restriction derived from operative paragraph 1(b) of the proposed Resolution went beyond the *Judgment* (*ibid.*, p. 17, para. 150), and Iceland supported Japan's argument that the proposal could be supported as it narrowed the scope of Article VIII of the Convention and paragraph 30 of the Schedule (*ibid.*, p. 17, para. 149).

117 *Ibid.*, p. 17, para. 151.

Extradite Case.[118] Compared with this case, the *Whaling in the Antarctic Case* may be seen as founding an extension of this jurisprudence of the Court owing to its endorsement of the obligations *erga omnes partes*, which recently may have provided encouragement to the Republic of the Marshall Islands to file with the Court, in April 2014, its Applications against nine States, accusing them of not fulfilling their obligations with respect to an early cessation of the nuclear arms race and of failure to develop international legal rules concerning nuclear disarmament, in particular under the Treaty on the Non-Proliferation of Nuclear Weapons, though also even as a matter of customary international law.[119] The touchstone in the *Obligations concerning Negotiations relating to Cessation of the Nuclear Arms Race and to Nuclear Disarmament Cases* is whether every State is entitled to appear before the Court to claim that a violation of the relevant rules of international law has occurred and should cease.

Thus, it is expected that the Court will have more opportunities in the near future to decide on the alleged violations of international obligations under a treaty by a State party to it, so that another State party, notwithstanding that it suffered no damage itself, may be awarded some remedy in relation solely to the violation of its own rights under the treaty. In such a case, the most important element for the Court is to clarify the requirements for the award of such remedies. Otherwise, the Court might very easily exercise a kind of judicial review in an arbitrary way, with the result that more States, which have declared their acceptance of optional clause jurisdiction, might run away from the jurisdiction of the Court, for example by revoking their declaration under Article 36, paragraph 2, of the Statute of the Court. Such a possibility arises because the Court was originally constituted to settle legal disputes between States in order to provide remedies for concrete damage, not to favour one party or category of parties to a treaty over another by downgrading the international adjudicatory process to one merely of review of violations of legal obligations.[120]

It is true that some international dispute settlement mechanisms, like the WTO, are ready for the implementation of public interests as represented in certain treaties or certain rules of customary international law, as well as for the solution of concrete disputes between parties. The *Whaling in the Antarctic*

---

118 *Questions relating to the Obligation to Prosecute or Extradite* (*Belgium v. Senegal*), *Judgment, I.C.J. Reports 2012*, p. 422.

119 *I.C.J. Press Release*, No.2014/18. Only three cases of nine are seen in the docket as of 1 January 2015: *Obligations concerning Negotiations relating to Cessation of the Nuclear Arms Race and to Nuclear Disarmament* (*Marshall Islands v. United Kingdom*) (*Marshall Islands v. Pakistan*) (*Marshall Islands v. India*) *Cases*.

120 Foster, *supra* note 84, p. 91.

Judgment, read together with the *Obligation to Prosecute or Extradite Judgment*, may encourage the Court to exercise its judicial control to decide on whether or not there has been a violation of legal obligations under particular treaties or customary international legal rules. Examples may include the Convention on International Trade in Endangered Species of Wild Fauna and Flora ('the CITES'), which, should the Court consider it to be a treaty for maintaining and enhancing the public interest of the international community as a whole, or at least the collective interest of the Contracting parties to it, could be used to claim suspension of Japan's whaling activities through the resort to the Court by any party to the CITES.[121] In this sense, the recent jurisprudence has led the Court to be endowed with a new function for ensuring the implementation of particular treaties or rules of customary international law, which would be applied to realize and to advance the public interest of the parties to the treaty, or of the international community as a whole, in addition to its classic function of resolving legal disputes between States to provide effective remedies for actual violations of the rights of the State party to the dispute which have given rise to actual damage.

Moreover, this issue also seems to be associated with the relationship between science and law, which was a central issue throughout the considerations in the *Whaling in the Antarctic Case*.[122] Whether or not the attempts to have the Court exercise its judicial control as a kind of compliance mechanism for certain rules of international law, which may reflect public interest, will be successful is dependent upon the validity of the criteria, according to which the Court may apply a 'standard of review' to the concrete case—that is, the test of reasonableness in the *Whaling in the Antarctic Case*. By using this test,

121 For the argument that Japan's research whaling programmes would be a breach of the CITES, Peter H. Sand, 'Japan's 'Research Whaling' in the Antarctic Southern Ocean and the North Pacific Ocean in the Face of the Endangered Species Convention (*CITES*)', *Review of European Community & International Environmental Law*, Vol. 17 (2008), pp. 56–71.

122 Cymie R. Payne, 'Australia v. Japan: ICJ Halts Antarctic Whaling', *ASIL Insight*, Vol. 18, Issue 9 (8 April 2014). Jacqueline Peel, 'Introductory Note to Whaling in the Antarctic (Australia v. Japan: New Zealand Intervening) (I.C.J.)', *International Legal Materials*, Vol. 54 (2014), p. 1, who states that 'the *Whaling Decision* demonstrates that the ICJ can be an effective and efficient forum for the resolution of contested law/science matters'. http://www.asil.org/insights/volume/18/issue/9/australia-v-japan-icj-halts-antarctic-whaling For more in details, see, *idem.*, 'ICJ Halts Antartic Whaling – Japan Starts Again', *Transnational Environmental Law*, Vol. 4 (2015), pp. 188–190. See also, Brendan Plant, 'Sovereignty, Science, and Cetaceans: The *Whaling in the Antarctic* Case', *The Cambridge Law Journal*, Vol. 74 (2015), pp. 43–44.

the Court was able to avoid entering into the details of science, for example, the question whether sample sizes were scientifically correct or necessary. A similar approach may provide it with 'a model for separating scientific matters and the non-scientific agenda in other complicated disputes involving science, society, and law'.[123] However, there would remain still some risk that such an application of 'standard of review' might impose on the Court—a body for dispute settlement in the light of international legal rules—the quite considerable burden to take into consideration criteria other than law. It remains unclear whether or not the extensive involvement of the Court with 'standard of review' is appropriate to its character and status in the international community, even though it is said that the Court has never refused to work as a kind of administrative tribunal for the application of 'standard of review'.[124]

Concluding Remarks

The *Whaling in the Antarctic Case*, in which Japan has appeared before the Court for the first time, either as an applicant or as a respondent, clearly shows Japan to have been lacking in, or short of, experience, in particular in the procedural aspects of the proceedings before the Court.[125] Apart from the questions on jurisdiction and admissibility, which have already been mentioned above, such procedural issues may be seen to have included the handling of the application for intervention by New Zealand pursuant to Article 63 of the Statute of the Court,[126] Japan's possible measures against a *de facto* joint

123 William de la Mare, Nick Gales & Marc Mangel, 'Applying Scientific Principles in International Law on Whaling. The Approach Might Address Disputes beyond Whaling and the Courtroom', *Science*, Vol. 345 (5 September 2014), p. 1126, which also states that 'the ICJ looked at the logic: If the sample sizes were purported to be necessary for achieving the objectives, then failing to collect the specified numbers must mean either that the objectives would not be attained or that the specified sample sizes were unnecessarily large'.

124 Robert Kolb, 'Chronique de la jurisprudence de la Cour international de Justice en 2014', *Revue Suisse de droit international et européen*, Tome 25 (2015), p. 155.

125 For Australia, this case is the fourth appearance before the Court, following the *Nuclear Test* case, the *Phosphate Lands in Nauru* case and the *East Timor* case. Australia also has been a party to the *Seizure and Detention of Documents* case against Timor-Leste since 2013. Surprisingly, the *Whaling* case is the first for Australia to proceed to the merits before the Court.

126 New Zealand sought to intervene in the proceedings, and the Court decided that the New Zealand's Declaration of Intervention is admissible. *Whaling in the Antarctic*

struggle of the Applicant and the Intervener,[127] as well as the refusal by the Applicant of a second round of written proceedings,[128] the utilisation of the experts,[129] and so on. As a matter of fact, the litigation strategy of Australia and New Zealand contributed to their obtaining successful findings from the Court in this case, which may have been some kind of compensation for their bad memories in relation to the *Southern Bluefin Tuna Cases.*[130]

(*Australia v. Japan*), *Declaration of Intervention of New Zealand, Order of 6 February 2013, I.C.J. Reports 2013*, p. 11,para. 23(1). Japan did not object to the admissibility of the Declaration. *Ibid.*, p. 9, para. 17. See also, Elena Geddis & Penelope Ridings, '*Whaling in the Antarctic*: Some Reflections by Counsel', *New Zealand Yearbook of International Law*, Vol. 11 (2013), pp. 145–146.

127 The intervention by New Zealand, an anti-whaling country, whose assertions were opposed to Japan's views, was decided to be admissible with the result that the Bench of the Court had a New Zealand's Judge as well as an Australian *ad hoc* Judge, which led Japan to raise its concerns relating to the equality of the Parties to the dispute. Written Observations of Japan on New Zealand's Written Observations, para. 2. http://www.icj-cij.org/docket/files/148/17388.pdf See also, Sir Geoffrey Palmer QC, 'A Victory for Whales: Sir Geoffrey Palmer QC Explains Aspects of the ICJ Decision', *Victoria University of Wellington Legal Research Papers*, Palmer Paper 42 (2014), p. 128. On this point, the Court confirmed that 'Australia and New Zealand cannot be regarded as being "parties in the same interest" within the meaning of Article 31, paragraph 5, of the Statute' because of the New Zealand's status of non-party to the proceedings, and thus concluded that 'the presence on the Bench of a judge of the nationality of the intervening State has no effect on the right of the judge *ad hoc* chosen by the Applicant to sit in the case pursuant to Article 31, paragraph 2, of the Statute'. *Whaling in the Antarctic (Australia v. Japan), Declaration of Intervention of New Zealand, Order of 6 February 2013, I.C.J. Reports 2013*, p. 10, para. 21. See also, Declaration of Judge Owada, *ibid.*, pp. 11–14, paras. 2–6; Penelope Riding, 'The Intervention Procedure in *Whaling in the Antarctic*: A Threat to Bilateralism?' *Australian Year Book of International Law*, Vol. 32 (2014), pp. 106–108.

128 Japan requested a second round of written proceedings, which Australia opposed, and the Court did not order a second round but instead decided to go to the oral proceedings. *Whaling in the Antarctic, I.C.J. Reports 2014*, p.235, para. 6. Separate Opinion of Judge Greenwood, *ibid.*, pp.418–419, paras. 32–38, who attempts to justify the decision of the Court on this point.

129 In this case, the experts who were appointed respectively by the parties to the dispute played an important role in influencing the Court's understanding of the scientific evaluations. As for the status of experts, the Court had made it clear in the *Pulp Mills Case* that they should not be included in the delegations of each party. *Pulp mills on the River Uruguay (Argentina v. Uruguay), Judgment, I.C.J. Reports 2010*, p. 72, para. 167. See also, Makane Moïse Mbengue, 'Between Law and Science: A Commentary on the *Whaling in the Antarctic* Case', *Questions of International Law*, Vol. 14 (2015), pp. 3–12.

130 In the *Southern Bluefin Tuna Cases*, the Arbitral Tribunal rejected the arguments of Australia and New Zealand, and instead held in favour of Japan in that 'it is without

Japan certainly 'lost' against Australia (and in fact, New Zealand) in the *Whaling* case in the sense that Japan's whaling activities under JARPA II were found to be in breach of the ICRW, which obliged Japan to revoke and to refrain from any fresh authorization in relation to JARPA II. However, according to the Court, the possibility of continuing whaling activities under a new plan pursuant to the ICRW is still open to Japan. This means that the Court rejected the argument that Japan's whaling activities should totally and finally be prohibited in the Antarctic, which must have been one of the main aims of Australia in this case. Australia also failed to persuade the Court to clarify that the object and purpose of the ICRW lies in the protection of whales.[131] If that is true, it may be added that it is a result of the Court's penetrating insights and prudent strategy, in that it has chosen solutions acceptable for both parties to the dispute. Whilst Japan will be able to keep engaging in whaling activities in the Antarctic 'for purposes of scientific research' under a new plan, in accordance with the ICRW, on the basis of the *Judgment* in the case, Australia will also attempt to persuade more Contracting parties to the ICRW to follow its own arguments, together with other anti-whaling countries, in the IWC, so that whales may be protected and preserved by the application of the severe standards for scientific research whaling, which the Court made clear in the *Judgment*.[132]

In any case, the Court's decision is expected to revitalize the political process of the IWC to fill a gap between the pro-whaling and the anti-whaling countries, with the result that the preservation of whales could be better managed within the framework of the ICRW; although there is no denying that there remains a risk of increasingly sharper conflicts between pro-whaling and

jurisdiction to rule on the merits of the dispute'. *Southern Bluefin Tuna case between Australia and Japan and between New Zealand and Japan. Award on Jurisdiction and Admissibility, Decision of 4 August 2000, Reports of International Arbitral Awards*, Vol. XXIII, p. 48, para. 72.

131 Shirley V. Scott & Lucia Oriana, '*Whaling in the Antarctic* (*Australia v. Japan: New Zealand Intervening*) Judgment of 31 March 2014: A Decisive Victory – but for Whom?' *The International Journal of Marine and Coastal Law*, Vol. 29 (2014), p. 557, who admits that '[o]n closer examination, the judgment represents somewhat less than a total victory either for the whales themselves or for the broader development of international environmental law'.

132 Richard Caddell, 'Science Friction: Antarctic Research Whaling and the International Court of Justice', *Journal of Environmental Law*, Vol. 26 (2014), p. 339, states that 'the ICJ has struck an effective balance: robust procedural obligations advanced in the judgment entrench both a broad right to conduct lethal sampling and a clearer requirement to objectively justify such a programme where it conflicts with recommended international standard'.

anti-whaling nations in the international forum, rather than the prospect of effecting reconciliation. These concerns stem from the undecisive and rather ambivalent nature of the *Judgment*, which only stated that Japan's activities under JARPA II did not fulfil the conditions of Article VIII of the ICRW, but which did not order Japan comprehensively to cease the use of lethal methods in scientific research whaling. Australia still might, entirely dependent on the progress of the discussions concerning the reasonableness of NEWREP-A within the IWC, request the Court to construe the meaning of the *Judgment* in accordance with Article 60 of the Statute of the Court, so that the *Judgment* would be available on a more favourable basis for Australia's arguments in the political process.

As to the lessons for Japan directly derived from the proceedings in this case, Japan has gained valuable and rewarding experience at first hand, in particular in relation to the procedure of the Court. Hopefully, the time will soon come when those lessons turn out to be useful and beneficial for Japan in its next appearance before the Court as well as in any further diplomatic negotiations.

Generally speaking, Japan has maintained friendly relations with Australia and New Zealand in almost all sectors, in particular in trade and security, despite the existence of some conflicts between them, including the differences of views on whaling. This is illustrated by Japanese Prime Minister Shinzo Abe's successful visits to Australia and New Zealand in July 2014, three months after the *Whaling in the Antarctic Judgment*[133] as well as by Australian Prime Minister Tony Abbott's official visit to Japan in April 2014, just a week after the *Judgment*.[134] On the one hand, it is necessary and essential for Japan,

133 During the official visit of Prime Minister Abe to Australia on 8–9 July 2014, he and Australian Prime Minister Tony Abott have signed the Agreement between Japan and Australia and for an Economic Partnership (http://www.mofa.go.jp/ecm/ep/page22e_000430.html) and the Agreement between the Government of Japan and the Government of Australia concerning the Transfer of Defence Equipment and Technology (http://www.mofa.go.jp/files/000044447.pdf). Before the visit to Australia, Prime Minister Abe also dropped by New Zealand to see its Prime Minister John Key on 7 July 2014 when these two Prime Minister have issued a Joint Press Release on Enhancing Japan-New Zealand Cooperation (http://www.mofa.go.jp/a_o/ocn/page3e_000194.html) to emphasize a good relationship between the two countries.

134 Ministry of Foreign Affairs of Japan, 'Visit to Japan of The Hon Tony Abbott, Prime Minister of the Commonwealth of Australia(Overview)' http://www.mofa.go.jp/a_o/ocn/au/page3e_000166.html, which confirms that '[i]n the series of conversations during Prime Minister Abbott's stay in Japan, Prime Minister Abe mentioned the difficult conditions in Japan following the court decision on the whaling case at the International Court

as well as for Australia and New Zealand, to develop and enhance such good relationships by their further diplomatic efforts. On the other hand, Japan should carefully note, in relation to the whaling issue, that the ICRW will require it to be accountable for the reasonableness and the transparency of a new plan for scientific research whaling pursuant to Article VIII if Japan wants to continue to engage in whaling activities within the framework of the ICRW. The decisions by the Court, as well as the actions by States based upon them, should also find their proper place in the diplomatic discourse.[135]

In this respect, the pledge by the Japanese Ministry of Foreign Affairs to comply with the *Judgment* of the Court after it had been delivered should be welcomed, not only in relation to the pursuit of its national interests but also for the conservation of whales and maintenance of the ecosystem in the Antarctic. Generally speaking, it is to be noted that respect for judgments and decisions by courts or tribunals must be beneficial to the parties in the long term. The *Whaling in the Antarctic Case* is no exception to this rule.

---

of Justice (ICJ) and the two leaders shared their views that they would make efforts so that this issue would not influence the overall bilateral relations'.

135 One of the important issues in dealing with standard of review during the political process is the question of who is entitled to apply the reasonableness test, since 'the decision on whether a problem is 'reasonable' or not inevitably depends on the reviewer's subjective judgment'. See, Dai Tamada, 'On the Way to Definitive Settlement of Dispute: Lessons from the *Whaling Case*', *Australian Year Book of International Law*, Vol. 32 (2014), pp. 122–123.

CHAPTER 11

Science in the Court! The Role of Science in 'Whaling in the Antarctic'

Anthony Press

> The trouble with the whaling industry today is that there are too many damned intellectuals mixed up in it![1]

Introduction

'The Court concludes that the special permits granted by Japan for the killing, taking and treating of whales in connection with JARPA II *are not "for purposes of scientific research" pursuant to Article* VIII*, paragraph 1, of the Convention'.*[2]

***

This paper deals with the way science featured in the case *Whaling in the Antarctic* (*Australia v. Japan: New Zealand intervening*). 'Science' was invoked by both Australia and Japan (and by New Zealand). The activities carried out under JARPA II and JARPA were portrayed as 'unscientific' by Australia and as 'science' by Japan. Scientists, as expert witnesses, were called upon to give written and oral evidence (two by Australia and one by Japan) and they were cross-examined in Court, and were questioned by the bench. Australia presented two Expert Witnesses to discuss science, Dr Nick Gales, Chief Scientist of the Australian Antarctic Program, and Professor Marc Mangel of the University of California, Santa Cruz. Japan presented one Expert Witness, Professor Lars Walløe from the University of Oslo.

It is uncommon for scientists to appear as witnesses in the International Court of Justice. In the *Case Concerning Pulp Mills on the River Uruguay* (*Argentina v. Uruguay*) the International Court of Justice observed:

1 Of Whales and Men, RB Robertson, 1956, The Reprint Society, London p. 191.

2 Judgment, Whaling in the Antarctic (Australia v. Japan: New Zealand intervening), International Court of Justice, 31 March 2013, para 246. At, http://www.icj-cij.org/docket/files/135/15877.pdf as accessed on 20 January 2015.

 | DOI 10.1163/9789004313828_013

> Regarding those experts who appeared before it as counsel at the hearings, the Court would have found it more useful had they been presented by the Parties as expert witnesses under Articles 57 and 64 of the Rules of Court, instead of being included as counsel in their respective delegations. The Court indeed considers that those persons who provide evidence before the Court based on their scientific or technical knowledge and on their personal experience should testify before the Court as experts, witnesses or in some cases in both capacities, rather than counsel, so that they may be submitted to questioning by the other party as well as by the Court.[3]

Besides being able to be cross-examined and/or questioned by the Court, expert witnesses (as opposed to 'experts' as counsel) are less likely to taken as 'advocates' for the position of parties to a dispute.[4] Foster (2014) in discussing the role of experts in international environment conflicts, wrote that '[d]isagreements over science feature with growing regularity in international disputes'.[5] Foster went on to consider '…the possibility that the Court could additionally decide to appoint its own experts to assist in the resolution of a case…'.[6]

While the Court did not appoint its own experts in the whaling case, the Judgment deals in detail with the evidence of the expert witnesses from both sides in the case. In doing so, it draws extensively on this evidence in delivering its Judgment. The Court examined the evidence before it about 'science' and 'scientific research' and ultimately drew its own conclusions as to the veracity of that evidence and its relevance to the case.

I deal with how 'science' was used by Australia, Japan and New Zealand to describe the 'Special Permit whaling' carried out by Japan in the Antarctic, and how the Court interpreted the 'science' before it. Like the Court, I do not attempt to describe what science is, although I will characterise the strength of some of the arguments used in the written and oral evidence in the proceedings.

---

3 Judgment, Case Concerning Pulp Mills on the River Uruguay (Argentina v. Uruguay), International Court of Justice, 20 April 2010, para 167.

4 Case Concerning Pulp Mills on the River Uruguay, Separate Opinion of Judge Greenwood, para 28. At, http://www.icj-cij.org/docket/files/135/15889.pdf as accessed 20 January 2015.

5 Foster, CE 2014, New Clothes for the Emperor? Consultation of Experts by the International Court of Justice, Journal of International Dispute Settlement, 2014, 5, 139–173, doi: 10.1093/jnlids/idt015p140.

6 Foster 2014, p. 141.

In general, I have followed the sequence of the Judgment (in so far as it relates to 'science') in order elucidate the reasoning of the Court.

JARPA II

JARPA II[7] was the whaling program authorised by Japan under Article VIII of the International Convention for the Regulation of Whaling and conducted in the Antarctic. It commenced in 2005, following the cessation of JARPA in 2004, and continued until the Judgment of the International Court of Justice (2014). For JARPA II, Japan authorised the killing of up to 935 Antarctic minke whales, 50 fin whales and 50 humpback whales per annum. The program was open ended (ie no forecast end date) and was to be 'reviewed'[8] every 6 years.

Japan asserted that JARPA II '...was launched in 2005 to meet specific research needs on the basis of the results of JARPA'[9] (author's emphasis). Australia, on the other hand argued '[i]t is clear that, in giving effect to ...overriding commercial considerations, JARPA II is in fundamental conflict with the requirements of established scientific process'[10] (author's emphasis). These arguments were, in effect, the basis of the dispute between Australia and Japan.

Japan had authorised JARPA II, as it had its predecessor JARPA, as Special Permit Whaling. Special Permit Whaling can be authorised under Article VIII of the International Convention for the Regulation of Whaling by a Party to the Convention. Argument in the case was whether the activities carried out in JARPA II were covered by Article VIII. As the Court said at para 42 of the Judgment:

> The present proceedings concern the interpretation of the International Convention for the Regulation of Whaling and the question whether

7 Japan 2005 Plan for the Second Phase of the Japanese Whale Research Program under Special Permit in the Antarctic (JARPA II) – Monitoring of the Antarctic Ecosystem and Development of New Management Objectives for Whale Resources, Government of Japan. At http://www.icrwhale.org/pdf/SC57O1.pdf as accessed on 23 January 2015.

8 'When the full-scale research program has been launched, a comprehensive report will be submitted to the IWC/SC after each six-year research period'. Japan 2005 pp. 20–21.

9 Japan 2012 Counter-Memorial of Japan, March 2012, p. 225 para 5.1. At, http://www.icj-cij.org/docket/files/148/17384.pdf as accessed 20 January 2015.

10 Australia 2011 Memorial of Australia, 9 May 2011, Volume 1 pp. 113–114, para 3.74, Government of Australia. At, http://www.icj-cij.org/docket/files/148/17382.pdf.

> special permits granted for JARPA II are for purposes of scientific research within the meaning of Article VIII, paragraph 1, of the Convention.[11]

Science is meant to play a central role in the way activities under the ICRW are managed. In 1950 the International Whaling Commission established a Scientific Committee.[12] The role of the Scientific Committee is to assist the International Whaling Commission '...in discharging its functions, in particular those relating to "studies and investigations relating to whales and whaling"'.[13]

The IWC, at its 'Scientific Committee Home Page'[14] As accessed 20 January 2015. states that '[a]n important feature of the [Whaling] Convention is the emphasis it places on scientific advice. The Convention requires that amendments to the Schedule "shall be based on scientific findings". To this end, the Commission has established a Scientific Committee'.

In current practice the Scientific Committee is made up of scientists nominated by States parties, and supplemented by '...advisers from intergovernmental organizations and scientists who have not been nominated by States parties [who] may be invited to participate in a non-voting capacity'.[15]

The functions of the Scientific Committee are underpinned by Article IV of the ICRW which provides that:

> The Commission may either in collaboration with or through independent agencies of the Contracting Governments or other public or private agencies, establishments, or organizations, or independently (*a*) encourage, recommend, or if necessary, organize studies and investigations relating to whales and whaling; (*b*) collect and analyze statistical information concerning the current condition and trend of the whale stocks and the effects of whaling activities thereon; (*c*) study, appraise, and disseminate information concerning methods of maintaining and increasing the populations of whale stocks.[16]

11 Judgment, Whaling in the Antarctic (Australia *v.* Japan: New Zealand intervening) 31 March 2013, Para 42.

12 The International Whaling Commission was established in under the provisions of the International Convention of the Regulation of Whaling.

13 Judgment, Whaling in the Antarctic (Australia *v.* Japan: New Zealand intervening) 31 March 2013, Para 47.

14 Scientific Committee of the International Whaling Commission, Homepage. At, https://iwc.int/scmain

15 Judgment, Whaling in the Antarctic (Australia *v.* Japan: New Zealand intervening) 31 March 2013, Para 47.

16 International Convention for the Regulation of Whaling 161 UNTS 72, Article VIII, para 3.

Under the ICRW, the International Whaling Commission can use the findings of the Scientific Committee in making amendments to the Convention's Schedule. Importantly, Article V of the ICRW provides that '...amendments of the Schedule...shall be based on scientific findings...'.[17]

The Scientific Committee also has a role with respect to Special Permit Whaling. Under paragraph 30 of the Schedule to the ICRW, '...the Scientific Committee reviews and comments on special permits before they are issued by States parties to their nationals for purposes of scientific research under Article VIII, paragraph 1, of the Convention'.[18] It should be noted that the Scientific Committee has no decision-making powers with respect to Special Permit Whaling. The Court observed that:

> [t]he Scientific Committee has not been empowered to make any binding assessment in this regard. It communicates to the Commission its views on programmes for scientific research, including the views of individual members, in the form of reports or recommendations. However, when there is a division of opinion, the Committee generally refrains from formally adopting the majority view.[19]

The reality of recent decades is that the Scientific Committee itself has been deeply divided in its views on Special Permit whaling, and mired in disputes about the scientific necessity of both JARPA and JARPA II (as reflected in the observation of the Court, above). Clapham (2014) wrote:

> [i]mportantly, the ICJ judgment represented the first time that an objective international body, independent of the IWC, had reviewed and assessed Japan's scientific whaling – and in doing so had found it wanting. By implication, the ruling also stood as a condemnation of the intractably mired process used by the IWC Scientific Committee to review and discuss scientific whaling programs; until recently, such reviews featured full participation by Japan in not only the review, but also in the writing of the subsequent report. Such reports invariably featured numerous statements along the lines of 'Some members felt... In contrast, others disagreed...', and debates often included charges by Japan that criticisms were politically motivated. Not surprisingly, the Scientific Committee

17 International Convention for the Regulation of Whaling 161 UNTS 72, Article V, para 2.

18 Judgment, Whaling in the Antarctic (Australia *v.* Japan: New Zealand intervening) 31 March 2013, Para 47.

19 Judgment, Whaling in the Antarctic (Australia *v.* Japan: New Zealand intervening) 31 March 2013, Para 47.

> could never reach agreement on either endorsement or condemnation of a scientific whaling program.[20]

In analysing how 'science' played a key role in the outcome of the case, I have generally followed the structure of the Judgment in order to illustrate the Court's majority thinking (re science), and the evidence it had before it.

Science in the Case, Part 1: Does the Court Need a Definition of Science?

This Chapter does not attempt to define science—the Court also declined to do so. But the Court's interpretations of the use of 'science' in the argument of the case are fundamental to understanding the outcome, and, ultimately Japan's response to it.[21]

Australia's argument (with respect to 'science') was that JARPA II '...is not a programme for purposes of scientific research within the meaning of Article VIII of the Convention...'.[22] Australia also argued that Japan had '...failed to comply with the procedural requirements set out in paragraph 30 of the Schedule for proposed scientific permits'.[23]

Japan, on the other hand, argued that JARPA II '...has been undertaken for purposes of scientific research and is therefore covered by the exemption provided for in Article VIII, paragraph 1, of the Convention' and also that they [Japan] had not breeched 'procedural requirements' as alleged by Australia.[24]

What is Scientific Research?

For JARPA II Special Permit Whaling to be legitimate, the activities authorised by Japan must be covered by Article VIII of the ICRW: Japan argued that JARPA II (and JARPA) were 'activities for the purposes of scientific research'; Australia

20 Clapham, PJ 2015 Japan's whaling following the International Court of Justice ruling: Brave New World – Or business as usual? Marine Policy 51, 238–241.

21 Judgment, Whaling in the Antarctic (Australia *v.* Japan: New Zealand intervening) 31 March 2013, Para 47.

22 Judgment, Whaling in the Antarctic (Australia *v.* Japan: New Zealand intervening) 31 March 2013, Para 48.

23 Judgment, Whaling in the Antarctic (Australia *v.* Japan: New Zealand intervening) 31 March 2013, Para 48.

24 Judgment, Whaling in the Antarctic (Australia *v.* Japan: New Zealand intervening) 31 March 2013, Para 49.

argued that those activities did not meet the necessary objective criteria to be science: 'Japan seeks to cloak its ongoing commercial whaling program in the labcoat of science',[25] Australia's Agent stated in opening its case before the Court.

In written and oral evidence, one of Australia's expert witnesses, Professor Marc Mangel, argued that:

> [c]onsistent with generally accepted scientific practice and with criteria identified by the Scientific Committee of the IWC, my opinion is that the essential characteristics of a program for the purposes of scientific research in the context of conservation and management of whales are that the program:
> (a) has defined and achievable objectives that aim to contribute knowledge that is important to the conservation and management of whale stocks;
> (b) employs appropriate methods that are likely to achieve the stated objectives, including:
> (i) lethal methods only where the objectives of the research cannot be achieved by any other means (for example, by the analysis of existing data and/or the use of non-lethal research techniques);
> (ii) setting sample sizes using accepted statistical methodology; and
> (iii) linking mathematical models to data consistently;
> (c) includes periodic review of research proposals and results and adjustment in response to such review; and
> (d) is designed to avoid adverse effects on the stocks being studied[26] (author's emphasis).

Article VIII of the ICRW says, in part:

> ...any Contracting Government may grant to any of its nationals a special permit authorizing that national to kill, take and treat whales for purposes of scientific research subject to such restrictions as to number and subject to such other conditions as the Contracting Government thinks fit, and the killing, taking, and treating of whales in accordance with the provisions of this Article shall be exempt from the operation of this Convention[27] (author's emphasis).

25 International Court of Justice, Verbatim Record CR2013/7.

26 Mangel, M 2011 An Assessment of Japanese Whale Research Programs Under Special Permit in the Antarctic (JARPA, JARPA II) as Programs for Purposes of Scientific Research in the Context of Conservation and Management of Whales, April 2011. In Australia 2011, Whaling in the Antarctic (Australia v. Japan), Memorial of Australia, Volume 1 Appendix 2. p. 237.

27 International Convention of the Regulation of Whaling, Article VIII.

The Court spent some time considering the 'meaning of the phrase "for purposes of scientific research"'.[28]

In presenting its case, Australia argued that 'scientific research' could be assessed against the objective criteria of Mangel (above). Australia also argued that the phrase 'for the purposes of scientific research' should be analysed in two parts, effectively: that which is 'scientific research'; and that which was authorised 'for the purposes' of that scientific activity.[29]

The Court reached the view that:

> ...the two elements of the phrase 'for purposes of scientific research' are cumulative. As a result, even if a whaling programme <u>involves</u> scientific research, the killing, taking and treating of whales pursuant to such a programme does not fall within Article VIII <u>unless</u> these activities are '<u>for purposes of</u>' scientific research[30] (author's emphasis).

The ICRW does not contain a definition of 'scientific research'—so what is scientific research in an international Convention? The answer to that question remains hypothetical, at least in this case. The Court did not provide a definition of 'science' or of 'scientific research', nor did it accept that the 'objective criteria' argued by Australia needed to be met in order for activities to be 'for the purpose of scientific research'. As Australia pointed out in its Memorial, '...of the 40 other international environmental agreements which currently contemplate the conduct of "scientific, research", none expressly define the concept'.[31]

In not accepting Australia's argument that, in the context of this case and the operation of the ICRW, scientific research has four objective and essential criteria, the Court:

> ...observes that the experts called by both Parties agreed that scientific research should proceed on the basis of particular questions, which could take the form of a hypothesis, although they disagreed about the level of specificity required of such a hypothesis. In short, the opinions of the experts reveal some degree of agreement, albeit with important nuances, regarding the role of hypotheses in scientific research generally.

---

28 'Meaning of the phrase "for purposes of scientific research"', Judgment, Whaling in the Antarctic (Australia *v.* Japan: New Zealand intervening) 31 March 2013, Part E.

29 Memorial of Australia 2011, paras 4.33 – 4.48.

30 Judgment, Whaling in the Antarctic (Australia *v.* Japan: New Zealand intervening) 31 March 2013, Para 71.

31 Memorial of Australia para 4.43.

To Kill a Whale: The Place of Lethal Whaling in Scientific Research

The use of lethal methods for scientific research in the context of JARPA II, was central to Australia's argument in this case. Mangel had addressed the issue of lethal take in scientific research in his expert opinion[32] and Japan's case rested strongly on its assertions that lethal take was not only required to obtain the data sought in JARPA II, but that non-lethal research methods could not provide the data required by JARPA II.

Australia argued strongly that the granting of permits under Article VIII to 'take, kill and treat' whales for scientific purposes should only be done when non-lethal methods were not available and that the scientific question was of merit. New Zealand, in its intervention, argued that 'whales may be killed only where that is necessary for scientific research and it is not possible to achieve the equivalent objectives of that research by non-lethal means'.[33]

In putting its case, Japan argued that, as Article VIII specifically authorised a Party to '...kill, take and treat whales for purposes of scientific research...', it was not required to refrain from issuing permits under Article VIII only if non-lethal methods were not available. Japan argued that it authorised lethal whaling only to the extent necessary to undertake the scientific research contemplated by JARPA II, even though it was not legally obliged to do so.[34] (See below for further discussion of lethal whaling).

The Court found that Article VIII of the ICRW did expressly allow the use of lethal methods.[35] The Court further found that there was not a '...requirement that lethal methods be used only when other methods are not available'.[36]

In further looking at Japan's ability to authorise lethal whaling, the Court observed, though, that '...the States parties to the ICRW have a duty to co-operate with the IWC and the Scientific Committee and thus should give due

---

32 Mangel, M 2011 An Assessment of Japanese Whale Research Programs Under Special Permit in the Antarctic (JARPA, JARPA II) as Programs for Purposes of Scientific Research in the Context of Conservation and Management of Whales, April 2011. In, Memorial of Australia Volume 1 Appendix 2. Memorial of Australia, Volume 1 Appendix 2 pp. 334–389.

33 Judgment, Whaling in the Antarctic (Australia *v.* Japan: New Zealand intervening) 31 March 2013, Para 81.

34 Judgment, Whaling in the Antarctic (Australia *v.* Japan: New Zealand intervening) 31 March 2013, Para 80.

35 Judgment, Whaling in the Antarctic (Australia *v.* Japan: New Zealand intervening) 31 March 2013, Para 83.

36 Judgment, Whaling in the Antarctic (Australia *v.* Japan: New Zealand intervening) 31 March 2013, Para 83.

regard to recommendations calling for an assessment of the feasibility of non-lethal alternatives'.[37] These observations would play a major role, ultimately, in the final deliberations of the Court (see below).

Does Special Permit Whaling Require Peer Review?

Australia had also argued, through Mangel's '...characteristics of a program for the purposes of scientific research in the context of conservation and management of whales...',[38] that peer review of JARPA and JARPA II was required to establish the legitimacy of the research and its methods (Mangel's criterion (c) above).

The Court found that, regardless of whether peer review was established practice in science generally, '...it does not follow that a programme can be said to involve scientific research only if the proposals and the results are subjected to peer review'.[39] The Court's view was that, while 'not precluding' peer review, the Convention had established mechanisms, including through Paragraph 30 of the Schedule and the 'Guidelines',[40] for the Scientific Committee to review proposed permits and to undertake reviews of ongoing and completed programs.

Avoiding Harm to Stocks

Australia (and New Zealand) also argued strongly that Special Permit Whaling must not adversely impact whale stocks. This was put forward by Mangel as his the fourth characteristic of 'a program for the purposes of scientific research'. Japan did not oppose this argument.

37 Judgment, Whaling in the Antarctic (Australia *v.* Japan: New Zealand intervening) 31 March 2013, Para 83.

38 Mangel, M 2011 An Assessment of Japanese Whale Research Under Special Permit in the Antarctic (JARPA, JARPA II) as Programs for Purposes of Scientific Research in the Context of Conservation and Management of Whales, p. 337., Appendix 2, pp. 334–389.

39 Judgment, Whaling in the Antarctic (Australia *v.* Japan: New Zealand intervening) 31 March 2013, Para 84.

40 Annex P to the International Convention on the Regulation of Whaling, Process for the Review of Special Permit Proposals and Research Results from Existing and Completed Permits. At https://iwc.int/private/downloads/u25vr6ymdasooo8w4040c4g0/Annex%20P%20updated.pdf as accessed 20 January 2015.

The Court observed '...the Parties and the intervening State appear to be in agreement in respect of this criterion. In the particular context of JARPA II, however, Australia does not maintain that meeting the target sample sizes would have an adverse effect on the relevant stocks, so this criterion does not appear to be of particular significance in this case'.[41]

So, What *is* Scientific Research

Australia failed to convince the Court that the criteria outlined by Mangel needed to be met in order for the JARPA II to be 'a program for the purposes of scientific research in the context of conservation and management of whales', or, in the context of the Court's finding, 'scientific research': 'The Court is not persuaded that activities must satisfy the four criteria advanced by Australia in order to constitute 'scientific research' in the context of Article VIII'.[42] The Court considered that Mangel's criteria constituted an expert's view of 'well-conceived scientific research' but that they did not necessarily apply to the ICRW.

While dismissing the arguments put forward by Australia on how to assess what is or is not 'scientific research' in this case, the Court did not substitute another, or its own, characterisation of scientific research. It concluded: '[n]or does the Court consider it necessary to devise alternative criteria or to offer a general definition of "scientific research"'.[43]

Science in the Case, Part 2: Whither 'Scientific Research'

The purpose of this section is to look at how the Court, without a definition of 'scientific research', proceeded to look at the arguments put forward about science and scientific research in the case. This is not an attempt to analyse the legal arguments.

The Court found that it did not need to '...pass judgment on the scientific merit or importance of...' a research program's objectives '...in order to assess

41 Judgment, Whaling in the Antarctic (Australia *v.* Japan: New Zealand intervening) 31 March 2013, Para 85.

42 Judgment, Whaling in the Antarctic (Australia *v.* Japan: New Zealand intervening) 31 March 2013, Para 86.

43 Judgment, Whaling in the Antarctic (Australia *v.* Japan: New Zealand intervening) 31 March 2013, Para 86.

the purpose of the killing of whales under such a programme'. It was also not for the Court to decide '...whether the design and implementation of a programme are the best possible means of achieving its stated objectives'.[44]

The Court then went to consider '...whether a programme's use of lethal methods is for purposes of scientific research'.[45] The Court had earlier said that it '...will consider if the killing, taking and treating of whales is "for purposes of" scientific research by examining whether, in the use of lethal methods, the programme's design and implementation are *reasonable in relation to achieving its stated objectives*'[46] (author's emphasis).

In order to do this, the Court took aspects of the arguments put forward by both sides in the case and considered the following 'elements':

- decisions regarding the use of lethal methods;
- the scale of the programme's use of lethal sampling;
- the methodology used to select sample sizes;
- a comparison of the target sample sizes and the actual take;
- the time frame associated with a programme;
- the programme's scientific output; and
- the degree to which a programme co-ordinates its activities with related research projects.[47]

The Court also considered the differences between the characteristics of commercial whaling and those of scientific research, saying:

> The Parties agree that the design and implementation of a programme for purposes of scientific research differ in key respects from commercial whaling. The evidence regarding the programme's design and implementation must be considered in light of this distinction. For example, according to Japan, in commercial whaling, only species of high commercial value are taken and larger animals make up the majority of the catch, whereas in scientific whaling 'species of less or no commercial value' may

44 Judgment, Whaling in the Antarctic (Australia *v.* Japan: New Zealand intervening) 31 March 2013, Para 88.

45 Judgment, Whaling in the Antarctic (Australia *v.* Japan: New Zealand intervening) 31 March 2013, Para 88.

46 Judgment, Whaling in the Antarctic (Australia *v.* Japan: New Zealand intervening) 31 March 2013, Para 67.

47 Judgment, Whaling in the Antarctic (Australia *v.* Japan: New Zealand intervening) 31 March 2013, Para 88.

> be targeted and individual animals are taken based on random sampling procedures.[48]

In dealing with this aspect of the case, I do not intend to look at the legal arguments regarding the characterisation of JARPA II as commercial whaling, except in so far as they relate to 'science' and 'scientific research'.

The Court found that motivations that a State may have which go beyond scientific research cannot be used to justify '…the granting of a special permit for a programme that uses lethal sampling on a larger scale than is reasonable in relation to achieving the programme's stated research objectives. The research objectives alone must be sufficient to justify the programme as designed and implemented'[49] (author's emphasis).

JARPA and JARPA II

JARPA

Central to the case were the arguments as to whether JARPA and JARPA II were programs for the purposes of scientific research under Article VIII of the ICRW. Australia had described JARPA and JARPA II (in various ways) as 'commercial whaling in disguise', whereas Japan characterised both as necessary scientific research required because of the adoption in 1982 of the moratorium on commercial whaling. The Moratorium on commercial whaling was established by setting the catch limits for all whale stocks at zero.[50] The Moratorium entered into force for Japan in 1987, and the JARPA Special Permit Whaling program began in the 1987–88 austral whaling season.

The JARPA Special Permit Whaling program was conducted from 1987 until the end of the 2004–2005 austral whaling season. JARPA II followed immediately, commencing in the 2005–2006 season.

48 Judgment, Whaling in the Antarctic (Australia *v.* Japan: New Zealand intervening) 31 March 2013, Para 89.

49 Judgment, Whaling in the Antarctic (Australia *v.* Japan: New Zealand intervening) 31 March 2013, Para 97.

50 Schedule to the International Convention on the Regulation of Whaling, paragraph 10. (e): 'Notwithstanding the other provisions of paragraph 10, catch limits for the killing for commercial purposes of whales from all stocks for the 1986 coastal and the 1985/86 pelagic seasons and thereafter shall be zero. This provision will be kept under review, based upon the best scientific advice, and by 1990 at the latest the Commission will undertake a comprehensive assessment of the effects of this decision on whale stocks and consider modification of this provision and the establishment of other catch limits'.

Japan's case centred on their argument that data on whales and whale stocks were required because the Moratorium meant that no data were available from commercial whaling and the lifting of the Moratorium required review of scientific information and a comprehensive assessment of the Moratorium. Paragraph 10(e) of the Schedule to the ICRW says that:

> ...this provision will be kept under review, based upon the best scientific advice, and by 1990 at the latest the Commission will undertake a comprehensive assessment of the effects of this decision on whale stocks and consider modification of this provision and the establishment of other catch limits.

It was Japan's contention that the purpose of JARPA was to provide the research and data to contribute to that 'best scientific advice' and therefore to the 'review' and the 'comprehensive assessment' of the Moratorium.

As established, JARPA was designed to lethally take 825 Antarctic minke whales and 50 sperm whales. At its inception, JARPA had three 'research objectives':[51]

- Objective 1: Estimation of the biological parameters required for the stock management of the Southern Hemisphere minke whale
- Objective 2: Elucidation of the role of whales in the Antarctic marine ecosystem, and
- Objective 3: Elucidation of the effect of environmental change on cetaceans

Japan later added a fourth objective:[52]

- Objective 4: Elucidation of the stock structure of Southern Hemisphere minke whales to improve stock management

After JARPA was established, but before its whaling program began, the sample size for the annual lethal take of minke whales was reduced to from 825 to 300 +/−10% (from 1987–1988 to 1994–1995).[53] From 1995–1996 the sample size for

51 Counter-Memorial of Japan, March 2012, p. 143 para 4.8.

52 In its Counter-Memorial, Japan stated that '... Objective 4 was originally implicit in Objective 1 but was separated during the research period'. Counter-Memorial of Japan, March 2012, p. 143 para 4.8.

53 The Judgment, at Para 107, states that the annual sample size of 300 minke whales went from 1987–88, to 1993–94, while the Counter-Memorial of Japan at page 145, paragraph

the annual lethal take of minke whales was set at 400 +/−10%. The proposed take of sperm whales was dropped from JARPA altogether.[54]

In total, between 1987 and 2005, more than 6,795 minke whales were killed in the JARPA Special Permit whaling program. The season by season take varied from a low of 241 in 1988/89 to a high of 440 (which was achieved annually for much of the time between 1995–1996 and the end of the JARPA).[55]

JARPA II

In early 2005, Japan presented its Research Plan for JARPA II to the IWC, and began Special Permit Whaling under JARPA II in the 2005–2006 austral whaling season. The JARPA II Research Plan was finalised following a Japanese-sponsored workshop (note: not an IWC workshop) held in January 2005. This workshop was convened by Japan to 'review' data obtained for JARPA. The IWC Scientific Committee review of JARPA was not held until December, 2006 – after JARPA II was underway.

The methodologies and kinds of data collected in JARPA II were very similar to those of JARPA, but the articulation of the research objectives for JARPA II was somewhat different.

The JARPA II Research Plan establishes its objectives as:

(1) Monitoring of the Antarctic ecosystem
   i. Monitoring of whale abundance trends and biological parameters
   ii. Monitoring of krill abundance and the feeding ecology of whales
   iii. Monitoring of the effects of contaminants on cetaceans
   iv. Monitoring of cetacean habitat
(2) Modelling competition among whale species and future management objectives
   i. Constructing a model of competition among whale species
   ii. New management objectives including the restoration of the cetacean ecosystem
      - Establishing future management objectives
      - Estimating surplus production (and hence allowable catch) by species, under some of the management objectives
      - Contribute towards a multi-whale-species management

4.14 states that 'Until the 1994/95 season, the sample size…was 300…and this was increased to 400…in the 1995/6 and 1996/7 seasons'.

54 Judgment, Whaling in the Antarctic (Australia *v.* Japan: New Zealand intervening) 31 March 2013, Para 104.

55 Table 4.78, p. 181, Counter-Memorial of Japan.

(3) Elucidation of temporal and spatial changes in stock structure
(4) Improving the management procedure for Antarctic minke whale stocks

- Improvement of MSYR (maximum sustainable yield rate) estimates for Antarctic minke whales
- Redefinition of appropriate management Areas
- Incorporation of effects arising from the inter-species relationships among the whale species[56]

As originally formulated, JARPA II was designed to sample, through lethal take, 850 +/–10% Antarctic minke whales, 50 humpback whales and 50 fin whales. It was designed to run indefinitely, and be reviewed every 6 years (6-year 'phases'[57]). History shows that JARPA II Special Permit whaling ceased with the handing down of the Judgment of the Court. The actual lethal take of whales under JARPA II was 3,364 Antarctic minke whales, zero humpback whales and 17 fin whales between 2005–2006 and 2013–2014.[58]

On Lethal Sampling

At the core of the case and its conclusion were the acts of lethal sampling undertaken in JARPA and JARPA II. Long, drawn out discussions in the IWC and its Scientific Committee, the vehement opposition of non-government organisations, the many diplomatic interchanges now spanning four decades, were all stimulated by the act of killing whales – in this case, framed in the pursuit of scientific knowledge. The case would never have gone to the ICJ if JARPA II had not involved the harvesting of whales.

Leaving aside some argument about experimental design, the case did not consider directly the non-lethal components of the JARPA II research program. Japan argued that the research could only be carried out by killing whales.

56 Government of Japan, 'Plan for the Second Phase of the Japanese Whale Research Program under Special Permit in the Antarctic (JARPA II) – Monitoring of the Antarctic Ecosystem and Development of New Management Objectives for Whale Resources', SC/57/O1 (2005), pp. 10–12[Annex 150]; also at Para 5.20, p. 232, Counter-Memorial of Japan.

57 Judgment, Whaling in the Antarctic (Australia *v.* Japan: New Zealand intervening) 31 March 2013, Para 119.

58 Counter-Memorial of Japan, Paras 5.72, 5.78 and 5.79, pp. 262–266. In addition to the 3624 minke whales reported killed up to 2010/11, a further 103 and 251 minke killed whales were reported by Japan to the IWC for the 2012/13 and 2013/14 seasons respectively.

Japan rejected Australia's assertions that the key research questions Japan purported to be investigating, could be better and more efficiently achieved through non-lethal means.

As noted above, the Court did not define science or scientific research. In the words of the Court 'The research objectives come within the research categories identified by the Scientific Committee... Based on the information before it, the Court thus finds that the JARPA II activities involving the lethal sampling of whales can broadly be characterized as "scientific research". There is no need therefore, in the context of this case, to examine generally the concept of "scientific research"'.[59]

The Court therefore looked at whether lethal sampling under JARPA II Special Permit whaling was justified—in its words '[w]hether the design and implementation of JARPA II are reasonable in relation to achieving the programme's stated research objectives'.[60]

As noted previously, Japan and Australia had diametrically opposing views on the necessity for lethal whaling to meet the stated objectives of JARPA II. Australia argued both that lethal whaling was unnecessary, and that it produced poor scientific information: '...the data collected by Japan through its continued whaling is neither useful nor reliable: killing whales is simply not the most effective way to obtain knowledge that is important for the conservation and management of whales'.[61] Interestingly, to emphasise its argument in this regard, Australia quoted a paper written by a long-term advisor to Japan's Institute for Cetacean Research, Dr D Butterworth, in which he said '[a]s far as the lethal components of these lethal research programs are concerned, the empirical evidence thus far of their value towards an improved basis for management is poor'.[62]

In its Memorial, Australia said:

> [i]t is the clear view of the IWC that scientific research should be conducted using non-lethal techniques where available. As noted by the Commission in Resolution 2003–2, 'Article VIII of the ICRW was drafted and accepted by States Parties in 1946, at a time when few alternatives to lethal investigations existed, a situation drastically different from today'. In the same Resolution, the Commission reaffirmed its clear view that

59 Judgment, Whaling in the Antarctic (Australia *v.* Japan: New Zealand intervening) 31 March 2013, para 127.

60 Judgment, Whaling in the Antarctic (Australia *v.* Japan: New Zealand intervening) 31 March 2013, p. 41.

61 Memorial of Australia, para 2.75.

62 Butterworth, D 'Science and sentimentality', *Nature* 357 (18 June 1992) 532, 532.

> 'non-lethal techniques available today will usually provide better data at less cost to both animals and budget', and it urged 'any country conducting or considering the conduct of Special Permit whaling to terminate or not commence such activities and to limit scientific research to non-lethal methods only'.[63]

Referring to the IWC's *Guidelines*,[64] Australia said that '...lethal methods may be utilised only where the objectives of the research cannot be achieved by any other means'[65] – an argument expanded on by Mangel in his role as Expert Witness for Australia.

In presenting its case against Japan, Australia emphasised that non-lethal methods including satellite tagging, biopsy sampling, faecal analyses and sighting surveys, could provide the critical data meant to be obtained through JARPA II. These methods, Australia said, were more effective ways of conducting whale research than lethal whaling, and technological developments for non-lethal research since the inception of JARPA made the killing of whales for JARPA II unnecessary.

Japan, in its Counter Memorial and oral evidence, dismissed Australia's arguments on the efficacy of non-lethal research techniques, saying that lethal research whaling was 'indispensable'.[66] Lethal whaling was required to determine: 'age composition' by analysing whale ear plugs;[67] 'pregnancy...and the reproductive cycle' and 'reproductive parameters' by '...observation of ovaries and testes'; and 'physical maturity' by '...examining dorsal vertebrae'. 'Because of the internal nature of these organs the only way to collect them is through lethal sampling',[68] Japan argued.

The collection of stomach contents from whales killed in JARPA and JARPA II was central to Japan's arguments in favour of lethal whaling, especially in relation to the objectives of 'monitoring of the Antarctic ecosystem' and 'modelling competition between whale species'. In their Counter Memorial, Japan dismissed the use of non-lethal methods '...for investigating diet of marine vertebrate predators':[69] '...none of them provides quantitative information on stomach contents, which is important for evaluating the impact of a whale's

63 Memorial of Australia, para 4.80.

64 Annex P to the International Convention on the Regulation of Whaling, Process for the Review of Special Permit Proposals and Research Results from Existing and Completed Permits.

65 Memorial of Australia, para 4.101.

66 Counter-Memorial of Japan, paras 4.56, 4.83 and 5.8.

67 Counter-Memorial of Japan, para 4.66 – 4.68 and 5.8.

68 Counter-Memorial of Japan, para 4.69.

69 Counter-Memorial of Japan, para 4.71.

prey consumption on the ecosystem'.[70] Similar arguments were used by Japan to reject the practicality of non-lethal techniques for studies of the effects of pollutants, and whale condition (in this case measured by 'blubber thickness').[71]

Whereas Australia had argued strongly that biopsy sampling could be used to provide the data necessary for stock assessment—a key objective for both JARPA and JARPA II—Japan responded by stating it was impractical for '...the Antarctic minke whale'.[72] Data from non-lethal methods, Japan argued, would be of lesser quality or reliability than data from lethal sampling and would be obtained at 'unrealistic' amounts of time and expense.

In the context of scientific research, Australia (and New Zealand) did not preclude the possibility that some research questions might require lethal research methods. Australia argued that the research 'exemption' provided by Article VIII of the ICRW was '...to be construed narrowly'[73] and that '[i]t is the clear view of the IWC that scientific research should be conducted using non-lethal techniques where available'.[74]

The Court ultimately formed the view that '...the Parties agree that non-lethal methods are not a feasible means to examine internal organs and stomach contents. The Court therefore considers that the evidence shows that, at least for some of the data sought by JARPA II researchers, non-lethal methods are not feasible'.[75]

How Good are the Data Obtained from Lethal Whaling and JARPA II?

In its Memorial, Australia had argued strongly that data obtained from lethal whaling in JARPA and JARPA II were scientifically both unnecessary and of such poor quality as to be unreliable for the purposes of the International Whaling Commission. Describing JARPA as 'a failure', and its objectives as '... irrelevant to the management procedure agreed by the IWC – the RMP [Revised Management Procedure] – and that were predicted, and ultimately proved, to be unachievable', Australia went on to say:

70 Counter-Memorial of Japan, para 4.72.

71 Counter-Memorial of Japan, para 4.77–4.81.

72 Counter-Memorial of Japan, para 4.78.

73 Memorial of Australia, para 4.71.

74 Memorial of Australia, para 4.80.

75 Judgment, Whaling in the Antarctic (Australia *v.* Japan: New Zealand intervening) 31 March 2013, para 133.

> ...the shift from JARPA to JARPA II is critical since Japan continues to collect the same data using the same methods that failed to produce useful or reliable scientific results in JARPA. Japan has simply framed a new purported scientific purpose for JARPA II in an attempt to justify its continued whaling.[76]

Australia dismissed the relevance of the science in JARPA by stating '[t]he main objective of JARPA was to collect data on the biological parameters of Antarctic minke whales that were central to the NMP [New Management Procedure], the management approach that was conclusively discarded by the IWC in 1994'.[77] Australia said that the aim of estimating natural mortality in Antarctic minke whales was '...practically unachievable, and that Japan's results would be so imprecise as to be useless'.[78]

The adoption of the Revised Management Procedure in 1994 made JARPA (and JARPA II) irrelevant, Australia argued: '...the RMP is a sophisticated management procedure that was deliberately designed to eliminate the need for data on biological parameters obtained through whaling'.[79] As Expert Witness for Australia, Mangel had written:

> The RMP thus eliminates the use of data obtained from whaling-dependent or other lethal-source data, which are often unreliable for purposes of management because they represent non-random samples of the population. Consistent with this, in 1995 the IWC adopted a Resolution (1995–9) that stated, among other things, 'that scientific research intended to assist the comprehensive assessment of whale stocks and the implementation of the Revised Management Procedure *shall be undertaken by non-lethal means*' (emphasis added).[80]

Australia also rejected the importance of the data in JARPA derived from analysis of whale stomach contents as '...well-established and uncontroversial information, which did not represent any advance in existing scientific knowledge and did not require that Japan kill one whale, let

76 Memorial of Australia, para 5.5.

77 Memorial of Australia, para 5.6.

78 Memorial of Australia, para 5.9, referring to W de la Mare, 'A Further Note on the Simultaneous Estimation of Natural Mortality Rate and Population Trend from Catch-at-Age Data', *Rep. int. Whal. Commn* 40, 1990, 489–492.

79 Memorial of Australia, para 5.12.

80 Mangel, M 2011, in Memorial of Australia, para 3.26.

alone 6,777'. 'Yet', Australia added, '...Japan relies on these "results" in support of JARPA II'.[81]

The 'seamless'[82] segue from JARPA to JARPA II, in the absence of a comprehensive review of JARPA, was also highlighted by Australia. In referring to the Scientific Committee's review of JARPA (concluded in 2006, after JARPA II had commenced), Australia said: 'For a long-term, large-scale "research" program, the Scientific Committee's statement [that JARPA had "potential" to improve management] was tantamount to a finding of abject failure. In short, JARPA had only managed to accumulate masses of data'.[83]

Australia dismissed as contrived, JARPA II's 'new overarching goal' to '... inform the development of "a new and improved management system for whales"'.[84] 'This proposed new management system will purportedly take into account interactions between baleen whale species in the Antarctic, and in particular possible competition between them for their single food source, krill'.[85]

Deeply embedded in Japan's justification for the JARPA II whaling program was the proposition that there was 'competition' for food (krill) between baleen whale species in the Antarctic,[86] and between baleen whales and other Antarctic krill predators (for example, penguins). Related to the proposition that there is competition for krill in the Antarctic, is the 'krill surplus hypothesis' which postulates that the demise of the populations of great whales during the heyday of the whaling industry created a surplus of krill.[87] This 'krill surplus' then allowed for other species to become more abundant – in effect, putting the Antarctic ecosystem 'out of balance'.[88]

Japan proposed that data from JARPA II could be used to build an 'ecosystem model' which took into account this proposed competition between whale species,[89] ultimately with the view that some species of whales could be

81 Memorial of Australia, para 5.12.

82 Counter-Memorial of Japan, para 4.99.

83 Memorial of Australia, para 5.15.

84 Memorial of Australia, para 5.18.

85 Memorial of Australia, para 5.18.

86 Japan 2014 'Whales and Whaling', Japanese Fisheries Agency. At, http://www.jfa.maff.go.jp/j/whale/pdf/140513english.pdf as accessed on 30 January 2015.

87 Murase, H., Nishiwaki, S., Ishikawa, H., Kiwada, H., Yoshida, T. and Ito, S., 2006. Results of the cetacean prey survey using echo sounder in JARPA from 1998/99 to 2004/2005. IWC Doc. SC/D06/J21.

88 Japan 2014 'Whales and Whaling', Japanese Fisheries Agency; Memorial of Australia paras 5.19 & 5.12; and Mangel 2011 para 5.36–7.

89 Government of Japan, 'Plan for the Second Phase of the Japanese Whale Research Program under Special Permit in the Antarctic (JARPA II) – Monitoring of the Antarctic

harvested (to reduce competition) in order to allow for the population growth in more commercially important whale species.[90]

The krill surplus hypothesis, and the proposition that the proposed 'ecosystem model' could be used to undertake ecosystem engineering for whaling ('multi-species management'[91]), was (and is) fiercely disputed in scientific circles, including the IWC Scientific Committee.[92]

Central to Japan's case was the argument that the Special Permit Whaling undertaken in JARPA and JARPA II was '...a legitimate scientific programme... [whose] objectives and methods, together with its valuable scientific outputs [is] designed to contribute to the working of the International Whaling Commission (IWC) and its Scientific Committee'.[93]

The JARPA Special Permit Whaling program '...was designed and implemented after the adoption of the Commercial Whaling Moratorium in order to provide the scientific data necessary for the IWC and its Scientific Committee to utilize for their work, and thus to contribute to the "review" and the "comprehensive assessment" envisaged in Paragraph 10(e) of the Schedule to the ICRW'.[94]

Ecosystem and Development of New Management Objectives for Whale Resources', SC/57/O1 (2005), pp. 10–12[Annex 150]; also at Para 5.20, p. 232, Counter-Memorial of Japan.

90 Government of Japan, 'Plan for the Second Phase of the Japanese Whale Research Program under Special Permit in the Antarctic (JARPA II) – Monitoring of the Antarctic Ecosystem and Development of New Management Objectives for Whale Resources', SC/57/O1 (2005), pp. 10–12[Annex 150]; also at Para 5.20, p. 232, Counter-Memorial of Japan; and Japan 2014 'Whales and Whaling', Japanese Fisheries Agency. At, http://www.jfa.maff.go.jp/j/whale/pdf/140513english.pdf.

91 Government of Japan, 'Plan for the Second Phase of the Japanese Whale Research Program under Special Permit in the Antarctic (JARPA II) – Monitoring of the Antarctic Ecosystem and Development of New Management Objectives for Whale Resources', SC/57/O1 (2005), pp. 10–12[Annex 150]; also at Para 5.20, p. 232, Counter-Memorial of Japan.

92 Nicol, S., Croxall, J., Trathan, P., Gales, N., and E. Murphy. 2007. Paradigm misplaced? Antarctic marine ecosystems are affected by climate change as well as biological processes and harvesting. *Antarctic Science* 19:291–295. See also, Report of the Scientific Committee, Report of the Standing Working Group on Scientific Permits, Appendix 2, 'Comments on the Government of Japan's Proposals for a Second Phase of Special Permit Whaling in Antarctica (jarpa ii)', *J. Cetacean Res. Manage. 8 (Suppl.)*, 2006; and Mangel 2011 para 5. 36–7.

93 Counter Memorial of Japan, para 12.

94 Counter Memorial of Japan, para 33. See also para 3.7.

JARPA II, most importantly for this case, '...was launched and conducted to collect scientific data contributing to the "review" and the "comprehensive assessment", and to follow up on the findings of JARPA which indicated that significant changes in the Antarctic ecosystem were taking place'.[95]

Further, Japan argued that Special Permit Whaling was required in order to 'overcome' the 'scientific uncertainties' associated with the IWC's management measures, in particular '...the need to improve the RMP [Revised Management Procedure] and to obtain data for the development of ecosystem modelling. The need to overcome these scientific uncertainties thus became one of the objectives of JARPA II'.[96]

JARPA II was also designed, Japan claimed, to take into account known changes in different whale population parameters, climate change impacts on cetacean populations, and 'a major shift' in the ecosystem.[97] Because of these changes, Japan said, '...we should have a better management tool to achieve appropriate utilization of more than one whale species with a better RMP, as the current RMP is a basically single species management model'.[98]

Japan ascribed arguments against JARPA II as unscientific: 'The whaling controversy at the IWC and elsewhere has drifted away from any basis in science and law for too long'.[99] Japan said that those that opposed whaling '... tended to underrate the importance of science-based management, the core principle of the ICRW'.[100]

Defending against strong criticism of the poor scientific outputs of its Special Permit Whaling programs, Japan said: 'JARPA and JARPA II have produced valuable scientific outputs, and Japan has provided and continues to provide these results to the IWC Scientific Committee. JARPA and JARPA II have made significant contributions to the deliberations at the Scientific Committee, and have been duly evaluated as such'.[101]

Japan's scientific expert, Professor Lars Walløe, disputed Australia's evidence on the value of data from JARPA and JARPA II,[102] especially as presented

95 Counter Memorial of Japan, para 33.

96 Counter Memorial of Japan, para 3.10.

97 Counter Memorial of Japan, para 5.17, and JARPA II Proposal.

98 Counter Memorial of Japan, para 5.17 and 5.54, and JARPA II Proposal.

99 Counter Memorial of Japan, para 19.

100 Counter Memorial of Japan, para 3.67.

101 Counter Memorial of Japan, para 34.

102 Walløe, L 2013 Scientific review of issues raised by the Memorial of Australia including its two Appendices, p. 7 At, http://www.icj-cij.org/docket/files/148/17418.pdf as accessed 20 January 2015.

by Mangel.[103] In doing so, Walløe chose to '...assume that the judges do not have in-depth scientific knowledge of the field. For this reason I have not found it necessary to set out detailed scientific evidence, but have tried to indicate in general terms which deliberations my views are based on'.[104] Walløe provided a general criticism of Mangel's 'characteristics of a program for the purposes of scientific research' describing it as '...an adequate – though somewhat idealistic and somewhat oversimplified – description of research in a fairly advanced biological field...'.[105] In positing rejection of Mangel's criticisms this way, Walløe went on to argue that '[e]xisting knowledge about the Southern Ocean ecosystem is very limited. This makes it difficult to tell which observations are likely to be important... [so] all...data that can be collected may be considered potentially valuable, not only data relevant to specific hypotheses'.[106]

Walløe also rejected Mangel's (and Australia's) argument regarding data necessary for the Revised Management Plan, stating, inter alia, that '...in the Japanese research programmes, the genetic information necessary to determine stock structure is obtained from samples of killed whales',[107] adding that while non-lethal means may be used [to collect genetic information] '...biopsy samples from minke whales...would be prohibitively expensive'.[108] He concluded his consideration on this matter by stating: '...it is only feasible to obtain a sufficiently large number of genetic samples from minke whales by lethal sampling'.[109] These arguments will be discussed further below.

The Court's view about these differences in view about science was that '[t]his disagreement appears to be about a matter of scientific opinion'.[110] The Court further went on to say, 'Taking into account the evidence indicating that non-lethal alternatives are not feasible, at least for the collection of certain data, and given that the value and reliability of such data are a matter of scientific opinion, the Court finds no basis to conclude that the use of lethal methods is per se unreasonable in the context of JARPA II'.[111]

103 Mangel, M 2013 Response to 'Scientific Review of Issues Raised by the Memorial of Australia Including its two Appendices' by Professor Lars Walløe [9 April 2013]. At, http://www.icj-cij.org/docket/files/148/17420.pdf as accessed 20 January 2015.

104 Walløe 2013 p. 3.

105 Walløe 2013 p. 3.

106 Walløe 2013 p. 5.

107 Wallow 2013 p. 11.

108 Wallow 2013 p. 11.

109 Wallow 2013 p. 11.

110 Judgment, Whaling in the Antarctic (Australia *v.* Japan: New Zealand intervening) 31 March 2013, para 134.

111 Judgment, Whaling in the Antarctic (Australia *v.* Japan: New Zealand intervening) 31 March 2013, para 135.

Non-Lethal Research and the Scale of Lethal Whaling in JARPA II

In looking at 'the scale' of lethal whaling established under JARPA II, the Court started with the following premise: lethal whaling, even where non-lethal methods were available, was not per se contrary to Article VIII of the ICRW (as discussed previously). The Court found, though, that the JARPA II Research Plan '... should have included some analysis of the feasibility of non-lethal methods as a means of reducing the planned scale of lethal sampling in the new programme'.[112]

The Court, in reaching this conclusion took into account, among other things, that Japan had stated that it does not use lethal methods '...more than it considers necessary',[113] and that expert evidence put forward by Australia[114] '...referred to significant advances in a wide range of non-lethal research techniques over the past 20 years'[115] which could be used to obtain the data sought under JARPA II.

The Court concluded '[i]t stands to reason that a research proposal that contemplates extensive lethal sampling would need to analyse the potential applicability of these advances in relation to a programme's design'.[116] In assessing whether Japan had undertaken this kind of analysis, the Court said:

> ...there is no evidence of studies of the feasibility or practicability of non-lethal methods, either in setting the JARPA II sample sizes or in later years in which the programme has maintained the same sample size targets. There is no evidence that Japan has examined whether it would be feasible to combine a smaller lethal take (in particular, of minke whales) and an increase in non-lethal sampling as a means to achieve JARPA II's research objectives.[117]

---

112 Judgment, Whaling in the Antarctic (Australia *v.* Japan: New Zealand intervening) 31 March 2013, para 137.

113 Judgment, Whaling in the Antarctic (Australia *v.* Japan: New Zealand intervening) 31 March 2013, para 137.

114 Gales, N 2013a Statement by Dr Nick Gales, 15 April 2013. At http://www.icj-cij.org/docket/files/148/17416.pdf; and Gales, N 2013b Statement by Dr Nick Gales in Response to the Expert Statement by Professor Lars Walløe, 31 May 2013. At, http://www.icj-cij.org/docket/files/148/17422.pd as accessed 20 January 2015.

115 Judgment, Whaling in the Antarctic (Australia *v.* Japan: New Zealand intervening) 31 March 2013, para 137.

116 Judgment, Whaling in the Antarctic (Australia *v.* Japan: New Zealand intervening) 31 March 2013, para 137.

117 Judgment, Whaling in the Antarctic (Australia *v.* Japan: New Zealand intervening) 31 March 2013, para 141.

JARPA II: Bigger and Bolder that JARPA

Japan was at pains to argue that JARPA II was not simply a continuation of JARPA: it had, they said, more sophisticated and complex objectives, and was more ambitious in its foreshadowed outputs, even though much of the methodology used in JARPA II was the same as in JARPA.

Japan explained these differences as follows: 'JARPA was focused on a one-time estimation of different biological parameters for minke whales, but JARPA II is a much more ambitious programme which tries to model competition among whale species and…detect changes in various biological parameters and the ecosystem'.[118] Further, JARPA II was also focussed on the impacts of climate change: 'a growing concern about climate change, including global warming, necessitated research whaling of a different kind from JARPA'.[119] Japan's argument was that the increased annual sample size for minke whales, and the inclusion of humpback and fin whales in JARPA II was required because of its focus on, especially, the first two objectives JARPA II, 'monitoring of the Antarctic ecosystem' and 'competition among whale species'.

The Court was not convinced of great material difference between the research objectives of JARPA II and JARPA. It concluded:

> Taken together, the overall research objectives of JARPA and JARPA II, as well as the subjects of study and methods used (i.e. extensive lethal sampling of minke whales) thus appear to have much in common, even if certain aspects differ. These similarities cast doubt on Japan's argument that the JARPA II objectives relating to ecosystem monitoring and multi-species competition are distinguishing features of the latter programme that call for a significant increase in the minke whale sample size and the lethal sampling of two additional species.[120]

Japan's decision to commence JARPA II, with its increased minke whale sample size and two additional whale species, immediately following the cessation of JARPA and before assessment of JARPA by the Scientific Committee, was also criticised by the Court (paragraphs 154 to 156 of the Judgment). The Court concluded:

---

118 Judgment, Whaling in the Antarctic (Australia *v.* Japan: New Zealand intervening) 31 March 2013, para 149.

119 Judgment, Whaling in the Antarctic (Australia *v.* Japan: New Zealand intervening) 31 March 2013, para 149.

120 Judgment, Whaling in the Antarctic (Australia *v.* Japan: New Zealand intervening) 31 March 2013, para 153.

> These weaknesses in Japan's explanation for the decision to proceed with the JARPA II sample sizes prior to the final review of JARPA lend support to the view that those sample sizes and the launch date for JARPA II were not driven by strictly scientific considerations. These weaknesses also give weight to the contrary theory advanced by Australia that Japan's priority was to maintain whaling operations without any pause, just as it had done previously by commencing JARPA in the first year after the commercial whaling moratorium had come into effect for it.[121]

How to Set a Sample Size (or Not)

The ICJ Judgment spends some time (paragraphs 157–198) examining the way Japan set the sample sizes for lethal whaling in JARPA II and the arguments put forward by both sides in the case. Japan had claimed that the sample sizes had been set 'on the basis of carefully selected parameters' and by using 'a standard scientific formula'.[122]

Japan argued that its final sample size of 850 +/–10% for minke whales was set by determining what data were required for the research (the research items), and what levels of change in research parameters needed to be detected; and then deciding that that number provided enough information on the various research items at 'a reasonable level of statistical accuracy overall'.

Australia on the other hand had argued that Japan could not justify the JARPA II sample sizes on rational statistical grounds.[123] The Court characterised Australia's case as '[i]n essence, Australia's contention is that Japan decided that it wished to take approximately 850 minke whales for purposes other than scientific research and then "retro-fitted" individual sample sizes to justify the overall sample size'.[124]

Cross examination of the expert witness for Japan, Professor Lars Walløe, on how Japan had established the figures for the lethal take of minke, fin and humpback whales, also provided a touch of drama in the proceedings when he said: 'The reason is, and that is one of the, as I state here, weaknesses of the

121 Judgment, Whaling in the Antarctic (Australia *v.* Japan: New Zealand intervening) 31 March 2013, para 156.

122 Judgment, Whaling in the Antarctic (Australia *v.* Japan: New Zealand intervening) 31 March 2013, para 159.

123 Memorial of Australia, paras 5.72 – 5.78; Mangel 2011, paras 5.38 – 5.48; and Mangel 2013 paras 5.1 – 5.3.

124 Judgment, Whaling in the Antarctic (Australia *v.* Japan: New Zealand intervening) 31 March 2013, para 158.

JARPA II documents, that I do not understand how they have calculated their sample sizes'[125] (author's emphasis).

But are the Numbers of Whales to be Killed Reasonable?

The Court made no assessment of the 'scientific merit' of the Special Permit whaling in JARPA II: '...the Court reiterates that it does not seek here to pass judgment on the scientific merit of the JARPA II objectives...'. It concluded that JARPA II could be '...broadly characterised as "scientific research"' and that it was '...not in a position to conclude whether a particular value for a given [research] variable has scientific advantages over another...'. The Court's role was to '...evaluate whether the evidence supports a conclusion that the sample sizes are reasonable in relation to achieving JARPA II's stated objectives'.[126]

The Court first looked at the fin and humpback whale sample sizes established for lethal take under JARPA II, which, as noted above, were set at an annual take of 50 for each species. Japan asserted that this number was set using a 12 year research period to provide data on two 'research items' age at sexual maturity and pregnancy rate. Japan had stated that the 12 year research period for fin and humpback whales (as opposed to the 6 year research period for minke whales) was precautionary and that there was not yet any consideration of implementing the RMP for these two species (paras 174–176 of the Judgment).[127]

The Court considered that the apparent inconsistency in the research periods set for fin and humpback whales, and for minke whales, as well as the differences in sample sizes, 'casts doubt on the centrality of the objectives that Japan highlights to justify the minke whale sample size of 850 (plus or minus 10 per cent)',[128] and, in particular, Japan's asserted emphasis on multi-species competition among whales, and its reason for including fin and humpback whales in the JARPA II Special Permit whaling.

The Court concluded also that, on the basis of the evidence presented by Japan, JARPA II would not have been able '...to produce statistically relevant information on at least one central research item to which the JARPA II

125 International Court of Justice, Court Record CR2013/14 p. 41.

126 Judgment, Whaling in the Antarctic (Australia *v.* Japan: New Zealand intervening) 31 March 2013, para 172.

127 Judgment, Whaling in the Antarctic (Australia *v.* Japan: New Zealand intervening) 31 March 2013, paras 174–176.

128 Judgment, Whaling in the Antarctic (Australia *v.* Japan: New Zealand intervening) 31 March 2013, para 178.

Research Plan gives particular importance'[129] ie change in age at sexual maturity (of fin and humpback whales).

Design, Numbers, and Random Sampling

Walløe had also provided evidence that, with respect at least to fin whales, the lethal sampling procedures used by Japan could not be considered 'random'. Japan had emphasised that 'random sampling' was an integral part of the methodologies used in JARPA II, and therefore critical to the efficacy of the data obtained under the Special Permit Whaling program. Only part of the fin whale population is found in the JARPA II research area. – the main population lying further to the north. There was also a technical obstacle: the larger of fin whales cannot be hauled aboard the whaling vessels, so only small fin whales can be sampled.[130]

Given the centrality of the Court's consideration of the setting of sample sizes in determining its final decision, it is worthwhile considering its findings at paragraph 181 of the Judgment:

> The Court finds that the JARPA II Research Plan overall provides only limited information regarding the basis for the decisions used to calculate the fin and humpback whale sample size. These sample sizes were set using a 12-year period, despite the fact that a shorter six-year period is used to set the minke whale sample size and that JARPA II is to be reviewed after each six-year research phase. Based on Japan's own calculations, the sample sizes for fin and humpback whales are too small to produce statistically useful results. These shortcomings, in addition to the problems specific to the decision to take fin whales, as noted in the preceding paragraph, are important to the Court's assessment of whether the overall design of JARPA II is reasonable in relation to the programme's objectives, because Japan connects the minke whale sample size…to the ecosystem research and multi-species competition objectives that, in turn, are premised on the lethal sampling of fin and humpback whales.[131]

129 Judgment, Whaling in the Antarctic (Australia *v.* Japan: New Zealand intervening) 31 March 2013, para 179.

130 Judgment, Whaling in the Antarctic (Australia *v.* Japan: New Zealand intervening) 31 March 2013, para 180.

131 Judgment, Whaling in the Antarctic (Australia *v.* Japan: New Zealand intervening) 31 March 2013, para 181.

In scientific terms, the Court's conclusion is that the research design of the JARPA II Special Permit Whaling program (sampling in relation to fin whales; duration of research periods; and sample sizes) was not sufficiently justified in the Research Plan, and the proposed research outputs, especially for a central component of the program, 'elucidating multi-species competition', could not be achieved (see also paras 191–195 of the Judgment).[132]

In ultimately setting the annual lethal take of minke whales at 850 +/–10%, Japan had said that this number provided a sample size to obtain sufficient information from most of the research items in the JARPA II Research Plan. These research items included were age at sexual maturity, pregnancy rate, blubber thickness, pathology, DNA mark-recapture, populations trends and the mixing of individuals from different stock in the study area. Figure 5.4 of the Japanese Counter-Memorial[133] (also reproduced at para 182 of the Judgment) shows the ranges of sample sizes calculated by Japan to provide 'sufficient data' for each of these items.

As detailed above, Australia's case was that the JARPA II Research plan provided insufficient information to determine how the final sample size for the lethal take of minke whales was arrived at, and why a large number (850) was chosen in preference to a smaller sample.

The evidence of the expert witnesses from both Australia and Japan was used by the court to conclude that '...the JARPA II Research Plan lacks transparency in the reasons for selecting particular sample sizes for individual research items'.[134] As Australia had argued, the Court considered that selecting different levels of change or accuracy to be measured, could have led to dramatically smaller number of whales to be killed annually in JARPA II. Evidence presented for Australia by Mangel was that the same level of accuracy could be achieved in research with smaller sample size, but a larger margin of error.[135]

Decisions such as these (ie about sample size, accuracy, precision, and margin of error) are the bread and butter of scientific research. Australia's case was that the JARPA II Research plan failed to address these trade-offs, and lacked transparency. The Court ultimately agreed: 'The absence of such evidence in connection with most of the sample size calculations described in the JARPA II

132 Judgment, Whaling in the Antarctic (Australia *v.* Japan: New Zealand intervening) 31 March 2013, para 195–196.

133 Counter-Memorial of Japan, Vol. I, p. 261.

134 Judgment, Whaling in the Antarctic (Australia *v.* Japan: New Zealand intervening) 31 March 2013, para 188.

135 Judgment, Whaling in the Antarctic (Australia *v.* Japan: New Zealand intervening) 31 March 2013, para 190.

Research Plan lends support to Australia's contention that a predetermined overall sample size has dictated the choice of the research period and the rate of change to be detected, rather than the other way around'.[136]

The Court's view was also that Japan had presented more material and reasons for determining what an adequate sample size might be to detect age at sexual maturity in minke whales, and the relationship between this 'research item' and the first two objectives of the JARPA II Research Plan, than it had for the other research to be conducted under JARPA II. 'This highlights the absence of evidence, at least in the JARPA II Research Plan, that could support a finding that the sample size for the lethal take of minke whales, a key component of the design of JARPA II, is reasonable in relation to achieving the programme's objectives'.[137]

In concluding its consideration of the selection of sample sizes for lethal take in the JARPA II Research plan, the Court concluded:

> Taken together, the evidence relating to the minke whale sample size, like the evidence for the fin and humpback whale sample sizes, provides scant analysis and justification for the underlying decisions that generate the overall sample size. For the Court, this raises further concerns about whether the design of JARPA II is reasonable in relation to achieving its stated objectives.[138]

Research to Reality: The Practice of JARPA II

As the evidence presented to the ICJ shows, the actual lethal take of whales in JARPA II fell far short of the sample sizes established in the Research Plan. Zero humpback whales were killed, 18 fin whales (instead of 400) and 3,618 Antarctic minke whales (instead of 6,800 or the potential a maximum of 7,480) were killed. In only the first year of the JARPA program did the annual take fall close to the target in the Research Plan (853 whales killed in 2005–2006). The reasons for the failure to achieve the stated annual sample size for minke whales was hotly disputed in the case. Australia claimed that the actual take

136 Judgment, Whaling in the Antarctic (Australia *v.* Japan: New Zealand intervening) 31 March 2013, para 195.

137 Judgment, Whaling in the Antarctic (Australia *v.* Japan: New Zealand intervening) 31 March 2013, para 189.

138 Judgment, Whaling in the Antarctic (Australia *v.* Japan: New Zealand intervening) 31 March 2013, para 205.

closely matched the economic target that Japan required for the program,[139] whereas Japan blamed the sabotage activities of Sea Shepherd and the 2007 fire aboard *Nishin Maru* for its failure to reach the projected sample size for minke and fin whales. Japan also said that it decided not to take humpback whales at the request of the Chair of the International Whaling Commission.

It is not the task of this Chapter to evaluate the veracity of these arguments, only to note the impact of the actual take on the goals of the JARPA II Research Plan. Following the logic of Japan's arguments with respect to setting sample sizes, failure to collect the nominated sample sizes, especially chronically, must have material consequences for achieving JARPA II's objectives. With respect to humpback whales, it is obvious that none of the species-specific research objectives which 'require' lethal sampling, could be achieved, and the small sample of small fin whales would only have produced basic data with little or no statistical robustness. With respect to Japan's research objectives of elucidating competition among whale species in the Antarctic, insofar as this objective required the lethal sampling of humpback and fin whales, as Japan asserted, JARPA II had failed. Australia described JARPA II's goal of producing a multi-species competition model as 'illusory'.

Japan conceded that its failure to meet its sample sizes with respect to minke whales had an impact on the length of time the research would need to be continue, and/or on whether they would need to accept '...a lower degree of accuracy' for some parameters.[140]

The Court was very critical of Japan's failure to modify the research program envisaged in the JARPA II Research Plan given the reality of Japan's whaling in the Antarctic, and its words are worthy of being reproduced in full here:

> The Court observes that, despite the number of years in which the implementation of JARPA II has differed significantly from the design of the programme, Japan has not made any changes to the JARPA II objectives and target sample sizes, which are reproduced in the special permits granted annually. In the Court's view, two conclusions can be drawn from the evidence regarding the gap between the target sample sizes and actual take. First, Japan suggests that the actual take of minke whales does not compromise the programme, because smaller numbers of minke whales can nonetheless generate useful information, either

139 Judgment, Whaling in the Antarctic (Australia *v.* Japan: New Zealand intervening) 31 March 2013, para 195.

140 Judgment, Whaling in the Antarctic (Australia *v.* Japan: New Zealand intervening) 31 March 2013, para 208.

> because the time frame of the research can be extended or because less accurate results could be accepted. The Court recalls, however, that the minke whale sample sizes for particular research items were based on a six-year research period and on levels of accuracy that were not explained in the JARPA II Research Plan or in these proceedings. Japan's statement that the programme can achieve scientifically useful results with a longer research period or a lower level of accuracy thus raises further doubts about whether the target sample size of 850 whales is reasonable in relation to achieving the stated objectives of JARPA II. This adds force to Australia's contention that the target sample size for minke whales was set for non-scientific reasons.[141]

The logic here is straightforward. Japan's case rested in large part on justifications for the scale of lethal take under the Special Permit granted for JARPA II. The methods used to set sample sizes, and the rationale behind them were disputed by Australia – but strongly defended by Japan. The reality of the implementation of JARPA II, was that only in its first year did the take of minke whales meet the stated objective of the program. This reality clearly must have had material consequences for the stated research objectives of JARPA II. This reality was ignored by Japan, leading to the Court's conclusion above that the target annual lethal sample size for minke whales 'was set for non-scientific reasons'.

Looking also at JARPA II's intended objective of 'elucidation of competition between species' and 'understanding changes in the Antarctic ecosystem', the Court also drew attention to the discrepancies between the Research Plan and its implementation:

> In the view of the Court, the gap between the target sample sizes for fin and humpback whales in the JARPA II Research Plan and the actual take of these two species undermines Japan's argument that the objectives relating to ecosystem research and multi-species competition justify the larger target sample size for minke whales, as compared to that in JARPA.[142]

The Court also drew attention to the statements by Japan in oral evidence that constructing an ecosystem model for the Antarctic did not require the

141 Judgment, Whaling in the Antarctic (Australia *v.* Japan: New Zealand intervening) 31 March 2013, para 209.

142 Judgment, Whaling in the Antarctic (Australia *v.* Japan: New Zealand intervening) 31 March 2013, para 210.

'missing' data from the humpback and fin whales not killed under JARPA II – at least some of these data could be obtained through non-lethal means. The Court's view was '…[Japan's own evidence] suggests there is no strict scientific necessity to use lethal methods in respect of this objective'.[143]

The Court's final conclusions in relation to sample sizes in JARPA II was that there was 'doubt' that JARPA II could be characterised as 'a programme for purposes of scientific research', and that the target lethal samples size for minke whales 'are larger than are reasonable'. The full text of paragraph 212 of the Judgment is reproduced here:

> Japan's continued reliance on the first two JARPA II objectives to justify the target sample sizes, despite the discrepancy between the actual take and those targets, coupled with its statement that JARPA II can obtain meaningful scientific results based on the far more limited actual take, cast further doubt on the characterization of JARPA II as a programme for purposes of scientific research. This evidence suggests that the target sample sizes are larger than are reasonable in relation to achieving JARPA II's stated objectives. The fact that the actual take of fin and humpback whales is largely, if not entirely, a function of political and logistical considerations, further weakens the purported relationship between JARPA II's research objectives and the specific sample size targets for each species—in particular, the decision to engage in the lethal sampling of minke whales on a relatively large scale.[144]

Australia had also argued strongly that the open ended timeframe for JARPA II cast further doubt on the scientific veracity of the Research Plan. The Court noted '…the open-ended time frame of JARPA II and observes that with regard to a programme for purposes of scientific research, as Annex P indicates, a "time frame with intermediary targets" would have been more appropriate'.[145]

The Court also pointed to two other elements of Australia's arguments concerning Japan's Special Permit Whaling as a legitimate scientific research program – its poor publication record, and failure to engage collaboratively. On research publication, the Court said: '… In light of the fact that JARPA II has

143 Judgment, Whaling in the Antarctic (Australia *v.* Japan: New Zealand intervening) 31 March 2013, para 211.

144 Judgment, Whaling in the Antarctic (Australia *v.* Japan: New Zealand intervening) 31 March 2013, para 212.

145 Judgment, Whaling in the Antarctic (Australia *v.* Japan: New Zealand intervening) 31 March 2013, para 216.

been going on since 2005 and has involved the killing of about 3,618 minke whales, the scientific output to date appears limited'.[146] On research collaboration '[t]he Court ...observes that some further evidence of co-operation between JARPA II and other domestic and international research institutions could have been expected in light of the programme's focus on the Antarctic ecosystem and environmental changes in the region'.[147]

Article VIII and JARPA II: Is the Special Permit Whaling 'for the Purposes of Scientific Research'?

The Court's ultimate consideration was that lethal take of whales in JARPA II 'is not unreasonable' per se, but that the '...target sample sizes in JARPA II are not reasonable in relation to achieving the programme's objectives'[148] (author's emphasis).

In reaching this conclusion, the Court reiterated its previous considerations and conclusions in four parts. First, that the JARPA and JARPA II had similar research objectives and Japan had provided little evidence to show how any differences between them had led to the '...considerable increase in the scale of lethal sampling in the JARPA II Research Plan'.[149] Second, that the sample sizes for the inclusion of fin and humpback whales in the Research Plan were, in effect, spurious. They were '...too small to provide the information that is necessary to pursue the JARPA II research objectives based on Japan's own calculations...';[150] and the random sampling of fin whales was unachievable. The third conclusion was that the Research Plan for JARPA II lacked transparency in the way that the sample size for the annual lethal take of minke whales was set.

The fourth conclusion turns on the lack of consideration by Japan of alternatives to lethal sampling. As shown above, Japan had provided evidence that

146 Judgment, Whaling in the Antarctic (Australia *v.* Japan: New Zealand intervening) 31 March 2013, para 219.

147 Judgment, Whaling in the Antarctic (Australia *v.* Japan: New Zealand intervening) 31 March 2013, para 222; See also para 226.

148 Judgment, Whaling in the Antarctic (Australia *v.* Japan: New Zealand intervening) 31 March 2013, para 224.

149 Judgment, Whaling in the Antarctic (Australia *v.* Japan: New Zealand intervening) 31 March 2013, para 225.

150 Judgment, Whaling in the Antarctic (Australia *v.* Japan: New Zealand intervening) 31 March 2013, para 225.

data derived from non-lethal research could be used to fill the gaps in the research program caused by the 'under-sampling' of humpback and minke whales, at least with respect to some of the research objectives. The Court concluded: '[t]he evidence...further suggests that little attention was given to the possibility of using non-lethal research methods more extensively to achieve the JARPA II objectives and that funding considerations, rather than strictly scientific criteria, played a role in the programme's design'.[151]

The Court, at paragraph 226 of the Judgment, was again very critical of Japan's failure to consider how the gap between the Research Plan the actual implementation of JARPA II affected the objectives of the program:

> Neither JARPA II's objectives nor its methods have been revised or adapted to take account of the actual number of whales taken. Nor has Japan explained how those research objectives remain viable given the decision to use six-year and 12-year research periods for different species, coupled with the apparent decision to abandon the lethal sampling of humpback whales entirely and to take very few fin whales.[152]

The Courts Conclusion?

'The Court concludes that the special permits granted by Japan for the killing, taking and treating of whales in connection with JARPA II are not "for purposes of scientific research" pursuant to Article VIII, paragraph 1, of the Convention'.[153]

In setting the 'remedies' in its Judgment, the Court ordered that '...Japan shall revoke any extant authorization, permit or licence to kill, take or treat whales in relation to JARPA II, and refrain from granting any further permits under Article VIII, paragraph 1, of the Convention, in pursuance of that programme'.[154]

151 Judgment, Whaling in the Antarctic (Australia *v.* Japan: New Zealand intervening) 31 March 2013, para 225.

152 Judgment, Whaling in the Antarctic (Australia *v.* Japan: New Zealand intervening) 31 March 2013, para 226.

153 Judgment, Whaling in the Antarctic (Australia *v.* Japan: New Zealand intervening) 31 March 2013, para 226.

154 Judgment, Whaling in the Antarctic (Australia *v.* Japan: New Zealand intervening) 31 March 2013, para 245.

Discussion

'It is to be expected that Japan will take account of the reasoning and conclusions contained in this Judgment as it evaluates the possibility of granting any future permits under Article VIII, paragraph 1, of the Convention'.[155]

***

In summing up Japan's case during the oral proceedings, Alan Boyle said: 'If JARPA II is not a program of scientific research, then neither are the primary research activities of institutions providing advice on sustainable catch levels for fisheries world wide'.[156] This statement and his subsequent commentary concerning international environmental law, reflect not only Japan's view of the 'rightness' of Special Permit Whaling, but also the implications perceived by Japan from any adverse decision. 'Science' cannot be defined by 'peer-reviewed academic publications, and supported by multiple hypotheses…',[157] Boyle went on to say.

The fact that the Court found that JARPA II was not consistent with Article VIII of the ICRW was obviously not what Japan had expected from the Judgment. Some elements of JARPA II could be categorised as 'science' but the Program and its Research Plan were 'not for the purposes of scientific research'. The Court had rejected Japan's arguments that if the Special Whaling Permit activities involved 'science' they were covered therefore by Article VIII.

While not accepting *in toto* Australia's arguments that scientific research should be defined through reference to objective criteria (Mangel's four elements, see above), the Court did reflect on whether the activities themselves were 'reasonable' to achieve the stated objectives of the JARPA II research plan.

This is a key finding for how future research activities could be construed. Australia's Expert Witness, Dr Nick Gales, had described in written and oral evidence the available non-lethal alternatives to the killing of whales for scientific research.[158] Japan's Expert witness, Lars Walløe, had argued that non-lethal alternatives to killing whales were not viable and too costly. Australia's evidence in relation to the use of non-lethal research methods in the Southern

155 Judgment, Whaling in the Antarctic (Australia *v.* Japan: New Zealand intervening) 31 March 2013, para 246.

156 ICJ Verbatim Record C2-13/22 para 13.

157 ICJ Verbatim Record C2-13/22 para 13.

158 Gales 2013a and b; see also International Court of Justice Verbatim Record CR 2013/10.

Ocean Research Program contradicted much of what Walløe (and Japan) had said in this regard, and provided an alternative perspective on the activities authorised in JARPA II.

The availability of alternatives to lethal whaling goes to the heart of the construction of any future research program authorised under Article VIII, or any challenge to it. The Court's use of the yardstick of 'reasonableness' requires an objective consideration of research objectives, research methodologies, sampling techniques and sample sizes – the mere statement that a sample size has been determined objectively, or that one research technique is justified over another, must be evaluated against alternative propositions. They must be, and be seen to be, 'reasonable in relation to achieving' the objectives of any authorised research program.

As noted previously, the Court did not provide a definition of 'science' or 'scientific research', and referred to 'differences of opinion' between scientists. But the Court did take into account some of the propositions put forward by Australia and its expert witnesses in relation to the connection between the objectives of JARPA II and the activities undertaken in it. For example, the Court's Judgment was that JARPA II had failed in its research objective of 'elucidating competition among whale species in the Antarctic', in so far as this research required lethal sampling of humpback and fin whales. The Court was also critical of the lack of transparency in setting sample sizes for lethal whaling; the poor publication record of the JARPA II program; and the lack of engagement of the JARPA II research program with other Southern Ocean research initiatives.

Can Science be Objectively Defined?

I am not critical of the Court for not defining 'science' or 'scientific research' in this case – to do so would have involved more than just reviewing the evidence of three expert witnesses and the competing opinions of counsel for Australia and Japan. Science is a complex body of work, defined not only by the common practices of communities of researchers and scholars, but also by the philosophical precepts of those same practitioners. To attempt to define *science* was beyond the realm of this case.

But science <u>can</u> be evaluated objectively. The usual practice is that of peer review through publication or of evaluation through panels of experts. Science presented in the public domain is open to criticism from others. Sometimes referred to as a 'self-correcting' process, the combination of peer review and publication (in the public domain) allows for commentary on methods, methodology and appropriateness of the research to meet its objectives, to be taken into account by the researcher or others in the field. It is therefore open to bodies such as the International Court of Justice to consider how a program of scientific research is viewed among the broad community of researchers in that field in order to assess whether the science is purposeful.

Future Lethal Whaling and the Advent of NEWREP-A

Many commentators in the media had interpreted the Court's ruling on the legality of JARPA II as an end to Antarctic whaling. History has now shown, though, that Japan, has produced a new proposal for whaling in the Southern Ocean and intends to commence its program in the 2015–16 austral summer. The advent of NEWREP-A[159] (a Special Permit Whaling program to replace JARPA II) will again focus global attention on whaling in the Antarctic.

Japan's new whaling program will be considered by the International Whaling Commission's Scientific Committee in May 2015. Mangel et al. have pointed to past 'politicised' review processes inside the Scientific Committee, stating that '...even though logical inconsistencies noted by the ICJ as flaws had been identified by many members of the scientific committee, these did not stand out in the committee's reviews as being so serious that they required attention'.[160] Clapham, in looking at how the IWC and its Scientific Committee will deal with a new Antarctic whaling program, has said '...the Scientific Committee will continue to be mired in disagreement'.[161]

While de la Mare et al. hold the view that '...the ICJ's approach represents a model for separating scientific matters and the non-scientific agenda in other complicated disputes involving science, society, and law',[162] Clapham's view, in so far as this sentiment relates to the International Whaling Commission, is

159 Japan 2014a Proposed Research Plan for New Scientific Whale Research Program in the Antarctic Ocean(NEWREP-A), The Government of Japan; Japan 2014b Outline of the Proposed Plan for the New Scientific Whale Research Program in the Antarctic Ocean, The Government of Japan.

160 Mangel et al. 2014.

161 Clapham 2014.

162 de la Mare, W, Gales, N and M Mangel 2014 Applying scientific principles in international law on whaling, Science Vol 345 pp. 1125–1126.

that '[s]uch change will not happen because of this judgement: Japan will continue business as usual in the coming years, while pursuing its long-term plan of attempting to obtain the votes necessary to lift the IWC moratorium and reinstate commercial whaling…'.[163]

In this author's view, the validity of NEWREP-A (regardless of whether it is ever the subject of any future challenge outside of the IWC) will depend not just on its formulation, but also its scientific relevance to its stated goals. In constructing NEWREP-A, Japan has procedurally addressed the major criticisms that the Court had of JARPA II[164] by explicitly detailing its justifications for lethal methods; scale of lethal sampling; sample size methodology; timeframe; collaboration and cooperation; scientific output, among others.

But the fact remains that NEWREP-A potentially relies on the lethal take of up to 333 Antarctic minke whales per annum (during its 12 year period), while still pursuing goals broadly similar to both JARPA and JARPA II.

In its publication 'Whales and Whaling', the Japanese Fisheries Agency emphasises both 'Marine Ecosystem Balance' wherein

> [w]ith the moratorium on commercial whaling protecting even abundant whale species, if their abundance continues to increase, the top layer alone of the food pyramid will expand, upsetting the balance of the marine ecosystem, since the fish available as food for so many whales would, by consequence, be reduced. The sustainable utilization of a certain number of whales based on scientific evidence is therefore important in terms of maintaining the marine ecosystem balance;

and 'Competition between Whales and Fisheries' wherein

> [t]he balance of the marine ecosystem will be affected if whales alone increase excessively in number.[165]

Like the 'Krill Surplus Hypothesis' discussed above these 'hypotheses' are open to scientific criticism from a theoretical point of view, let alone from the

163 Clapham 2014.

164 Hodgson-Johnson, I and Jabour J 2014 Japan's new whaling program is a small win for whales, but…. At, https://theconversation.com/japans-new-whaling-program-is-a-small-win-for-whales-but-34420 as accessed 30 January 2015.

165 Japan 2014c Whales and Whaling, at (http://www.jfa.maff.go.jp/j/whale/pdf/140513english.pdf accessed 30 January 2015.

perspective of justifications for 'ecosystem management' (ie the utilisation of whales to maintain ecosystem balance), or the commencement of a new lethal whaling research program. The Court's observation that future Special Whaling Permits must to take into account the reasoning behind its Judgment significantly raises the bar for lethal whaling. Merely asserting that the Court's views have been taken into consideration may not be seen as 'reasonable' if subsequently challenged.

Conclusion: The Judgment, Its Implications and Prospects

Akiho Shibata

1 A Diverse Face of the Judgment

A famous statute of Ashura at Kōfukuji-temple in Nara, Japan, has three faces: the left face is representing Anger; the right face, Despair; and the front face, Hope. The reaction to the *Whaling* judgment, rendered by the International Court of Justice (Court) on 31 March 2014, was like the faces of Ashura. Some angrily criticized the Court for not treating adequately the law and facts in ordering Japan to suspend its special permit whaling in the Antarctic.[1] Others expressed despair that the Court did not stop once and for all the killing of whales in the name of science.[2] The preceding chapters in this Book, however, represent Hope, the hope of academic critical examination that discloses the significance of the judgment for the law and policy relating to whaling and for international law generally.

The preceding chapters demonstrate that the *Whaling* judgment is full of nuances and subtleties, the academic examination of which leads to different interpretations as to their significance and implications. The *Whaling* judgment is one of those judgments from which law school casebooks will have tremendous difficulty in selecting excerpts. It would not do justice, in trying to properly present the judgment, to select just a patchwork of paragraphs, leaving out the nuances emanating from between the lines, and the things that are not said expressly in the judgment. Thus, the law school professor would have to say to his or her students: you need to read the whole judgment in order to properly understand it. This integrative nature of the judgment might have been the reason why, according to **Morishita** (Chapter 8), Australia and Japan could not produce a joint paper introducing the *Whaling* judgment at the 65th meeting of the International Whaling Commission (IWC) held in Slovenia in September 2014. Australia considered it better to let the delegations read the

1 Hiroshi Hatanaka, 'Problems regarding the *Whaling in the Antarctic* judgment: A perspective from a scientist who drafted JARPA II', *Geiken Tsushin*, No. 463 (September 2014) (in Japanese).

2 Phillip J. Clapham, 'Japan's whaling following the International Court of Justice ruling: Brave New World or business as usual?' *Marine Policy*, Vol. 51(2015), pp. 238–241.

 | DOI 10.1163/9789004313828_014

whole judgment by themselves, whereas Japan did produce a paper with excerpts of the judgment.[3]

As **Tams** (Chapter 6) attests, the *Whaling* judgment is 'noteworthy for its silences, its omissions, its lacunae'. **Sakai** (Chapter 10) agrees: the judgment was 'not decisive but rather ambivalent'. Many of the preceding chapters have drawn inferences allowed by the Court's silence or ambivalence and tried to close the *lacuna* by their suggested reasoning and justifications. It is in such inferences and suggestions many of the interesting nuances and differences of opinion appear.

A good example of such subtleties is the Court's interpretation of Article VIII, paragraph 1, of the International Convention for the Regulation of Whaling (ICRW), a crucial provision in the case. In particular, it relates to whether the Court saw it as an exception or not. It is true, as **Foster** (Chapter 1) confirms, the Court made no finding on whether Article VIII, paragraph 1, provided for an exception, despite the fact that the Parties to the case made extensive arguments on the issue. **Christakis** (Chapter 4) considers that, from the beginning, the Court interpreted Article VIII as an exception or derogation from the State Parties' obligations under ICRW, and, consequently, the Court was able to utilise its usual interpretative tools to address 'discretion' as provided in such derogation clauses. In this context, he concludes that the outcome would not have been different even if the Court applied, instead, the concept of 'margin of appreciation' developed and elaborated under the jurisprudence of the European Court of Human Rights. **Hayashi** (Chapter 7), from a regime theory perspective, considers Article VIII as a flexibility clause introducing the soft and dynamic nature of the ICRW regime and, indeed, as an exception from the moratorium obligation. The Court as well might have considered it as a dynamic regime when it declared ICRW as 'an evolving instrument'. **Sakai**, on the other hand, considers that the Court inconspicuously comes to support the Australian argument that special permit whaling is to be treated as exceptional, by setting up a standard of an objective nature to be complied with by State Parties, and by introducing a relatively higher standard of review 'for purposes of scientific research'.

The Japanese understanding of Article VIII as being an exemption from the Convention was based on the logic that the ICRW itself is an exception from States' freedom of high seas fisheries (whaling), and the special permit whaling in Article VIII reserves the freedom of marine scientific research on the high seas by exempting it from the operation of ICRW.[4] The requirements

3 IWC, *Chair's Report of the 65th Meeting* (2014), subsection 15.1.1, paras. 235–248.

4 Lowe, public sitting held on Thursday 4 July 2013, at 10 a.m., CR 2013/15, p. 15, paras. 7–9.

as enunciated in Article VIII must therefore be interpreted in light of this general freedom of high seas, and not within the context of the ICRW. From this understanding, Article VIII should be characterized as a free-standing provision. Consequently, the discretion the State Parties enjoyed under Article VIII must be wide. The Court, however, cleverly disconnected the first half of the Japanese argument on the freedom of high seas (paragraph 57) from the latter part on the legal nature of special permit whaling under the ICRW (paragraphs 52 and 59). The Court's conclusion, in paragraph 55, on the interpretation of Article VIII, was actually decisive in situating the Article VIII exemption in relation to the 'obligations under the Schedule'. This is a typical characterization of an exception clause, as **Christakis** confirms. **Morishita**, on the other hand, reads this paragraph as the Court accepting 'the current or traditional basic framework and interpretation of Article VIII'. Whatever the 'traditional interpretation of Article VIII' would be, the Court was no longer interested in the issue of whether or not there is discretion in light of the legal nature of Article VIII. Thus, the Court, in paragraph 61, was able to conclude that Article VIII, paragraph 1, does indeed gives discretion to a State Party to ICRW, but the crucial issue has become *how* that discretion be assessed by the Court, namely the standard of review.

2 An Innovation: The Standard of Review

The Court's innovative approach to the introduction and application of the concept of standard of review attracted much attention in several of the chapters, because, as **Rothwell** (Chapter 9) argues, it might have 'created a benchmark for how States may seek to legally challenge the conduct of whaling in international law'. He therefore suggests the judgment established not only the criteria for judicial review, but also general rules of conduct for future special permit whaling. In a more nuanced way, **Hayashi** examines the idea that the standard pronounced by the Court consists of 'a list of missing actions and explanations', and neither the Convention nor the Court name these missing actions and explanations as generally required under Article VIII of the Convention. Nevertheless, **Hayashi** acknowledges that they are practically the requirements that should have been fulfilled in order to render the Japanese special permit whaling (JARPA II) reasonable. This demonstrates how subtle the line is that distinguishes the standard of judicial review applicable in a particular case under Article VIII and the rules of future conduct under the same article.

As to its provenance, **Hamamoto** (Chapter 2) carefully examines the arguments of the parties and the intervener to demonstrate how the Court came to adopt its version of standard of review. **Rothwell** also sees these standards as 'blended aspects of the arguments presented by the parties'. **Sakai**, however, questions whether the Court provided sufficient justification in introducing and applying the standard of review as to its legal basis, its scope of application, its content, and its methodology of application, including, in particular, the burden of explanation imposed on the respondent. **Fitzmaurice** (Chapter 3) concurs that the standard of review employed by the Court, with its key concepts such as reasonableness and objectivity, cannot be derived from just an interpretation of Article VIII of ICRW, and it must have been based also on other factors outside the realm of Article VIII, without explicitly specifying what these factors were.

Foster, with piercing clarity, analyzes the important function the standard of review played in the *Whaling* judgment. As the Court focused on the 'purpose' criteria of Article VIII, paragraph 1, it was inevitably drawn into an examination of the inherently subjective element of motivation. In this context, **Sakai**'s criticism, paraphrasing the dissenting opinion of Judge Abraham, that the Court should have respected the intention of Japan, seems convincing. However, according to **Foster**, the standard of review the Court introduced and applied, despite its novelty, was in fact effective in potentially closing the gaps and reconnecting the seemingly ambiguous logic of the Court. For example, the application of a standard of review by the Court enabled it to indirectly assess the motivation of Japan, employing inferences from circumstances and 'reasons', without becoming entangled in an inquiry whether science was the sole or the preponderant motivation, or only one amongst other motivations Japan could have been conceiving in conducting JARPA II. Consequently, the Court was able to conclude, in paragraph 97, that

> a State often seeks to accomplish more than one goal when it pursues a particular policy. Moreover, an objective test of whether a programme is for purposes of scientific research does not turn on the intentions of individual government officials, but rather on whether the design and implementation of a programme are reasonable in relation to achieving the stated research objectives.

There are interesting differences in the reading of the judgment as regards the issue of burden of proof within the context of the application of standard of review. Where **Tamada** (Chapter 5) is of the opinion that the Court, in paragraph 68, 'makes clear its intention to examine the standard of review…on the basis of a burden of proof imposed on Japan', **Foster** is more nuanced. 'The Court's application of a standard of review in the *Whaling* case seems to have been connected with a relatively informal and malleable approach to allocation of the burden of proof'. The Court did not refer to the burden explicitly, and, according to **Foster**: 'Inference appears to have been determinative in the Court's consideration of the seven elements of the JARPA II's design and implementation'. **Hamamoto**, on the other hand, argues that by transforming the reasonableness test within the process of standard of review into an obligation to explain with detailed reasons, the Court was able to examine without placing the burden of proof on either side. This is because whether Japan stated reasons or not is a purely objective question, so that Australia does not have to indicate each point where Japan did not provide reasons.

A few chapters went further in drawing some policy inferences from the Court's consequently effective, but legally precarious, employment of the standard of review in the case. Despite his criticisms, **Sakai** accepts 'the Court's penetrating insights and prudent strategy that it has chosen' to bring about a solution acceptable for both parties to the dispute. **Foster** concurs:

> We can understand, given all the dimensions of the case, why the Court turned to the idea of applying a standard of review in this case. This protected Japan against allegations beyond the four corners of the identified standard, and can be seen as an important effort to bring about peaceful resolution of the parties' dispute.

Whether the judgment in fact has brought the disputing parties closer to a resolution will be briefly examined in Section 4 below. **Foster**, however, is more restrained when she predicts the potential for the Court's move from 'adjudication' to 'review', with associated move from requiring 'proof' to applying a 'test'.

3 Unsettled Implications

The preceding chapters in this Book demonstrate convincingly that there are in fact several agenda items emanating from the *Whaling* judgment that may have potential implications and ramifications for the understanding of the ICRW and the work of IWC and its Scientific Committee; for the role of ICJ in dispute settlements; and for the understanding of international law generally. However, like **Foster**, the general tone of the authors is decidedly cautious in identifying the extent and the reach of such potential. One of the reasons could be the judgment's policy-oriented nature, as described above.[5] Another

---

5 *See also* Mari Koyano, 'The Significance of the Whaling Judgment from a Perspective of International Administrative Law', *Kokusai Mondai*, No. 636 (November 2014), pp. 43–58 (in Japanese).

could be the recognition that, as **Rothwell** indicates, the case submitted to the Court by Australia from the beginning had a limited scope, and, accordingly, the judgment should be read in that light.

On the procedural issues, **Tamada** highlights the winning and losing litigation strategies of the parties and the intervener in the *Whaling* case as regards the Statute's Article 63 non-party intervention, and its Article 43 party-appointed expert examination, and draws many important lessons for future litigating countries, including Japan. However, he is careful not to extend his critical analysis to the legal institutions themselves, concluding that most of the unfavourable results for Japan were 'unavoidable' under the existing institutions of intervention and expert examination. **Tams**, on the other hand, carefully reads 'important messages' from the Court's side-stepped treatment on the jurisdictional and standing issues. On standing, **Tams** suggests a 'more healthy reading' of the judgment: 'the Court, after decades of equivocation, has now embraced the idea of public interest standing, at least on the basis of multilateral treaties protecting collective interest'. **Sakai**, however, is avowedly more cautious in establishing its precedential value, pointing out the specific circumstances of the case in which the Court did not address the issue, and the Court's decision (intentional, according to **Sakai**) not to refer to obligations *erga omnes*, or to Article 48, paragraph 1 (a), of the Articles on State Responsibility. **Tams**, expecting such counter-argument, did not forget to add an admonition: If the Court wishes to shape the conduct of parties on crucial questions such as jurisdiction and standing, silence might not have been its best approach.

On the law of treaties, particularly on the issue of treaty interpretation, a comprehensive examination by **Fitzmaurice** attests that the *Whaling* judgment raised more questions than it provided answers to in relation to many of the contemporary issues of treaty interpretation: evolutionary interpretation, interpretation based on systemic integration, and the role of subsequent agreements and practice in treaty interpretation. Her analysis on the possible employment by the Court of an evolutionary interpretation is particularly illuminating, as it touches upon the very core of the dispute at hand: the object and purpose of the ICRW. According to **Fitzmaurice**, ICRW has a multiple object and purpose, and, therefore, 'the best way forward is to analyse the issue on a case-by-case basis, by the usual means of interpreting a treaty on the basis of its object and purpose'. In this sense, she questions whether it was necessary for the Court to characterise the ICRW as 'an evolving instrument' (paragraph 45). Paragraph 56, read together with paragraph 45, should then mean that 'the IWC may evolve the focus of the Convention through amendments to the Schedule, but may not re-define and redesign its object and purpose fundamentally'. This restrained reading of paragraphs 45 and 56 is consonant with that of **Morishita** and, indeed, that of the Japanese government when deciding to continue the special permit whaling in the western North Pacific in 2014 and resume it from the end of 2015 in the Antarctic as NEWREP-A.[6] **Hayashi**, instead, looks to international relations theory in explaining the Court's logic: 'Certain parts of soft law in treaty regimes cannot adequately be analysed in the light of the law of treaties, for the simple reason that the law of treaties has developed, and is conceived, as an instrument to deal with static treaties'. According to **Hayashi**, the Court in fact 'found ways to take into account certain aspects of a dynamic regime that do not squarely fall within any concept of the law of treaties, and that only the examination of the operation of the treaty regime can reveal'. On the role of subsequent agreement and practice in treaty interpretation, **Fitzmaurice** considers the judgment as 'very illuminating'. According to her, in this case, it was the subsequent agreement and practice of the Parties, members of the IWC, and not of the IWC itself, that was under scrutiny, and the Court was very strict and cautious in its assessment of the relevance of Article 31, paragraph 3 (a) and (b), regarding the evidentiary value of the resolutions of the IWC in this regard.

On the duty of cooperation, again, there were different nuances in assessing its significance in the reasoning of the Court. On the one hand, **Fitzmaurice**, at least within the specific context of this case, considers that the duty of cooperation is 'a very weak obligation, lacking well defined legal content'; and, thus, it does not contribute to the interpretation of Article VIII to mean that the use of lethal methods must be modified or restricted. **Hayashi**, on the other hand, sees 'a twist' function in the duty of cooperation in transforming soft law to hard law. She considers that

> once a Contracting Government acknowledges, and consents to, its duty to co-operate with the IWC in general, it is difficult to delineate those recommended acts which it is obliged to carry out, on one hand, and, on the other hand, those which it can, after a consideration in good faith, nevertheless decline to carry out.

This reasoning leads to what **Hamamoto** calls a far-fetched obligation to explain with detailed reasons why a State's actions are in conformity with norms reflected even in non-legally binding instruments. In his recent article, Naoya Okuwaki also sees the fundamental role played by the duty of cooperation in

6 Government of Japan, *Proposed Research Plan for New Scientific Whale Research Program in the Antarctic Ocean* (*NEWREP-A*), November 2014, <http://www.jfa.maff.go.jp/j/whale/pdf/newrep--a.pdf>.

the reasoning of the Court to reach its ultimate conclusion that Japan should revoke any extant permit granted in relation to JARPA II, and refrain from granting any further permits in pursuance of that programme.[7] Certainly, the potential broader implications of the Court's succinct pronouncement on the duty of cooperation awaits in-depth academic examination.[8]

Finally, at a policy level, a few authors touched upon a possible implication of the judgment on the future of the IWC. Although **Morishita** is generally pessimistic in relation to its impact in changing the basic positions of the IWC members, he does recognise a possibility of the judgment promoting 'some dialogue among [the disputing parties] because some legal views are presented by the authoritative ICJ on some aspects of the disputes'. **Press** (Chapter 11) agrees by saying that 'the solution to the on-going whaling war does not lie in the interpretation of the judgment, in court, or even at sea. Instead, it almost certainly lies with the Whaling Commission, the IWC Scientific Committee and their founding laws'.[9] It is interesting to note that the two opposing policy-makers have a shared perception of the judgment that the Court did make its own scientific findings, but whether the judgment 'further jeopardized the *raison d'être* of the IWC' must be carefully examined in light of the actual practice under the Commission and its Scientific Committee. As will be briefly discussed in Section 4 below, the judgment seems to revitalize the scientific discussion in the Scientific Committee in reviewing the proposed NEWREP-A.

One implication of the judgment seems to garner consensus amongst the authors. The judgment 'significantly raised the bar for lethal whaling'. In what sense is this so? **Press** explains:

> The availability of alternatives to lethal whaling goes to the heart of the construction of any future research program authorised under Article VIII, or any challenge to it. The Court's use of the yardstick of 'reasonableness' requires an objective consideration of research objectives, research methodologies, sampling techniques and sample sizes – the mere statement that a sample size has been determined objectively, or that one research technique is justified over another, must be evaluated against alternative propositions.

7 Naoya Okuwaki, 'Lessons of Whaling Judgment of ICJ: From the Perspective of the Duty to Cooperate', *Nihon Kaiyō Seisaku Gakkaishi* [*Journal of Japan Society of Ocean Policy*], No. 4 (November 2014), p. 18 (in Japanese).

8 *See* Akiho Shibata, 'ICRW as an Evolving Instrument: Potential Broader Implications of the *Whaling* Judgment'", *Japanese Yearbook of International Law*, Vol. 58, Year 2015 (in press).

9 Citing Indi Hodgson-Johnston and Julia Jabour, 'Japan's new whaling program is a small win for whales, but…', *The Conversation*, November 20, 2014.

Hamamoto elaborates: With the judgment, a scientific common sense can no longer be taken for granted as regards the lethal scientific whaling; it has to be explained, and unless it is explained (as the Court put it in paragraph 195 of the judgment) it 'lends support to [the] contention that a predetermined overall sample size has dictated the choice of…the rate of change to be detected, rather than the other way around'. Referring to paragraph 138 of the judgment, he also sees the possibility that the proponent country will be required to review and justify the sample sizes of its scientific whaling programme every year, without awaiting for the review by the Scientific Committee. Indeed, the Court, when rendering its judgment in March 2014 that impugned the appropriateness of sample sizes of JARPA II, plainly ignored the fact that the review of JARPA II by the Expert Panel and then the Scientific Committee was already scheduled in February and May 2014.[10]

The 'significantly higher bar' established by the Court for lethal scientific whaling represents the clearest practical implication of the judgment, the subsequent development of which has brought an unexpectedly significant impact on the regime of special permit whaling under the ICRW. The prospect of the judgment having a real impact seems greatest in this respect, which I will examine in the next section.

4 A Hope after 15 Months from the Judgment

After all, the *Whaling* judgment did not intend to untangle the Gordian knot of the whaling controversy once for all.[11] Instead, it confined itself, wisely according this author's view, to pronounce the legal parameters for the exercise of the

10 During the oral proceedings, Japanese counsels in many occaions informed the Court of this scheduled review. Akhavan, ICJ, public sitting held on 2 July 2013, CR 2013/12, p. 58, para. 68; Hamamoto, ICJ, public sitting held on 3 July 2013, CR 2013/13, p. 22, paras. 39, 68–71.

The Expert Panel and the Scientific Committee (with several members disengaged in the discussion) reviewing JARPA II in 2014 made several comments that are much more favourable for Japan than the Court. For example, with regard to the objective of improving the management procedure for Antarctic minke whales, 'the Panel agreed that the information from JARPA and JARPA II, particularly with respect to stock structure and abundance will greatly improve any future *Implementation Simulation Trials*, should these ever be requested by the Commission'. *Report of the Scientific Committee, Bled, Slovenia, 12–24 May 2014*, IWC/65/Rep01 (2014), p. 70.

11 Malgosia Fitzmaurice, 'The International Convention for the Regulation of Whaling and International Whaling Commission: Conservation or Preservation – Can the Gordian Knot be Cut (or Untangled)?', *Yearbook of Polar Law*, Vol. 5 (2013), p. 489.

authority to issue special permits under Article VIII of the ICRW.[12] Moreover, at least for the parties to the case and the intervener, the Court succeeded in setting parameters that came to be acceptable to all of them. This positive aspect of the judgment should be particularly acknowledged. Here, this author sees a hope of an emerging solution to the pugnacious element of the IWC regime, namely, the special permit whaling. The judgment succeeded in circumscribing the dispute regarding special permit whaling to a narrowly defined technical framework and established signposts for a possible future resolution of the controversy. This framework and signpost will best be applied and implemented by the Scientific Committee based on robust scientific discussion,[13] rather than by the Commission based on political maneuver and majority voting. It will take time, but a beam of hope is clearly visible.

The cumulative grounds for such an emerging solution that could be observed during the 15 months from the judgment up to the time of writing this paper is: (1) Japan's acceptance of the judgment as a whole; (2) the adoption of IWC resolution 2014–5 in September 2014, particularly its first operative paragraph, and the Japanese acquiescence in the said paragraph; (3) the submission by Japan of a new special permit whaling programme in the Antarctic, the NEWREP-A, in November 2014 and a robust review exercised by the Expert Panel in February 2015 and by the Scientific Committee in May 2015, based on the 'new' criteria provided in resolution 2014–5; and, finally, (4) the expectation that Japan will sincerely cooperate with this 'revitalized' Scientific Committee. Let me briefly examine these four grounds.

(1) *Japan's Acceptance of the Judgment as a Whole*

All this process started with the Japanese declaration, minutes after the judgment was read out by the Court in The Hague, that 'Japan will abide by the judgment of the Court', although expressing at the same time its disappointment and regret over the Court's finding.[14] Since the primary authority to interpret the judgment, and to determine the ways to implement it, lies with the parties to the case,[15] a more important development was the April 18 statement of

12 *See also* Sir Geoffrey Palmer, 'A Victory for Whales', *New Zealand Law Journal*, May 2014, pp. 124–128.

13 Australian government is also of the view that Australian scientists will provide 'robust scrutiry' of Japan's whaling plan. *The Gurdian*, 'Australian government to scrutinise new Japanese whaling plan', 19 November 2014.

14 Statement by Chief Cabinet Secretary, the Government of Japan, on International Court of Justice (Australia v. Japan: New Zealand intervening), 31 March 2014.

15 Constanze Schulte, *Compliance with Decisions of the International Court of Justice* (2004), pp. 29–32.

the Japanese Minister for Agriculture, Forestry and Fisheries[16] that indicated how Japan interpreted the judgment and intended to implement it. First, according to this statement, Japan interpreted that the Court had reaffirmed that one of the purposes of ICRW is the sustainable utilisation of whale resources, and, therefore, Japan has confirmed its basic policy of pursing the resumption of commercial whaling by conducting special permit whaling in order to gather scientific data essential for the management of whale resources. Second, in responding to the expectation of the Court expressed in paragraph 246 of its judgment, Japan, taking into account the advise of the Court, decided to modify the sample sizes of its whale research programme in the western North Pacific, the JARPN II,[17] despite the fact that JARPN II was outside the purview of the judgment. The reduction of the sample sizes for 2014 programme was indeed significant.[18] Third, for 2015 and onward, Japan decided to submit a new whale research plan in the Antarctic 'reflecting the criteria mentioned in the judgment' and ensuring an 'internationally open and highly transparent process' in drafting the plan. This was indeed what happened when Japan submitted draft NEWREP-A in November 2014. In the proposal, Japan praised the judgment as it provided 'meaningful guidance on the issuance of future permits by Japan'.[19] The plan appeared 'fastidiously tailored to address

16 Statement by Minister for Agriculture, Forestry and Fisheries on Policy towards the future Whale research Programs, 18 April 2014. <http://www.jfa.maff.go.jp/e/pdf/danwa.pdf>.

17 *Report of the Scientific Committee, Bled, Slovenia, 12–24 May 2014, supra* note 10, p. 73.

18 Table 1: 2014 JARPN II modification as to sample sizes

Original proposal			2014 Modification
Offshore component	Common Minke	100	0
	Sei	100	90
	Bryde	50	25
	Sperm	10	0
Coastal component			
Miyagi	Common Minke	120	51 (actual catch 30)
Kushiro	Common Minke		51

The scientific justifications for those modifications are provided by Japan in 2015, after recommended by the Scientific Committee in 2014 to 'develop a more comprehensive document for review at the 2015 Annual Meeting that details how the above recalculations of sample size and changes in allotments of lethal versus non-lethal methods fit with achieving the overarching programmatic objectives'. Government of Japan, *Responses to SC 65b recommendation on Japan's Whale Research Program under Special Permit in the Western North Pacific (JARPN II)*, SC/66a/SP10 (2015), pp. 1–5.

19 NEWREP-A proposal, *supra* note 6, Section 1.3, p. 7.

the complete scope of the Court's judgment'.[20] Japan now acknowledges that there are 'certain conditions' for special permit whaling to be 'legally permissible',[21] the legal authority under Article VIII of the ICRW that Japan had previously characterized as an 'absolute right'.[22] This demonstrates Japan's determination to comply with the judgment as a whole, including its reasoning and the standard of review. This was a crucial policy decision, considering the fact, at least amongst academia, the gaps and silences in the Court's reasoning were identified and criticized, as the preceding chapters demonstrated.

(2) *IWC Resolution 2014–5 and Japan's Acquiescence in its Paragraph 1*

An unexpected bonus for the Court, perhaps, was that, through the adoption of IWC resolution 2014–5,[23] the legal parameters set by the Court have now been transformed into the operational criteria to be implemented by the Scientific Committee within the IWC regime. The proponent of the resolution, New Zealand, must have had the intention of strengthening the special permit review process, which had been criticized by some scientists as 'weak' and 'politicized'.[24] Basically, the proposal had two objectives. First, it tried to establish substantive criteria that reflected the Court's standard of review, that will be implemented by the Scientific Committee when reviewing special permits. Second, it tried to establish a new procedure for such review, requesting the Commission to make recommendation on the proposal. This second element in effect tried to amend paragraph 30 of the Schedule under ICRW. Japan cast a negative vote on this resolution because of this element.[25] However, for the purpose of this paper, the first element is more significant because, first, the criteria clearly reflected the Court's stringent and negative attitude towards utilizing lethal methods in scientific whaling. This was not quite the case for the existing criteria as enunciated in Annex P, which are neutral to the

---

20 Hodgson-Johnston and Jabour, *supra* note 9. *See* NEWREP-A proposal, *supra* note 6, Section 1.3, pp. 7–11, where the proposal explicitly refers to several paragraphs of the judgment which are acrimonious to Japan.

21 NEWREP-A proposal, *supra* note 6, Section 1.3, p. 11.

22 Report of the Scientific Committee (2004), *J. Cetacean Res. Management*, Vol. 7 (suppl.), 2005, p. 45.

23 IWC Resolution 2014–5: *Resolution on Whaling under Special Permit*, adopted by the vote of 30 in favor, 20 against, and 5 abstaining.

24 William de la Mare and Nick Gales, Marc Mangel, 'Applying scientific principles in international law on whaling', *Science*, Vol. 345, Issue 6201 (2 September 2014), p. 1126.

25 IWC, *Chair's Report of the 65th Meeting*, *supra* note 3, para. 149.

so-called lethal versus non-lethal methodology.[26] The first operative paragraph of IWC resolution 2014–5 enumerates five criteria as subparagraphs (a) to (e). The most significant is its subparagraph (d), which is almost a carbon copy of paragraph 144 of the judgment. The Court, in paragraph 144 of its judgment, demanded from the proponent of lethal whaling the consideration of the possibility of making more extensive use of non-lethal methods 'in order to reduce or eliminate the need for lethal sampling'. Subparagraph (d) of the resolution instructs the Scientific Committee to evaluate whether 'non-lethal alternatives are feasible to either replace or reduce the scale of lethal sampling proposed'. This criterion clearly reflects the underlying conservationist approach to whales.[27] During the Commission discussion, Norway objected that subparagraph (b) of the resolution went beyond the judgment, as it instructed the Scientific Committee to evaluate whether the elements of the research that rely on lethally obtained data are likely to lead to improvements in the conservation and management of whales.[28] As will be explained below, the criterion in subparagraph (b) will have a significant impact on the way the Scientific Committee actually review the special permits.

Despite these criticisms, the first operative paragraph of resolution 2014–5 is significant because it was acquiesced in by Japan to be applied by the Scientific Committee in its review of NEWREP-A proposal in 2015. In its preliminary response to the Expert Panel Report on NEWREP-A, Japan stated that, '(w)hile (Japan is) aware that the Review Panel also referred to the first operative paragraph of IWC resolution 2014–5, which Japan voted against, specific areas referred to in the said paragraph are generally covered by the five items under the Terms of Reference (shown in Annex P)'.[29] As explained above, this author does not consider the criteria provided in the first operative paragraph of resolution 2014–5 are 'covered' by the existing criteria enunciated in Annex P, because the 'new' criteria in the resolution provide cognizable discouragement

26 Annex P on Process for the Review of Special Permit Proposal and Research Results from Existing and Completed Permits, Report of the Scientific Committee (2008), *J. Cetacean Res. Management*, Vol. 11 (suppl.), 2009, pp. 398–401. As to lethal methods, Annex P refers to 'an assessment of why non-lethal methods been considered insufficient', 'advise and suggestions on components of the programme that might be achieved using non-lethal methods', and 'evaluation of the utility of the lethal techniques used by the Special Permit Programme compared to no-lethal techniques'. This document was adopted by consensus in the IWC.

27 Shibata, *supra* note 8.

28 IWC, *Chair's Report of the 65th Meeting*, *supra* note 3, para.150.

29 Government of Japan, *Proponents' preliminary response to the Report of the Expert Panel to review the proposal for NEWREP-A*, SC/66a/SP/1 (2015), p. 2.

to using lethal methods in scientific whaling, based on the underlying conservationist outlook towards whales. This is far beyond what Japan accepts as a scientific policy, namely 'it does not use lethal means more than it considers necessary' (paragraph 80 of the judgment). This being said, with the acquiescence by Japan, the Scientific Committee definitely felt comfortable in applying the criteria and its underlying outlook, as provided in the first operative paragraph of IWC resolution 2014–5 within the context of review of NEWREP-A proposal.

(3) *Robust Review by the Scientific Committee on NEWREP-A Proposal*

This led to a decidedly more confident Expert Panel in reviewing the proposed NEWREP-A in February 2015. This change of attitude is clearly visible if one compares the tone of its 2015 report with that of the one submitted a year ago, just before the ICJ judgment, in February 2014 with regard to the review of JARPA II. Seven out of ten experts on both Panels are the same persons. The conclusion of the 2015 Expert Panel Report is forthright: 'the Panel was not able to determine whether lethal sampling is necessary to achieve the two major objectives; therefore, the current proposal does not demonstrate the need for lethal sampling to achieve those objectives'.[30] This author cannot summarize the 35-page scientific reasoning for the Panel to reach this conclusion, but the following should be noted in order to highlight the Panel's logic that clearly reflects the underlying reasoning of the *Whaling* judgment, being transformed into the Scientific Committee's review criteria through IWC resolution 2014–5. First, with regard to the primary objective of NEWREP-A, of improving the performance of Revised Management Procedure (RMP) and its relation to the need of lethal sampling, the Panel, applying the criterion in subparagraph (b) of resolution 2014–5, concludes that the proposal does not yet evaluate 'the level of improvement' that might be expected in RMP performance by improved precision in biological parameters,[31] namely the Age at Sexual Maturity (ASM) of Antarctic minke whales, to detect an annual increase of 0.1 per year, by lethally sampling 333 whales per year.[32] It is significant that the Panel, interpreting subparagraph (b), which requires an evaluation of the 'likelihood' of improvements in the management of whales by lethal research, considered it necessary for the proponent to demonstrate the 'degree' of improvement. Second, applying the criteria in subparagraphs (c) and (d), the Panel recommended an examination

30 *Report of the Expert Panel to review the proposal by Japan for NEWREP-A, 7–10 February 2015, Tokyo, Japan*, SC/66a/Rep/6 (2015), Section 8, p. 35.

31 *Id.*, Section 2.1.2, p. 6.

32 NEWREP-A proposal, *supra* note 6, Section 2.1.2, pp. 13–14 and Section 3.3, pp. 30–31.

into non-lethal methods that will provide 'essential information in terms of evaluating lethal versus non-lethal techniques and thus in determining lethal sample size'.[33] Such examination should be given high priority and be completed before there is a final conclusion on lethal techniques and sample size, without which the Panel cannot complete its full review.[34] Here, faithfully following the reasoning of the Court, the Panel imposed the burden of explanation on the proponent proposing to use lethal sampling, and required it to demonstrate that the use of non-lethal methods cannot reduce any further the size, and possibly replace the use, of lethal sampling proposed in the programme. According to the Panel, that has not been demonstrated yet in the NEWREP-A proposal.

In response, Japan decided to sincerely respond to all 29 recommendations made by the Panel,[35] faithfully complying with the duty to cooperate with the Scientific Committee and the consequent obligation to give due regard to its recommendations as a State Party to the ICRW (paragraph 83 of the judgment). Japan even expressed appreciation to the Panel, saying that many of those recommendations are 'useful advice' to further improve the NEWREP-A proposal.[36] Japan also started to tackle the thorny request to evaluate the 'level of improvement' that might be expected in RMP performance by improving the ASM data by sampling 333 Antarctic minke whales. It preliminarily concluded that the availability of age data from NEWREP-A will provide substantially improved estimates of key aspects of the minke whale population's dynamics, although recognizing that further time is necessary for completing the necessary simulations.[37] As to the lethal versus non-lethal methods, Japan put forward its own view, rebutting the Panel's assumption that 'the necessity of lethal sampling cannot be proven unless the feasibility studies of all the conceivable non-lethal research techniques, both current and future ones, are completed and the conclusion is reached that none of the non-lethal techniques is a feasible alternative to lethal sampling'.[38] Instead, according to Japan, a more 'reasonable approach' would be to 'determine the feasibility of

33 *Report of the Expert Panel on NEWREP-A*, *supra* note 30, Section 3.1.3.1, p. 11.

34 *Id.*, Section 3.1.3.1, p. 11 and Section 8, p. 35.

35 Fisheries Agency of Japan and Ministry of Foreign Affairs of Japan, *Response to the Report of the Expert Panel of the International Whaling Commission's Scientific Committee (IWC SC) to review the proposal for a New Scientific Whale Research Program in the Antarctic Ocean (NEWREP-A)*, 14 April 2015. <http://www.jfa.maff.go.jp/j/whale/pdf/review2.pdf>.

36 Government of Japan, *Proponents' preliminary response*, *supra* note 29, p. 2.

37 Government of Japan, *Proponents additional responses to the report of the Expert Panel to review the proposal for NEWREP-A*, SC/66a/SP/8 (2015), pp. 18–22.

38 Government of Japan, *Proponents' preliminary response*, *supra* note 29, p. 5.

non-lethal methods based on the currently available scientific and technical knowledge, and if deemed unfeasible, to initiate lethal sampling in the meantime while continuing feasibility studies on non-lethal methods on an ongoing basis'.[39] In addition, Japan proposed its own tests for assessing the feasibility and practicability of non-lethal methods.[40] This might be a nuanced but important difference of interpretation between the Panel and Japan on subparagraph (d) of first operative paragraph of IWC resolution 2014–5, which in effect means a difference of interpretation on paragraph 144 of the *Whaling* judgment.[41] At the same time, it is important to note the recognition by Japan that the feasibility assessment of all *current* non-lethal methods be made before deciding to use lethal sampling and that such assessment should be an ongoing one while the lethal programme continues.

The full Scientific Committee's review on NEWREP-A in May 2015 was again robust and discreet.[42] It established a small working group to evaluate the additional explanation by Japan,[43] and the Committee *concurred* with the conclusions of this working group.[44] As to the improvement of RMP performance, the small working group was of the view that 'it is still to be determined whether a reduction in uncertainty in year-class strength has any appreciable

---

39 *Id.*

40 The four major tests proposed by Japan are (i) whether the same data sought (e.g. age) can be obtained by a non-lethal method; (ii) whether enough data can be obtained in terms of quantity for statistical analysis; (iii) whther it is of sufficient quality for analysis (e.g. accuracy); and (iv) whether the cost to obtain the data is realistic and reasonable. Applying these tests, Japan concludes that it is 'clear that feasibility assessment of non-lethal methods for age data based on currently available scientific knowledge has been completed and it has demonstrated the need for lethal sampling at this juncture'. *Id.*

41 The Court, in paragraph 83 of its judgment, denied the existence of 'a requirement that lethal methods be used only when other methods are not available'. The Court, in paragraph 135, also found that there is 'no basis to conclude that the use of lethal methods is per se unreasonable in the context of JARPA II'. The nuanced difference of interpretation on paragraph 144 seemed to emerge when one leans more on those two paragraphs of the judgment when reading the Court's requirement, in paragraph 144, to consider 'more extensive use of non-lethal methods in order to reduce or eliminate the need for lethal sampling'.

42 *Report of the Scientific Committee, San Diego, CA, USA, 22 May–3 June 2015*, IWC/66/Rep01 (2015), subsection 17.1, pp. 88–97.

43 *Annex Q-1: Report of the Small Group Established to Review SC/66A/SP8*, Annex Q: Matters related to Item 17: special permits, *Report of the Scientific Committee (2015)*, pp. 1–2. The small working group was composed of the following 7 members: Punt (Chair), Butterworth, Cooke, de la Mare, Kitakado, Matsuoka, and Palka.

44 *Report of the Scientific Committee (2015)*, *supra* note 42, subsection 17.1.4, p. 92.

effect on performance statistics for *Implementation Simulation Trials*'. It also concluded that allowing for time-varying ASM has little impact on the majority of the results from the Statistical Catch at Age model (SCAA). At the same time, it considered that, because of additional variations, the sample size of 333 Antarctic minke whales proposed in NEWREP-A may be too small and suggested a management-related (or biologically-based) justification for the sample sizes.[45]

Based on these reports, the Committee *agreed* that, although the preliminary results indicated that collection of age data will reduce uncertainty in estimates of future recruitment, whether this is likely to lead to 'substantial improvements' in conservation and management is yet to be demonstrated.[46] This agreement by the Scientific Committee is significant, as it now interprets subparagraph (b) of first operative paragraph of the IWC resolution 2014–5 as requiring from the proponent of lethal whaling an explanation as to not only 'a degree' but 'the substantial degree' of improvements in the proposed objective of the research using lethal sampling. The Committee therefore agreed that additional work needs to be done to evaluate the level of improvements. However, the Committee could not agree to the view, as expressed by the Expert Panel and some members of the Committee, that the proposal does not demonstrate the need for lethal sampling to achieve the objectives.[47] As to lethal versus non-lethal methods, the Committee could not agree beyond what the Expert Panel concluded, namely it needs further information on the field experiments, laboratory work and analysis on using non-lethal methods to determine whether non-lethal means can be used further to reduce sample sizes or to possibly replace the lethal methods altogether in the NEWREP-A proposal.[48] In conclusion, the Committee '*agrees* that the analyses recommended by the Panel and further specified in Annex Q-1 should be completed, and that progress should be reviewed again next year'.[49] From this conclusion,

45 *Annex Q-1*, *supra* note 43, Table 1, p. 2.

46 *Report of the Scientific Committee* (*2015*), *supra* note 42, subsection 17.1.4, p. 92.

47 *Id.*, subsection 17.1.5.1, p. 94. 42 members of the Committee expressed the view that 'they do not consider that there is any scientific justification for collecitng lethal samples until it is demonstrated that the information collected is necessary and reasonable in accordance with the guidelines in Resoltuion 2014–5 adopted by the Commission'. *Annex Q-5: Statement concerning Review of the NEWREP-A Proposal*, Annex Q: Matters related to Item 17: special permits, *Report of the Scientific Committee* (*2015*), p. 9. At the 2015 Scientific Committee annual meeting, there was a total of 122 members attending from national delegations.

48 *Report of the Scientific Committee* (*2015*), *supra* note 42, subsection 17.1.5.2, p. 95.

49 *Id.*, subsection 17.1.5.5, p. 96.

the Committee seemed to implicitly endorse the approach proposed by Japan as to lethal versus non-lethal methods, evaluating vigorously the feasibility of currently available non-lethal methods while implementing the lethal programme, and that its lethal component be reviewed continuously taking into account the results of such evaluation.

(4) *Japan's Cooperation with the Scientific Committee*

Thus, the 2015 Scientific Committee reviewing Japan's new special permit whaling programme in the Antarctic (NEWREP-A) did not endorse a simple 'agree to disagree' policy of an IWC member, nor did Japan try to scorn the Committee's review process as just a ceremonial passage for its predetermined plan to hunt whatever number of whales it deems necessary. First, the Committee, in implementing the first operative paragraph of IWC resolution 2014–5, has now clearly established an extremely high standard by which a programme involving lethal sampling should demonstrate 'the substantial improvements' in the performance of RMP—the key objective of the NEWREP-A, especially in relation to the calculation of sample sizes. Japan decided to accept and surmount this standard, and now faces a formidable task of demonstrating how the more precise data that will be obtained from NEWREP-A would substantially improve the management performance. Second, the Committee, again accepted by Japan, requires from the proponent of lethal whaling a demonstration that all currently available techniques of non-lethal methods be actually tested to evaluate whether such non-lethal methods can reduce and possibly replace altogether the lethal sampling in the programme. Japan has revised its NEWREP-A proposal, and expanded significantly its non-lethal research component,[50] including two dedicated vessels for sighting survey.[51] Japan now faces a tremendously costly programme, if it decides to implement NEWREP-A. Finally, the Scientific Committee, again accepted by Japan, now sees the review on special permit whaling as a continuous, annual endeavour, which heretofore was only a periodic one.[52] The Scientific Committee, from its 2014 meeting after the *Whaling* judgment, had become forthright in requesting from the proponent of a lethal programme more information at the next annual meeting, because the information

---

50 Government of Japan, *Addendum to the proposed research plan for New Scientific Whale Research Program in the Antarctic Ocean* (NEWREP-A), SC/66a/SP/2 (2015), pp. 4–6.

51 *Report of the Scientific Committee* (2015), *supra* note 42, subsection 10.18.1.2, pp. 52–53.

52 See Revised Annex P: Process for the review of Special permit Proposal and Research Results from Existing and Completed Permits (2012). Available at <https://iwc.int/document_3100>.

available was not sufficient to evaluate the justification for the sample sizes.[53] The same conclusion was reached as to the NEWREP-A proposal. These requests are not explicitly endorsed by paragraph 30 of the Schedule or Annex P, but Japan, under the obligation to give due regard to the recommendations of the Scientific Committee, has so far faithfully responded to them. Considering the Committee's approach to require continuous evaluation of lethal versus non-lethal methodology, favouring the latter, it is almost certain that the proponent of lethal whaling would submit annually to the Committee detailed reports on the examination on the feasibility of non-lethal methods and on the evaluation, based on such examination, of whether lethal methods be reduced or replaced by non-lethal ones. These reports will be scrutinized by the Committee every year, with possible further recommendations from the Committee. The administrative cost for proponents of lethal whaling to prepare for, and to engage in, the annual meeting of the Scientific Committee will become unimaginably high.

5 Conclusion

It may be true that 'no great statement of law or resounding principle appear in the judgment',[54] if we take the cautious reading of the judgment as advocated by the preceding chapters of this Book. On the other hand, as this author argued above, the *Whaling* judgment, with its standard of review, had a significant practical impact on the way the special permit review is conducted within the IWC regime. After 15 months from the judgment, it can now be said that the remaining controversy surrounding the special permit whaling in the Antarctic is confined to the following scientific and technical issues. First, whether the proponent can convincingly explain how substantially the data obtained from lethal methods contribute to the improvements in the performance of whale management schemes such as RMP and its *Implementation Simulation Trials* for Antarctic minke whales. Second, whether the proponent can convincingly explain the currently available non-lethal methods cannot reduce or replace the lethal sampling in order to achieve the research objective. This involves

53 *Report of the Scientific Committee* (*2014*), *supra* note 10, subsection 17.3, p. 74. Previously, when reviewing the JARPN II in 2009, the Expert Panel did make a similar determination that it could not complete a full review without further information, but the Scientific Committee was able only to express concern over it. Report of the Scientific Committee (2009), *J. Cetacean Res. Manage.*, Vol. 11 (suppl.2), 2010, pp. 74–78.

54 Palmer, *supra* note 12, p. 128.

actual testing of those non-lethal methods by the proponent of lethal whaling and her continuous effort, based on the results from such testing, to reduce the size of lethal sampling.

The immediate practical consequences of the *Whaling* judgment should first be noted. With the Court establishing the high standard of review for special permit whaling and the Scientific Committee operationalizing it in a rigorous manner, for Japan, the special permit whaling in the Antarctic has now become an even more costly endeavour to maintain. It will soon become an issue for serious debate among the Japanese people whether it is worth continuing this endeavour, pumping millions of taxpayer's money.[55] On the other hand, for the vast majority of ICRW members, especially for developing countries, the special permit whaling has become a non-viable option, because of its extremely demanding scientific and technical justification and the consequent human and financial resources required. This is a curious consequence, as one is required already to have a vast scientific knowledge and state-of-the-art technology before considering commencing a scientific activity to increase its scientific knowledge.

As to medium to longer-term consequences, it is expected that the discussion on the above two issues will continue in the Scientific Committee for some more years, because there are still no decisive scientific answers to them. However, as scientific knowledge increases, and the relevant technologies improve, there can only be a convergence, and not a divergence, of scientific opinions that can emerge from such discussion. As to the former, there will come a point where no further improvements can be foreseen, or even necessary, in the management scheme, so that no additional 'scientific' whaling is necessary, meaning, in effect, that a decision on commercial whaling is possible, at least from the scientific point of view. As to the latter, there will come a point where data from advanced non-lethal methods would be considered sufficient to maintain the functional precision of the relevant management schemes. Whichever comes earlier, scientific whaling with the objective of improving whale management procedure is on the wane, as it should be, because it is only a means to an end—contributing to a scientifically informed decision as to the resumption or continued suspension of commercial whaling of Antarctic minke whales.

The *Whaling* judgment could also be read as an admonition to ICRW members to seriously prepare for this eventuality. Although this author does not concur with the interpretation that the *Whaling* judgment in its paragraph 56

55 It is reported that the Japanese special permit progammes are subsidized by the Ministry of Agriculture in the amount of 700 to 1100 million JYN per year from 2007 to 2014.

'reaffirmed' that one of the purposes of the ICRW is the sustainable utilization of whale resources, the Court indeed left the door open for the Commission members by consensus to redefine and possibly reorient the Convention most suitable for the contemporary international community. Without such political efforts on the part of all members of the IWC, the ingenious signposts provided by the Court towards possible resolution of the special permit whaling controversy would not come to its fruition.

Conference Report

Yuri Takaya

The Symposium on 'Whaling in the Antarctic: The ICJ Judgment and its Implications' was organized by Kobe University in Japan from May 31st to June 1st of 2014, only 2 months after the Court's decision. With 13 distinguished speakers, the following 7 sessions were held: (1) contemporary whaling and its overview; (2) the Judgment and its significance; (3) procedural issues concerning whaling in the Antarctic; (4) whaling in the Antarctic as an international scientific dispute; (5) ICRW, its Article VIII, and the law of treaties; (6) the Judgment's implications: IWC and whaling in future; and (7) litigation strategy and tactics. The *Chatham House Rules*[1] were applied to the symposium as well as to this Report.

Opening Remarks

The Symposium was opened with welcoming remarks by Prof. *Shigeki SAKAMOTO* (Doshisha University, Japan), referring that this Symposium was organized as the final outcome of the project '*Jurisdictionalization* of International Law' supported by the Japan Society for the Promotion of Science. A brief summary of the case and its background was given to highlight the core issues.

Session 1: Contemporary Whaling and Its Overview

This session was opened with a presentation entitled 'Contemporary Whaling and Overview' introducing: (1) the 1946 International Convention on the Regulation of Whaling (ICRW); (2) types of whaling; (3) NAMMCO – the North Atlantic Marine Mammal Commission; (4) whaling outside the ICRW; and (5) possible whaling in future. A brief explanation was given that the 1946

1 When a meeting, or part thereof, is held under the Chatham House Rule, participants are free to use the information received, but neither the identity nor the affiliation of the speaker(s), nor that of any other participant, may be revealed. Texts are available at: http://www.chathamhouse.org/about/chatham-house-rule#sthash.E2fT1muv.dpuf (last accessed on 31 December 2014).

ICRW was preceded by the 1931 Geneva Convention and the 1937 London Convention and the Protocol, and that it consists of the text of the Convention and the Schedule which includes a list of whale types, and catch limits for commercial and aboriginal subsistence whaling. It highlighted the following areas of disagreement: (i) the generic definition of what a whale is in the Schedule (small and medium sized whales such as narwhal and beluga); (ii) the territorial scope of the ICRW application; and (iii) the appropriate international organization stipulated in Article 65 of the UNCLOS. The purpose of concluding the 1946 ICRW was defined in its Schedule as being to provide for the proper conservation of whale stocks and thus make possible the orderly development of the whaling industry. In the ICRW, consisting of 88 state parties, Japan is the only state that conducts scientific whaling, while Iceland appended a reservation to the moratorium. It was pointed out that aboriginal subsistence whaling has a potential of becoming a very contentious issue for the following reasons: (i) inconsistency in its definition; (ii) demands for increases in quotas (e.g. Denmark for Greenland); (iii) commercialization of aboriginal whaling; (iv) the Makah tribe issue in the US; (v) aboriginal whaling of medium and small size whales (narwhal and beluga); (vi) aboriginal whaling outside the IWC (Canada, Indonesia); and (vii) a need to reassess Japan's small coastal whaling in the light of aboriginal whaling.

Session 2: The Judgment and Its Significance

This session focused on Australia's and Japan's perception of the *Whaling in the Antarctic* case. The presentation was provided from a scholar's point of view in terms of the background, the objectives before the Court and the reaction of Australia to the Judgment. In Australia, protracted efforts had been made to halt whaling since 1978 when the then Liberal/National Coalition government of Malcolm Fraser banned all whaling in Australian waters. While Australia initiated the moratorium and a number of the IWC resolutions in 1986–87, before JARPA in 2001–2003, the International Fund for Animal Welfare (IFAW) played a core role in developing international legal strategies that pro-conservation governments could pursue through international courts against the conduct of JARPA II. Based on the IFAW panel assessment that the whaling failed to meet the requirements of paragraph 30 of the Schedule as so was unlawful, it was clarified that the Australia's objectives before the ICJ was to contest, as a party of the ICRW, Japan's interpretation of the ICRW via its conduct in relation to issuing Article VIII special permits. In the end, the Court's decision was evaluated from Australia's perspective as a vindication of the politically and legally risky strategy of challenging the international legal validity of JARPA II and as a precedent for the interpretation of Article VIII of the ICRW in terms of the conduct of special permit whaling especially in the Southern Ocean.

Further, the case was evaluated from a Japanese scholar's viewpoint, not as a Party at the bar, who brought up nine legal questions. The 1st question raised was why Japan did not submit any preliminary objections before going on to the proceedings on the merits. Japan chose not to raise the issue of jurisdiction as a preliminary matter, but to set out its arguments on jurisdiction in its Counter-Memorial by invoking the Judgment in *Application of the Interim Accord of 13 September 1995 (the former Yugoslav Republic of Macedonia v. Greece)*. The answer was explored in relation to the list of pending cases before the Court (10 cases as of 31 March 2014) and four possible reasons were suggested as follows: (1) less likelihood of Japan's objections being accepted by the Court; (2) greater likelihood of its winning the case on the merits; (3) preference for good relations with Australia; and (4) cost for the case. In the end, it was concluded that Japan preferred the Court to resolve the case quickly. The 2nd question was whether or not Australia established its *Locus Standi* in the case. Australia did not base its claim on damage incurred as a result of Japan's violations of international obligations, but on the obligation *erga omnes partes* to all contracting parties to the ICRW, though the Court evaded the arguments on standing. The answer was explored by considering the impact of the *Questions relating to the Obligation to Prosecute or Extradite (Belgium v. Senegal)* case and the introduction of the standard of review. The 3rd question was whether or not the Court adopted an appropriate method for interpreting the ICRW. While describing the ICRW as an 'evolving' instrument, the Court interpreted it in accordance with the VCLT and recognized the IWC's recommendations as non-binding on any or all of the contracting governments except 'when these are adopted by consensus or by a unanimous vote'. The Court considered whether or not a recommendation 'adopted by consensus' was equivalent to an agreement of the contracting parties to the convention. As for the object and purpose of the ICRW, the Court rejected both Parties' arguments and observed that 'neither a restrictive nor an expansive interpretation of Article VIII is justified'. This raises the next question, whether or not the Court so construed principal obligations and exceptions in the ICRW regime so that the moratorium on the commercial whaling is a principal obligation of the regime while scientific research under special permits is an exception. The 4th questions was related to the term 'for the purposes of scientific research' in Article VIII, para. 1 of the ICRW, namely, from where the 'two elements'—'for purpose of' and 'scientific research'—came, and why they should be examined separately. The Court seems to have avoided stepping into defining 'scientific' problems; however, it left it unclear whether the distinction is artificial or decisive and whether 'for purpose of' is, rather, subjective. The 5th question was whether or not the Court applied the standard of review too strictly to the conduct of the Japanese authorities. Considering that there should be a strong presumption that a contracting party has made the determination to grant special permits for scientific research in light of careful consideration, maximum deference should be accorded, in the IWC to the determination of the national authorities. The 6th question was whether or not the *actori incumbit probatio* is still valid. By asking 'the authorizing state, which has granted special permits, to explain the objective basis for its determination' (Judgment, para. 68), the Court placed the burden of proof on the State granting a special permit under Article VIII, in opposition to the *actori incumbit probatio*. The 7th question was why the duty to co-operate with the IWC and the scientific committee was imposed on Japan by IWC resolutions and guidelines. The 8th question was whether or not Japanese whaling activities in the North Pacific Ocean would continue after the Judgment, considering that the Judgment may have certain legal implications in relation to the Japanese Whale Research Program under Special Permit in the North Pacific Phase II (JARPN II). The 9th question was whether or not the Court would actively exercise its judicial control in subsequent cases on violations of international legal rules. After examining those nine questions, the need was emphasized to continue the dialogues between the parties concerned to arrive at a solution in any cases in the future.

Session 3: Procedural Issues in the Whaling in the Antarctic Case

This session was started with a presentation entitled 'Unfavourable but unavoidable procedures: Procedural aspects of the *Whaling* case' examining potential effects of the litigation strategy of the applicant states on the Court's Judgment. It consisted of two parts: (1) the intervention of New Zealand; and (2) the examination and cross-examination of experts. First, it analyzed the intervention by New Zealand as new and well-planned litigation strategy for the purposes of: (a) adding Australia's Judge *ad hoc*; (b) doubling Japan's work; and (c) involving a 'non-party'—New Zealand—but being almost the same as a party in the litigation. It explored the possible means for Japan to have avoided this 'unfavorable' situation by requesting the Court to apply: (a) Article 47 of the ICJ's Rules of Court for joinder of the cases; or (b) Article 31(5) of the Statute, to have the applicant states treated as one party. Second, by categorizing the case as a 'scientific dispute', it considered the examination and cross-examination of experts in the case to identify the function of experts' opinions under Article 43(5) of the Statute. Although the negative impact of expert opinions on Japan were affirmed, it was left unclear whether experts served as counsel or advocates of the States, and emphasized the need for independence and impartiality of expert's opinions. In conclusion, it described Japan's position as slightly unfavourable but unavoidable from the procedural point of view; however, lessons remained from the applicant states' litigation strategy.

The following presentation addressed the topic of 'Roads Not Taken, Opportunities Missed: Procedural and Jurisdictional Questions Sidestepped in the *Whaling* judgment'. Focusing on the implications of silences in the Court's reasoning (insofar as it related to procedural and jurisdictional matters), this presentation addressed three issues in particular: (1) the Court's rather rushed treatment of question of jurisdiction; (2) the absence, from the judgment, of language reflecting the public interest or '*erga omnes*' character of the application; and (3) the Court's decision to restrict the written proceedings to one round. First, it considered that the Court unanimously upheld jurisdiction on the basis of the two parties' optional clause declarations even though it might have been argued that the ICRW embodied a 'special dispute settlement agreement' excluded from the scope of the Australian optional clause. Second, the presentation examined legal implications of the Court's decision not to discuss the applicant's *Locus Standi*. This silence was perhaps surprising, given that Australia claimed to be acting in defense of a public concern and to invoke Japan's responsibility '*erga omnes partes*' under the ICRW. The presentation argued that the Court's silence could be read in two ways: the Court may have simply taken for granted that States have standing to bring claims in defense of obligations *erga omnes partes*; or it may have preferred silence so to avoid an express recognition of such a right. On either explanation, the case remained as the first one that admitted an avowed public interest claim. Third, the presentation briefly commented on the reasoning leading the Court to proceed to the oral hearings after only one round of written proceedings under Article 45 para. 2 of the Rules of Court. It considered two institutional aspects: first, with only one round of written argument, the proceedings could be completed relatively quickly; and second, the oral hearings in the case became far more relevant than in other proceedings, considering that both parties' written memorials were not overbearing as in other cases and Japan did not cover the material in its entirety. Against that background, the move towards one round of written proceedings may well be perceived as a useful change in the Court's procedure; however, if only for reasons of procedural fairness, it needed to be communicated to the Court's clients and implemented in a principled manner with some explanation in the Judgment.

Session 4: Whaling in the Antarctic as an International Scientific Dispute

At Session 4, a presentation entitled 'Motivations and Methodologies: Was Japan's Whaling Programme for Purposes of Scientific Research?' was made which had three goals: (1) to investigate the Court's decision to adopt a methodology in the case based on employing a standard of review; (2) to identify apparent consequences including those related to the burden of proof and its discharge; and (3) to ask whether the Court was likely to use such a methodology again in future. It evaluated how well the standard of review offered practical benefits to the Court in exercising its adjudicatory function, and interfacing between science and motivations through the medium of an expectation that Article VIII research program's design and implementation be reasonable in relation to achieving its stated observes.

The above-presentation was complemented by a presentation entitled 'Standard of Review in the *Whaling* Judgment' to clarify the distinction between an obligation to act reasonably and an obligation to state reasons. The following issues were examined: (1) the genesis of the concept of standard of review in the case; (2) the standard of review adopted by the Court; (3) the precise content of the obligation to give due regard to Recommendations; (4) the requirement of reasonableness as an obligation to state reasons in detail; and (5) the requirement of reasonableness as an obligation to state reasons within a short period of time. Focusing on the terms 'standard of review', employed by the Court for the first time, and 'the margin of appreciation', used by Japan, it was pointed out that the Court did not follow the standard of review proposed by either of the parties but adopted a two-step approach (Judgment, para. 67) to prove Japan's violation. While accepting the existence of a duty to co-operate with the IWC and to give duty regard to recommendations adopted by consensus, the presentation pointed out as being unusual, the precise content of the 'obligation to give due regards to recommendations' that the Court held Japan to have violated. In addition, it explored the reasons why the Court ignored two factors: Japan's reason for a gap in sample sizes and the actual catches, due to the sabotage activities of anti-whaling NGO; and Japan's intention to modify the JARPA II research programme if the scientific committee scheduled in 2014 revealed the necessity of any modification. In the end, the presentation concluded by posing another question, whether or not the 'obligation to state reasons' be extended to other fields of international law, though it answered this in the negative.

Session 5: ICRW, Its Article VIII, and the Law of Treaties

Session 5 was devoted to the Law of Treaties. It started with a presentation entitled 'The Whaling Convention and Thorny Issues of Interpretation' with a focus on two aspects of the case in the law of treaties: (1) the object and purpose of the ICRW; and (2) the 'treaty as an evolving instrument'. Although both parties heavily relied on the Preamble to clarify the object and purpose of the ICRW, Australia explored an evolutionary interpretation of the ICRW, while Japan was the proponent of a more static approach. Clarifying the existence of two original purposes of the ICRW—the orderly regulation of the whaling industry and the protection of whales—the presentation first examined the interpretation of the 'object and purpose' of the ICRW in relation to its Article VIII. As to (1), Australia claimed that a special permit under Article VIII was an exception to the general obligations under the ICRW and that Article VIII needed to be read by reference to Articles 31 and 32 of the VCLT. On the other hand, the Court regarded it as an integral part of the ICRW and denied that Japan's whaling was not 'for the purpose of' scientific research'. As to (2), while clarifying that the ECtHR concept of 'living instrument' was not applicable to the ICRW, the presentation held that the notion of the ICRW as a 'living instrument' was not fully explained in the Judgment and that para. 45 of the Judgment was misleading. Additionally it considered the question of subsequent practice of the parties (Art. 31 (3) (a) (b) of the 1969 VCLT). While Australia and New Zealand argued that the Court should rely on certain resolutions by the IWC offering a restrictive view of the permissible scope of lethal means in scientific research, the Court regarded those resolutions as neither subsequent agreement nor subsequent practice in relation to the interpretation of Article VIII. It concluded that the Judgment proved to be fertile ground for reflections on the law of treaties, but at the same time left many questions unresolved and still awaiting clarification.

Further at this Session, there followed a presentation on 'The Margin of Appreciation in the Use of Exemptions in International Law: Comparing the ICJ Whaling Judgment and the case law of the European Court of Human Rights'. Considering the Judgment as the first ICJ case that expressly referred to this concept created by the ECtHR, the presentation introduced the definition and function of the 'margin of appreciation' (MoA) in relation to exceptions by comparison to the concept of residual liberty of states emphasized in the

Lotus case. In the whaling case, MoA was argued by both parties, Japan claiming it to be an 'axiom of international law', while Australia and New Zealand denied its existence in international law. Although the Judgment did not endorse the term MoA, it assessed Article VIII as allowing a 'narrow' MoA, and extended the arguments into the ECtHR cases. For example, the MoA concept slipped into the concept of 'standard of review', and MoA was considered as a tool to respect pluralism and cultural relativism.

Session 6: Judgment's Implications: IWC and the Whaling in Future

This session included a presentation on 'Very Like A Whale: What the *Whaling* Judgment Reveals about the ICRW Regime' consisting of two parts: (1) the power of soft-law instruments in the ICRW regime that were, the product of—were 'yielded by'—organs of the regime set up under hard law provisions; and (2) the role of two factors in a regime, namely, the 'duty to cooperate' with the supervisory organs of the regime and the duty of good faith of a state in participating in an international regime. The two factors provide examples, when a case regarding the regime is addressed by a judicial organ such as the ICJ, of the possibility for soft law aspects to be transformed into hard law by a binding judgment, though this effect could only materialize in a judicial proceeding if the regime lacks its own dispute settlement mechanism.

The next presentation in this Session was entitled 'IWC and the ICJ Judgment' considering the implications of the Judgment for future developments at the IWC by posing three questions. It first considered whether or not the Judgment changed the basic legal framework and/or the interpretation of the ICRW. Quoting from the Judgment (paras. 55, 83, and 94), the presentation clarified that the object and purpose of the ICRW remains literally as written in the Preamble and the interpretation of Article VIII is not affected/restricted, and the Australian 'evolving' interpretation is denied. Second, it explored whether or not the Judgment affected the international policy making system regarding whales and whaling. While the IWC made efforts to arrive at a package compromise proposal named the 'Future of the IWC' project, Australia made a clear political decision to submit a case to the Court in line with its Prime Ministerial election campaign. The presentation concluded that it was too early to foresee the effect of the Judgment on the IWC's internal system, considering that, beside the commercial whaling moratorium adopted in 1982, the IWC had neither dispute a settlement mechanism nor the ability to make legally binding decisions, resulting into its dysfunction. The presentation added that the Court produced Judgments on 'reasonableness' of scientific activities which was not inconsistent with the views of the IWC scientific committee. Third, it focused on Japan's course of action after the Judgment by quoting a policy statement of the Ministry of Agriculture, Forestry, and Fisheries of 18 April 2014, setting out Japan's basic policy on the whaling issue. As the Judgment did not deny the possibility of a future special permit research program under Article VIII of the ICRW, Japan's policy is consistent with the Judgment, even though anti-whaling countries and NGOs regarded it as a milestone on the way to total cessation of whaling. In the end, the presentation concluded that the 65th IWC meeting would be important for the future of the IWC.

Session 7: Litigation Strategy and Tactics

Finally, Session 7 was devoted to litigation strategy and tactics. It included presentations on 'Some remarks on the Whaling litigation: Japanese perspective' providing the background of Japan's arguments and a presentation entitled 'An observer's view of the presentation of Australia's case to the ICJ in "Whaling in the Antarctic"' to provide different perspectives from the point of view of policy and science.

Concluding Remarks

Categorizing the implications given by all presentations into three: (1) policies; (2) proceedings; and (3) general international law, the Concluding Remarks highlighted critical points discussed during symposium and ended with words of appreciation for invited speakers as well as participants.

Index

www.ingramcontent.com/pod-product-compliance
Lightning Source LLC
LaVergne TN
LVHW010623300926
849162LV00001B/46